Energy

Energy

Crises, Challenges and Solutions

Edited by

Pardeep Singh
Department of Environmental Studies, PGDAV College,
University of Delhi, New Delhi, India

Suruchi Singh
Department of Botany, Sunbeam College for Women,
MGKVP University, Bhagwanpur, Varanasi, India

Gaurav Kumar
Department of Environmental Studies, PGDAV College,
University of Delhi, New Delhi, India

Pooja Baweja
Department of Botany, Maitreyi College,
University of Delhi, New Delhi, India

WILEY Blackwell

This edition first published 2022

Registered Offices
John Wiley & Sons, Inc., 111 River Street, Hoboken, NJ 07030, USA
John Wiley & Sons Ltd, The Atrium, Southern Gate, Chichester, West Sussex, PO19 8SQ, UK

Editorial Office
9600 Garsington Road, Oxford, OX4 2DQ, UK

For details oft our global editorial offices, customer services, and more information about Wiley products visit us at www.wiley.com.

Wiley also publishes its books in a variety of electronic formats and by print-on-demand. Some content that appears in standard print versions of this book may not be available in other formats.

Library of Congress Cataloging-in-Publication Data

Names: Singh, Pardeep, editor. | Singh, Suruchi, 1987– editor. | Kumar, Gaurav, 1984– editor. | Baweja, Pooja, 1977– editor.
Title: Energy : crises, challenges and solutions / edited by Pardeep Singh, Suruchi Singh, Gaurav Kumar, Pooja Baweja.
Description: Hoboken, NJ : Wiley-Blackwell, 2022. | Includes bibliographical references and index.
Identifiers: LCCN 2021007246 (print) | LCCN 2021007247 (ebook) | ISBN 9781119741442 (cloth) | ISBN 9781119741510 (adobe pdf) | ISBN 9781119741558 (epub)
Subjects: LCSH: Renewable energy sources. | Energy development. | Energy policy.
Classification: LCC TJ808 .E565 2022 (print) | LCC TJ808 (ebook) | DDC 333.79/4–dc23
LC record available at https://lccn.loc.gov/2021007246
LC ebook record available at https://lccn.loc.gov/2021007247

Cover Design: Wiley
Cover Image: © Funny Solution Studio/Shutterstock

Set in 9.5/12.5pt STIXTwoText by Straive, Pondicherry, India

10 9 8 7 6 5 4 3 2 1

Contents

Preface *xiv*
List of Contributors *xvi*

1 Energy Crisis and Climate Change: Global Concerns and Their Solutions *1*
Sandeepa Singh
1.1 Introduction *1*
1.2 Energy Crisis *2*
1.3 Role of Renewable Energy in Sustainable Development *3*
1.4 Climate Change and Energy Crisis *5*
1.5 Climate Change *6*
1.5.1 Environmental and Social Consequences of Climate Change *7*
1.5.2 Process and Causes of Global Warming *9*
1.6 Cleaner Alternatives to Coal to Alleviate Climate Change *10*
1.6.1 Carbon Sequestering and Clean Coal *10*
1.6.2 Natural Gas and Nuclear Energy *11*
1.6.3 Hydrogen *11*
1.7 Climate Change and Energy Demand *12*
1.8 Mitigation Measures for the Energy Crisis and Global Warming: Reduce Emissions of Greenhouse Gases (IPCC) *12*
1.9 Conclusion *13*
1.10 Future Considerations *14*
References *15*

2 Advances in Alternative Sources of Energy: Opening New Doors for Energy Sustainability *18*
Jyoti Tyagi
2.1 Introduction *18*
2.2 Need of Novel Research in Alternative Sources of Energy *19*
2.3 Recent Advances in Renewable Sources of Energy *20*
2.3.1 Solar Energy *21*
2.3.1.1 Solar Photovoltaic *21*
2.3.1.2 Solar Power Generation *24*
2.3.1.3 Photovoltaic/Thermal (PV/T) Collectors *24*
2.3.2 Wind Energy *25*

2.3.2.1 Onshore Wind Energy Technology *25*
2.3.2.2 Offshore Wind Energy Technology *27*
2.3.3 Hydropower *29*
2.3.3.1 Flow Control Technologies *30*
2.3.3.2 Digitalization of Hydropower Plants *30*
2.3.3.3 Evolution in Hydroelectric Energy Storage *31*
2.3.3.4 Technology Evolution: Small-Scale Hydropower Plants *31*
2.3.3.5 Gravity Hydropower Converters *32*
2.3.3.6 Pump as Turbines (PAT) *32*
2.3.3.7 Developments in Fish-Friendly Hydropower *32*
2.3.4 Geothermal Energy *33*
2.3.4.1 Direct Dry Steam Plants *34*
2.3.4.2 Flash Power Plants *35*
2.3.4.3 Binary Plants *35*
2.3.4.4 Combined-Cycle or Hybrid Plants *35*
2.3.4.5 Enhanced Geothermal Systems (EGS) *35*
2.3.5 Bioenergy *37*
2.3.5.1 Biopellets and Biogas *38*
2.3.5.2 Bioethanol and Biodiesel *38*
2.3.5.3 Advanced or 2G Biofuels *39*
2.3.6 Ocean Energy *40*
2.3.6.1 Wave Energy *40*
2.3.6.2 Tidal Energy *41*
2.3.6.3 Ocean Thermal Energy Conversion (OTEC) *41*
2.3.6.4 Salinity Gradient Energy *41*
2.4 Future Fuel: Hydrogen *42*
2.4.1 Hydrogen Production Methods Using Renewable Sources *43*
2.4.1.1 Renewable Electrolysis *43*
2.4.1.2 Biomass Gasification *43*
2.4.1.3 Thermochemical Water Splitting *43*
2.4.1.4 Bio-Hydrogen Production *44*
2.5 Challenges *44*
2.5.1 Efficiency *44*
2.5.2 Large-Scale Production *45*
2.5.3 Cost-Effective Production *46*
2.6 Future: Alternative Sources of Energy *46*
2.7 Conclusions *47*
References *48*

3 Recent Advances in Alternative Sources of Energy *55*
Pradeep Pratap Singh, Ambika, and Maya Verma
3.1 Introduction *55*
3.2 Different Innovations Employed in Major Types of Alternative Sources of Energy *56*
3.2.1 Solar Energy (Semiconductor Technology to Harness Solar Power) *56*

3.2.2 Hydropower *60*
3.2.3 Wind Energy *61*
3.2.4 Geothermal Energy *61*
3.2.5 Biomass Energy *62*
3.2.6 Hydrogen as a Fuel *63*
3.3 Environmental Impacts *64*
3.4 Future Prospects *65*
3.5 Conclusions *65*
References *66*

4 Energy and Development in the Twenty-First Century – A Road Towards a Sustainable Future: An Indian Perspective *72*
Shikha Menani and Kiran Yadav
4.1 Introduction *72*
4.2 Energy Consumption and Economic Development *73*
4.3 Environmental Issues – A Corollary of Economic Development *76*
4.4 Air Quality – Deterioration Leading to Development of another Mars *77*
4.5 Carbon Footprints – Gift of Mankind to Mother Earth *78*
4.6 Sustainable Development *80*
4.6.1 Problems Faced by the Country in Implementing Sustainable Development Goals (SDGs) *81*
4.6.1.1 Financial Resources *81*
4.6.1.2 Social Issues Not Covered *82*
4.6.1.3 Natural Calamities and Pandemics *82*
4.6.1.4 Illegal Activities Barring the Achievement of the SDGs *82*
4.6.2 Paris Accord *82*
4.6.3 Steps Taken by India to Reduce the Carbon Emission *84*
4.6.3.1 Sustainability Index *84*
4.6.3.2 Mandatory CSR *85*
4.6.3.3 Innovative Schooling Ideas *86*
4.6.3.4 Solar Powered Transportation System *86*
4.7 Coronavirus Pandemic and its Impact on the Carbon Emission *87*
4.8 Conclusion *88*
References *89*

5 Energy Development as a Driver of Economic Growth: Evidence from Developing Nations *91*
Md Rashid Farooqi, Akhlaqur Rahman, Md Faiz Ahmad, and Supriya
5.1 Introduction *91*
5.2 Energy and Economic Development *92*
5.2.1 The Impact of Economic Development on Energy *94*
5.2.2 Economic Development and Fluctuations in Energy Consumption *95*
5.2.3 Energy Consumption in Developing Nations *96*
5.2.4 The Price of Energy and Management of Demand *97*
5.3 Energy Services in Developing Nations *99*

5.4 Energy Supplies in the Developing Nations *100*
5.5 Energy and the Environment in Developing Nations *102*
5.6 Conclusion *103*
References *104*

6 Pathways of Energy Transition and Its Impact on Economic Growth: A Case Study of Brazil *108*
Pooja Sharma
6.1 Introduction *108*
6.2 The Rationale for Public Investment in Research and Development in Energy Sector *112*
6.3 Overview of the Electricity Sector in Brazil *113*
6.3.1 Energy Policies in Brazil *113*
6.3.1.1 Energy Sources and Associated Policies *113*
6.3.1.2 The First Phase of Reforms in the Electricity Sector: 1990s *114*
6.3.1.3 Second Reform of the Electricity Market: 2004 *114*
6.3.2 Climate Change: National Policy 2009 *114*
6.3.3 Prioritization of Policies in Choice of Energy Mix (International Atomic Energy Agency, 2006) *114*
6.4 Market Structure *115*
6.4.1 Government Players *115*
6.4.2 Private and Public Players *116*
6.5 Programmes and Laws Under the Government of Brazil *116*
6.6 An Overview of the Sources of Finance in the Energy Sector: Brazil *116*
6.6.1 The Regime for Funding Agency (World Energy Outlook 2013) *118*
6.6.1.1 Regime Structure and Legal Regulatory: Key Takeaways *119*
6.6.2 Source of Funding and Trends in Research and Development *120*
6.6.2.1 Finance and Innovation in Renewable Energy: Key Takeaways *120*
6.7 Climate-Resilient Growth: Environmental Consequences *121*
6.7.1 Environmental Consequences: Key Takeaways *121*
6.8 Social Consequences: Availability, Affordability and Accessibility *122*
6.8.1 Social Consequences: Key Takeaways *122*
6.9 The Political Economy of Energy Transition: A Brazilian Experience *123*
6.10 Interlinking Economic Growth and Energy Use: A Theoretical Construct *123*
6.10.1 Renewable Energy Consumption, per Capita GDP Growth, CO_2 Emissions, Research and Development Expenditure: A Comparison of BRICS *124*
6.11 Conclusion *125*
References *126*
Websites *127*
Appendix A *128*

7 Renewable Energy: Sources, Importance and Prospects for Sustainable Future *131*
Shachi Agrawal and Renu Soni
7.1 Introduction *131*
7.2 Sources of Renewable Energy *132*

7.2.1 Solar Energy *133*
7.2.1.1 Active Solar Energy Technology *133*
7.2.1.2 Passive Solar Energy Technology *133*
7.2.2 Wind Energy *134*
7.2.3 Hydropower *136*
7.2.4 Geothermal Energy *137*
7.2.5 Biomass *138*
7.2.6 Tidal Energy *138*
7.3 Advantages and Disadvantages of Various Renewable Energy Resources *139*
7.4 Importance of Renewable Energy *140*
7.5 Benefits of Renewable Energy Production to the Society *141*
7.6 Renewable Energy and Sustainable Development Goals *142*
7.7 Limitations in Renewable Energy *143*
7.8 Current Status and Future Perspectives *143*
7.9 Conclusion *144*
References *145*

8 Clean Energy Sources for a Better and Sustainable Environment of Future Generations *151*
Aparna Nautiyal and Ayyagari Ramlal
8.1 Introduction *151*
8.2 Conventional Sources of Energy *152*
8.2.1 Hydro Energy *153*
8.2.2 Wind Energy *154*
8.2.3 Geothermal Energy *155*
8.2.4 Solar Energy *155*
8.2.5 Ocean Energy *156*
8.3 Environmental Impacts of Renewable Resources *156*
8.4 Mitigation Strategies and Sustainable Development of Renewable Resources *157*
8.5 Biomass and Microorganisms-Derived Energy *157*
8.6 Alternative Energy Resources *160*
8.6.1 Biodiesel from Bioengineered Fungi *160*
8.6.2 Microbial Fuel Cells (MFCS) *161*
8.6.3 Waste-to-Energy Technology *161*
8.6.4 Hydrogen as a Fuel *162*
8.6.5 Fuel Cell *163*
8.6.6 Radiant Energy *163*
8.7 Challenges: Implementation to the Usage of Renewable Energy *164*
8.7.1 Social Barriers *164*
8.7.2 Ecological and Environmental Issues *164*
8.7.3 Commercialization and Scalability *165*
8.7.4 Material Requirement *165*
8.8 Conclusion *165*
References *165*
Suggested Readings *168*

9 **Sustainable Energy Policies of India to Address Air Pollution and Climate Change** *169*
Prem Lata Meena, Vinay, and Anirudh Sehrawat
9.1 Introduction *169*
9.2 Energy Sector of India *170*
9.2.1 Energy Reserves *170*
9.2.1.1 Coal and Lignite *170*
9.2.1.2 Petroleum and Natural Gas *170*
9.2.1.3 Renewable Energy Sources *170*
9.2.2 Production of Energy *170*
9.2.3 Consumption of Fossil Fuel and Electricity *171*
9.2.3.1 Coal and Lignite *171*
9.2.3.2 Crude Oil and Natural Gas *171*
9.2.3.3 Petroleum Products Consumption *171*
9.2.3.4 Consumption of Electricity *171*
9.2.4 Energy Sector and Greenhouse Gases Emission *172*
9.3 India's Potential and Policies to Exploit Renewable Sources *172*
9.3.1 Solar Energy *172*
9.3.2 Wind Energy *172*
9.3.3 Hydropower *173*
9.3.4 Biomass Energy *173*
9.4 National Strategies to Promote Renewable Energy: Policy Framework with Their Objectives *174*
9.4.1 India's Electricity Act *174*
9.4.2 National Electricity Policy (NEP), 2005 *175*
9.4.3 NAPCC-National Action Plan on Climate Change, 2008 *175*
9.4.4 Copenhagen Accord *176*
9.4.5 India's Intended Nationally Determined Contribution (INDC) *176*
9.5 Financial Instruments to Promote Renewable Sources in India *176*
9.5.1 Coal Tax *176*
9.5.2 Subsidy Cuts on Fossil Fuels *178*
9.5.3 Renewable Energy Certificates (RECs) *178*
9.5.4 Perform, Achieve and Trade Scheme *178*
9.5.5 Other Government Policies, Their Budget and Status *179*
9.6 Conclusion *179*
References *180*

10 **A Regime Complex and Technological Innovation in Energy System: A Brazilian Experience** *182*
Pooja Sharma
10.1 Introduction *182*
10.2 Brazil: Its Changing Role in Global Governance *183*
10.3 Brazilian Energy: A Regime Complex *184*
10.3.1 Role of Brazil and Regime Complex for Climate Change *185*
10.4 Implications of Climate Regime on Brazilian Energy Regime *186*
10.5 A Shift in Energy Regime: Technological Innovations in Energy Sector *187*
10.6 Conclusion *188*

References 188
Websites 189
Appendix A 190

11 Opportunities in the Living Lights: Special Reference to Bioluminescent Fungi *191*
Pramod Kumar Mahish, Nagendra Kumar Chandrawanshi, Shriram Kunjam, and S.K. Jadhav
11.1 Introduction *191*
11.2 History of Bioluminescence *192*
11.3 Bioluminescence in Terrestrial Organisms *193*
11.4 Bioluminescence Molecules *194*
11.5 Bioluminescent Fungi *196*
11.5.1 Diversity *196*
11.5.2 Mechanism of Bioluminescence in Fungi *197*
11.5.3 Significance *198*
11.6 Opportunities in Fungal Bioluminescence *198*
11.6.1 Glowing Tree *198*
11.6.2 Bioassay of Toxicity *199*
11.6.3 In-Vivo Imaging *200*
11.6.4 Animal Model Study *201*
11.6.5 Bioactive Secondary Metabolites *201*
11.7 Conclusion *201*
References *202*

12 Production of Liquid Biofuels from Lignocellulosic Biomass *208*
Manoj Kumar Singh, Sumit Sahni, and Anita Narang
12.1 Introduction *208*
12.2 Ethanol from Lignocellulosic Biomass *210*
12.2.1 Pretreatment of LCB *211*
12.2.2 Detoxification *212*
12.2.3 Hydrolysis *212*
12.2.3.1 Acid Hydrolysis *213*
12.2.3.2 Enzymatic Hydrolysis *213*
12.2.4 Fermentation *214*
12.2.5 Product Recovery *214*
12.3 Bio-gasoline from Lignocellulosic Biomass *214*
12.3.1 Hydrolysis to Monosaccharides *215*
12.3.2 Hydrogenation of Monosaccharides to Polyols *215*
12.3.3 Conversion of Polyols and Carbohydrates to C_5/C_6 Alkanes *216*
12.3.3.1 Monosaccharides *216*
12.3.3.2 Cellulose and Biomass *217*
12.4 Jet Fuels from Lignocellulosic Biomass *217*
12.4.1 Production of Jet Fuels from Sugars and Platform Molecules *217*
12.4.2 Production of Oil to Jet Fuels *219*
12.4.3 Production of Gas to Jet Fuels *220*
12.4.4 Production of Alcohol to Jet Fuels *221*

12.5 Conversion of Lignin to Hydrocarbons *221*
12.6 Conclusion *223*
References *224*

13 Sustainable Solution for Future Energy Challenges Through Microbes *231*
Sumit Sahni, Manoj Kumar Singh, and Anita Narang
13.1 Introduction *231*
13.2 Importance of Energy and Energy Statistics *232*
13.3 Brief History of Biofuels *233*
13.4 Classification of Biofuels *234*
13.4.1 First Generation (1G) *234*
13.4.2 Second Generation (2G) *236*
13.4.2.1 Enzymatic Pretreatment Process *240*
13.4.2.2 2G Biodiesel *243*
13.4.3 Third Generation (3G) *243*
13.4.4 Fourth Generation (4G) *245*
13.5 Conclusions *245*
References *246*

14 Fungal Microbial Fuel Cells, an Opportunity for Energy Sources: Current Perspective and Future Challenges *250*
Sudakshina Tiwari, Deepali Koreti, Anjali Kosre, Pramod Kumar Mahish, S.K. Jadhav, and Nagendra Kumar Chandrawanshi
14.1 Introduction *250*
14.2 General Introduction of Microbial Fuel Cells (MFCs) *251*
14.2.1 FCs *252*
14.2.2 Electrode of MFCs *252*
14.2.3 Proton Exchange Membrane *252*
14.2.4 Microorganisms and Their Electron Transfer Mechanism *253*
14.3 Factor Affecting the MFCs' Performance *253*
14.3.1 Configuration of Reactor *253*
14.3.1.1 Single-Chamber MFCs *253*
14.3.1.2 Dual-Chamber MFCs *253*
14.3.2 Buffer *254*
14.3.3 Substrate *254*
14.3.4 Electrolyte Resistance *255*
14.4 Fungal Microbial Fuel Cells *255*
14.4.1 Saccharomyces cerevisiae *256*
14.4.2 Candida melibiosica *257*
14.4.3 Hansenula anomala *257*
14.5 Other Fungi Used as a Biocatalyst in Microbial Fuel Cells *257*
14.6 Batteries Design with the Use of Fungal Electrode *258*
14.6.1 Batteries Design *258*
14.6.2 Structure and Composition of Lithium-Based Batteries *259*
14.6.3 Lithium–Sulphur (Li-S) Batteries *259*
14.6.4 Lithium-Ion Batteries *260*

14.6.5 Lithium-Air Batteries *260*
14.6.6 Role of Fungi in Batteries Design *260*
14.7 Application of MFCs *261*
14.7.1 Bioelectricity Production *262*
14.7.2 Biohydrogen Production *263*
14.7.3 Biosensor *263*
14.7.4 Wastewater Treatment *264*
14.7.5 Bioremediation *264*
14.7.6 Dye Decolorization *264*
14.8 Challenges and Future Prospective *265*
14.9 Conclusion *266*
Acknowledgements *266*
References *266*

15 Current Perspective of Sustainable Utilization of Agro Waste and Biotransformation of Energy in Mushroom *274*
Anjali Kosre, Deepali Koreti, Pramod Kumar Mahish, and Nagendra Kumar Chandrawanshi
15.1 Introduction *274*
15.2 Sustainable utilization of Agro waste Through Mushroom Cultivation Technology *276*
15.3 Lignocellulosic Biomass *278*
15.3.1 Characteristics of Lignocellulosic Biomass *279*
15.3.2 Cellulose *279*
15.3.3 Hemicelluloses *279*
15.3.4 Lignin *280*
15.4 Spent Mushroom Substrate (SMS) *280*
15.4.1 Biotechnological Importance of Lignocellulosic Biomass *281*
15.4.2 Applications of Spent Mushroom Substrate (SMS) *282*
15.4.2.1 Fertilizers *282*
15.4.2.2 Wastewater Treatment *283*
15.4.2.3 Enzyme Recovery *283*
15.4.2.4 Energy *283*
15.5 Biotransformation of the Spent Mushroom Substrate (SMS) Into Energy *284*
15.5.1 Biohydrogen Production from SMS *285*
15.5.2 Biogas Production from Spent Mushroom Substrate (SMS) *287*
15.5.3 Bioethanol from Spent Mushroom Substrate (SMS) *288*
15.5.4 Biobutanol from Spent Mushroom Substrate (SMS) *289*
15.5.5 Bio-Coke *290*
15.5.6 Electricity Generation Using Mushroom Technology *291*
15.5.7 Solar Steam Generation Device *292*
15.6 Challenges *293*
15.7 Conclusion *293*
References *294*

Index *303*

Preface

Energy is an indispensable component of every aspect of development, wealth, health, nutrition, infrastructure, and education. Energy is a necessary element in development that should be a fundamental right. Many development indicators are strongly related to per-capita energy consumption. The economic development of many countries has come at the cost of the environment. It should not be presumed that a reconciliation of the two is not possible. There is a need to take enhanced global actions to address emission problems.

Fossil fuel is the most conventional energy source, but its usage is full of dichotomy as its utilization has increased during economic development, but that also increased greenhouse gas emissions. Also, fossil fuel conservation will include finding a way to tap into the Earth's supply so that the commonly used oil fields are not drained completely. What will pave the way for natural recovery? The depletion also creates an enormous destructive waste product that then impacts the rest of life. The nexus concept is the interconnection between energy, water, food, land, and climate. Such interconnections enable us to address trade-offs and seek synergies among them. Putting pressure on one component will affect other components as well. Energy, water, food, land, and climate are essential resources of our natural environment and support our quality of life. Competition between these resources is increasing globally and is exacerbated by climate change. Improving resilience and securing resource availability would require improving resource-use efficiency.

Many policies and programmes are announced nationally and internationally to replace the conventional mode and emphasize the conservation of fossil fuels and reuse of exhausted energy, so a gap in implications and outcomes can be broadly traced by comparing the data.

The book 'Energy Crises, Challenges, and Solutions' aims to highlight problems and solutions related to conventional energy utilization, formation, and multitudes of ecological impacts and tools for the conservation of fossil fuels. The book also discusses modern energy services as one of the sustainable development goals and how the pressure on resource energy disturbs the natural flows. The book covers holistic issues related to energy and its contribution in triggering climate change and replenishing fossil fuel, emphasizing fossil fuel conservation and thus nature recovery. The compilation also highlights direct and indirect implications on different sectors. Many policies and legislations have been documented, but still, energy-related problems cannot be checked. This book helps identify these gaps, especially in the developed region. It benefits researchers and all other sectors and stakeholders, students, industries, and governmental agencies directly or indirectly associated with energy research.

We are highly delighted and express our gratitude to all the authors for their outstanding cooperation towards the compilation of this book. We also extend our sincere thanks to all the reviewers for their valuable suggestions and comments, which have helped us tremendously prepare this book. We also thank Wiley Publication, Andrew, Rosie, and Shiji for their generous support and efforts.

Editors
Pardeep Singh, Suruchi Singh, Gaurav Kumar, and Pooja Baweja

List of Contributors

Ambika
Department of Chemistry
Hansraj College
University of Delhi
New Delhi, India

Shachi Agrawal
Department of Botany
Gargi College
University of Delhi
New Delhi, India

Md Faiz Ahmad
School of Management
Malla Reddy University
Hyderabad, Telangana, India

Nagendra Kumar Chandrawanshi
School of Studies in Biotechnology
Pt. Ravishankar Shukla University
Raipur (C.G.)
Raipur, Chhattisgarh, India

Md Rashid Farooqi
Department of Commerce and Management
Maulana Azad National Urdu University
Hyderabad, Telangana, India

S.K. Jadhav
School of Studies in Biotechnology
Pt. Ravishankar Shukla University
Raipur (C.G.)
Raipur, Chhattisgarh, India

Deepali Koreti
School of Studies in Biotechnology
Pt. Ravishankar Shukla University
Raipur (C.G.)
Raipur, Chhattisgarh, India

Anjali Kosre
School of Studies in Biotechnology
Pt. Ravishankar Shukla University
Raipur (C.G.)
Raipur, Chhattisgarh, India

Shriram Kunjam
Department of Botany
Govt. V. Y. T. Autonomous P.G. College
Durg (C.G.)
Durg, Chhattisgarh, India

Pramod Kumar Mahish
Department of Biotechnology
Govt. Digvijay Autonomous P.G. College
Rajnandgaon (C.G.)
Rajnandgaon, Chhattisgarh, India

Prem Lata Meena
Department of Polymer Science
Bhaskaracharya College of Applied Sciences
University of Delhi
Dwarka, New Delhi, India

Shikha Menani
Department of Commerce
PGDAV College
University of Delhi
New Delhi, India

Anita Narang
Department of Botany
Acharya Narendra Dev College
University of Delhi
New Delhi, India

Aparna Nautiyal
Department of Botany
Deshbandhu College
University of Delhi
New Delhi, India;
i-4 Centre
Deshbandhu College
University of Delhi
New Delhi, India

Akhlaqur Rahman
District Institute of Education and Training
Bihar Education Service
Patna, Bihar, India

Ayyagari Ramlal
Division of Genetics
ICAR – Indian Agricultural Research Institute (IARI)
Pusa, New Delhi, India

Sumit Sahni
Department of Botany
Acharya Narendra Dev College
University of Delhi
New Delhi, India

Anirudh Sehrawat
University School of Environment Management
Guru Gobind Singh Indraprastha University
Dwarka, New Delhi, India

Pooja Sharma
Daulat Ram College
University of Delhi
New Delhi, India

Sandeepa Singh
Department of Botany
Maitreyi College
University of Delhi
New Delhi, India

Manoj Kumar Singh
Department of Botany
Acharya Narendra Dev College
University of Delhi
New Delhi, India

Pradeep Pratap Singh
Department of Chemistry
Swami Shraddhanand College
University of Delhi
New Delhi, India

Renu Soni
Department of Botany
Gargi College
University of Delhi
New Delhi, India

Supriya
Disaster Management Professional
Sitamarhi, Bihar, India

Sudakshina Tiwari
School of Studies in Biotechnology
Pt. Ravishankar Shukla University
Raipur (C.G.)
Raipur, Chhattisgarh, India

Jyoti Tyagi
Department of Chemistry
Zakir Husain Delhi College
University of Delhi
New Delhi, India

Maya Verma
Department of Physics
Hansraj College
University of Delhi
New Delhi, India

Vinay
University School of Environment Management
Guru Gobind Singh Indraprastha University
Dwarka, New Delhi, India

Kiran Yadav
Department of Commerce
PGDAV College
University of Delhi
New Delhi, India

1

Energy Crisis and Climate Change

Global Concerns and Their Solutions

Sandeepa Singh

Department of Botany, Maitreyi College, University of Delhi, New Delhi, India

1.1 Introduction

The energy system needs substantial decarbonization to combat climate change (Fawzy et al. 2020). However, climate change itself is often affected by energy system components that include long-term variations in climate parameters, instability and extreme weather situations (Field et al. 2014). Energy security and climate change are also genuine global issues, which also surface worldwide public policy debates. Significant international attempts to examine climate change policies have been undertaken in the past, in order to make them more conducive to the cause (Helm and Hepburn 2009; Giddens 2011; Held et al. 2011). In this ever-changing research area, it is vital to review literature that quantifies impacts and evaluates how this information is used to develop energy systems models.

Multiple-source energy use has always been a primary factor in human survival and civilizational growth. When comparing our ancestors' energy consumption quantity and trend with modern society, significant changes can be observed. Hunter-gatherer men's daily energy needs were about 2500 kcal before the arrival of modern agricultural techniques to sustain their nutrition, storage and reproduction, while modern man's requirements may exceed more than 100 times their ancestors, depending on their carbon footprint (Dias 2006; WWF 2006). As his key energy source for around 500,000 years, the prehistoric man primarily used food and fire from dry biomass burning. Besides providing heat, light and fuel for cooking, fire became the main defence and cold protection device. Subsequently, people started to spend their resources on land production, and energy use became more complex so that long farming cycles dominated short hunting and gathering times. Hence, a significant advance towards civilization and urban formation was the age of energy that started with fire and firewood and extended into food energy production developments. Both fire and wood were renewable biomass sources during this time.

Moreover, at the beginning of the seventeenth and eighteenth centuries, the invention of steam engines by James Watt allowed the industrial revolution to flourish by using non-renewable energy resources, especially 'fossil fuels', which included natural gas, oil and

Energy: Crises, Challenges and Solutions, First Edition. Edited by Pardeep Singh, Suruchi Singh, Gaurav Kumar, and Pooja Baweja.

coal. Fossil fuels have been the key drivers of twenty-first-century economic growth from an energy perspective. However, the heavy use of fossil fuels has created some severe environmental issues since the start of the industrial revolution, including global warming, photochemical smog and catastrophic air pollution, many of which are widely documented in the scientific literature.

It is a founded reality that if the greenhouse gas emissions rise is not reversed, events leading to catastrophic changes in the global environment will result in a cascade with a consequential effect on human society and economy. Study efforts to seek alternative and sustainable energy fuels were triggered by a steady increase in energy usage and environmental pollution. Materials and techniques for the efficient use of alternative fuel resources are already being developed by many nations worldwide. These alternative fuels, also known as advanced or non-traditional fuels, are compounds that can be used as fuels rather than conventional fuels. The word traditional fuel and nuclear materials such as uranium apply to petroleum (oil), coal, propane and natural gas. Biodiesel, bio alcohol (ethanol, methanol), hydrogen, chemically stored electricity (batteries and fuel cells), non-fossil natural gas, non-fossil methane and vegetable oils are common alternatives and well known. According to recent research, livestock, forestry and other land activities account for 23% of global human-origin greenhouse gas emissions. Such emissions are caused by land-use changes, such as deforestation, allowing space for crops, houses, and factories. Another 44% of the potent greenhouse gas methane comes from human-driven agriculture, peatland degradation and other land-based sources (IPCC 2019).

Climate policy security writers have stressed that adequate discursive action has not been taken to counter climate change. There appears to be a broad gap between public policy discourses and their final discourses on climate change and energy security. While less attention has been paid to the relationship between the two major policy areas of climate change and energy protection itself, much attention has been paid to this in recent past. For two reasons, this is a big research field. Since energy accounts for about 60% of global emissions (Baumert et al. 2005), managing energy emissions would be critical for climate reduction goals.

Moreover, in many countries, energy is viewed as a priority policy area, as it is an essential channel for economic development, progress and prosperity. Furthermore, it is relevant not only from a domestic viewpoint, but energy management also functions as a strategic foreign policy (Giddens 2011). Although climate change is bound to affect the energy market, the implications of policies designed to control climate change are expected to be immediate and potentially broader. Climate change issues pervade modern energy policy and resource renewability and energy protection problems that eventually contribute to economic growth. Increasingly strict environmental legislation must be placed on electricity providers to make major investments in lowering pollution, using renewable energy supplies, transmission infrastructure, replacing outdated technology and upgrading the grid.

1.2 Energy Crisis

The energy crisis refers to the world's rising need for energy to feed the growing population. With neither improving or reducing greenhouse gas (GHG) emissions, over-dependence on efficiency and carbon trading has been demonstrated as a phenomenal mistake.

According to the energy industry calculations, several thousand billion tonnes of coal are currently underground and could well be adequate for a 150-year demand at current extraction rates. Therefore, it becomes apparent that the bulk of coal reserves should stay on the ground if humans hope to avoid any climate catastrophe. In the United States, Russia, China, Australia and India, three-quarters of the world's coal reserves are concentrated in five countries. Therefore, human life and society's fate depends in no small degree on the coal-related decisions of these five major nations. Therefore, being the historical environmental polluters, the nations that are big emitters of greenhouse gases must compensate for harming developing countries by paying and promising not to burn coal and oil on the ground and keep their forests undisturbed and planting trees. To this end, a new Fund (Forest Carbon Partnership Facility, FCPF) was introduced by the World Bank in 2007, which could help achieve this goal. Providing significant rights to historical polluters implies partnership in ordinary individuals' ongoing theft and promoting scarcity rentals to private organizations rather than harnessing them to raise public revenue. Carbon emitters should be required to pay higher prices in relation to their emissions, and trading should be permitted to open up a source of funding for the poor to be transferred. The formulation of a global carbon tax will be a feasible solution to this issue. A sustainability strategy that contributes to productivity needs to be formulated in order to achieve the wider objectives of energy protection and climate reform.

Developing such a strategy will lead to decreased carbon usage, deter the expansion of less carbon-intensive technology and restructure revenue gradually. In all subsequent stages of the supply chain, higher input prices on carbon content, and fossil fuels will encourage higher efficiency, reducing depletion and eventually reducing pollution (Veritas 2004).

The need to obtain fossil fuels from politically turbulent or aggressive countries that have long been a cause of concern in the United States, or over-dependence on a few selected suppliers, has been a cause of concern for the European Union in Russia. There are growing demands for energy 'dependence' on others and promoting renewables, which can achieve the desired goals and help combat climate change, as such conditions can be used for strategic gains (Friedman 2005). An urgent transition from traditional to renewable energy is required. It has been estimated that the United States could supply enough power to the entire United States by generating wind energy in the Dakotas and with the aid of 254×254 km of the Sahara desert, the world's electricity demand could be met efficiently. Thus, desert-rich nations should be financially encouraged to produce and export solar energy to the world. Likewise, several nations might become a tidal force, offshore wind, wave and current (Helweg-Larsen and Bull 2007).

1.3 Role of Renewable Energy in Sustainable Development

Renewable energy production has undeniably proved to be an effective and realistic approach to achieving sustainable development. Not only has renewable energy emerged as a sustainable alternative to the clean energy system with recent and rapid energy technology advances, but it is also a way to meet other socio-economic needs, including improvising energy stability (Wei et al. 2012), mitigating climate change (Wang et al. 2014b) and minimizing environmental impacts associated with fossil fuels (REN21 2017). A total of 176 countries made efforts to realign the world's renewable energy targets in 2016.

According to available figures, renewable power accounted for around 30% of the installed capacity and 24.5% of the international level. According to projections, renewable energy accounts for 20% of the world's energy supply (Rees 1999). In the future global energy scenario, given renewable energy's huge technological and economic potential, it is realistic to expect a golden opportunity for renewable energy. There are real worries regarding the global energy crisis, as only small amounts of oil and gas are available and can be recovered. Natural forces generated millions of years ago non-renewable coal, oil and natural gas reserves, and no new supplies are being developed now. So the globe will eventually also run out of those provisions, but it is still unclear when it will happen in the future. For hopeful, who believe the energy peak still exists, the challenge is to find ways to conserve the currently available energy resources to potentially postpone the upcoming peak age of oil-gas-fuel and find new energy sources can effectively substitute fossil fuels. Exploring renewable options available for biofuels in solar, wind, nuclear and geothermal energy is now critical. Renewables will also allow new ways of transforming the economy and efficiently replacing carbon-free electricity. This move would take decades, but it would help restore businesses, power grids and lives associated with new energy sources.

Economic development, industrial society and the developed world's overall lifestyle rely heavily on the energy provided by oil and gas supply. Accessible food production powered by industrial farming needs low-cost supplies of natural gas and oil for goods and downstream purposes including pesticides, fertilizers, farming, harvesting, packaging, transport and marketing. All these big shifts in global industrial society made cheap oil energy products possible (Newton 2012). The availability of stable, skilled and reliable energy supply is essential for sustaining modern societies. Renewable energy can effectively support energy security problems at both ends, including supply and demand. Due to the finite and exhaustible state of fossil fuels, renewables can effectively meet the enormous energy demand in an increasingly important way. According to the Intergovernmental Panel on Climate Change (IPCC) study, 2.5% of renewable capacity would address a substantial 80% of world energy demand in 20505 (IPCC 2011). On the demand side, renewables could solve energy protection, fulfil sustainability needs and be environmentally friendly. The evolving relationship between renewable energy production and energy protection has been extensively studied (Valentine 2011), evaluating the evolution of energy security concept and validating symbiosis.

Renewable energy can positively affect the climate and other beneficial social benefits, including expanded opportunities and jobs in education, reduced energy scarcity and gender disparity (Tsai and Chou 2006). Consumption of renewable energy will generate far fewer carbon emissions and harmful environmental contaminants than fossil fuels, a long way towards combating climate change and reducing environmental pollution (Varun et al. 2009; Wang et al. 2014a). Jobs figures in this energy sector grew to 9.8 million in 2016, with most countries opting for the transition to renewable energy resources, a substantial 1.1% rise over 2015 (REN21 2017). Furthermore, since access to modern and advanced energy resources is a road to sustainable growth, distributed renewable energy technologies provide unusual opportunities to alleviate energy insecurity in underdeveloped and rural areas (Wei et al. 2014).

As mentioned in the latest Organisation for Economic Co-operation and Development (OECD)/International Energy Agency (IEA) book, almost one-fifth of the world's total

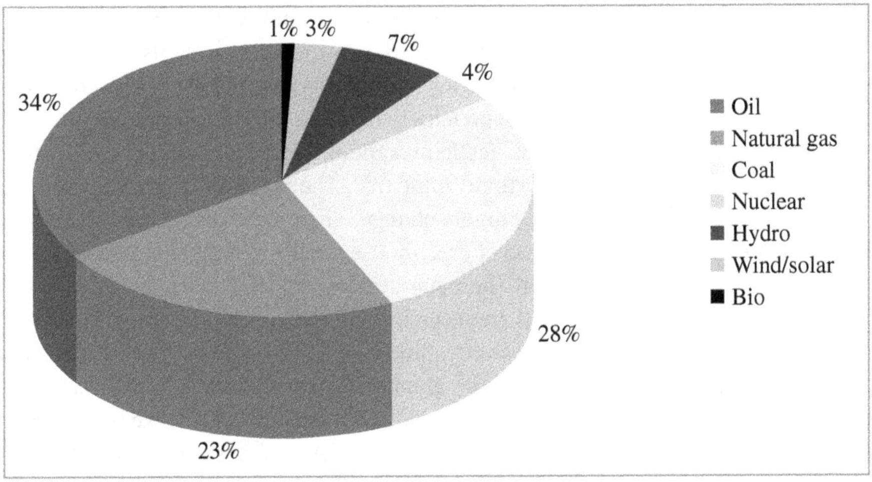

Figure 1.1 World's primary energy consumption in 2017. *Source:* Paul Homewood, https://notalotofpeopleknowthat.wordpress.com/2017/06/19/bp-energy-review-2017/.

electricity produced comes from renewable sources (OECD/IEA 2004). Renewable energy for power generation: status and prospects 'emphasises that renewable energy, after oil (34%), coal (28%), nuclear energy (4%) and natural gas (23%) is the second most efficient energy supply around the world' (Figure 1.1). Renewable energy growth took place from 1973 to 2000 at 9.3% annually and is expected to increase at 10.4% annually in 2010 and beyond. Wind turbines have risen to 52% most rapidly and are bound to increase further, overtake biopower and curb GHGs by developing environmentally friendly technologies (wind, solar and fuel cells). The crucial task is to minimize GHG and renewable energy generation by building broad research and development capacity in environmentally sustainable technologies. Little more than 50% of the world's land is listed as arid and part of the rural and desert environment with no water and power grids. Here, water pumps based on diesel engines are used to supply borehole water to the inhabitants. In exchange, diesel engines are impaired by maintenance and high running costs, contributing to emissions from the atmosphere. Energy obtained from wind systems can be an acceptable alternative process.

1.4 Climate Change and Energy Crisis

Climate change and climate change policies are two big energy sector contributors. Climate change policies adopted and enforced by different countries already influence planning, development and investment decisions at their locations. The transition from conventional fossil fuels to renewable energies, such as wind, geothermal, bioenergy and hydroelectricity has important implications for the goals of stable, clean and affordable energy, so that organizations and institutions responsible for achieving these goals must come forward and contribute.

The energy sector is highly responsible for climate change and is associated with greenhouse gas (GHG) emissions from fossil-fuel-based manufacturing plants, particularly power plants. Consequently, the industry's heavy reliance on fossil fuels requires this sector to be a priority of government remedial policies to regulate and mitigate greenhouse gas emissions. This could include a rigorous licencing system, strict emission quotas, carbon capture technologies and renewable portfolio standards. The industry is to be blamed for the heavy use of fossil fuels for causing climate change. The process begins with coal, gas and oil combustion, leading to the release of greenhouse gases that trap heat in the atmosphere, resulting in global warming. Even, there has been an ongoing debate on this subject as scientists have long tried to distinguish between human-induced changes and those that may be attributed to natural climate variability. Since developed nations have the highest emissions levels, they must therefore bear the greatest responsibility for global warming. Furthermore, as a precautionary measure, developing countries must also take measures to mitigate possible emission increases as their economies grow and populations increase, as the Kyoto Protocol clearly emphasizes (United Nations 2001).

Importantly, human activities in the form of carbon dioxide (CO_2), the most important contributors to future climate change, occur primarily through fossil fuel production. As a result, attempts to regulate CO_2 emissions could negatively impact worldwide people's economic growth, investment, trade, employment and living standards. The energy sector is highly responsible for climate change and is associated with greenhouse gas (GHG) emissions from fossil-fuel-based manufacturing plants, particularly power plants. Consequently, the industry's heavy reliance on fossil fuels requires this sector to be a priority of government remedial policies to regulate and mitigate greenhouse gas emissions. This could include a rigorous licencing system, strict emission quotas, carbon capture technologies and renewable portfolio standards. The industry is to be blamed for the heavy use of fossil fuels for causing climate change. The process begins with coal, gas and oil combustion, leading to the release of greenhouse gases that trap heat in the atmosphere, resulting in global warming. Even, there has been an ongoing debate on this subject as scientists have long tried to distinguish between human-induced changes and those that may be attributed to natural climate variability. Since developed nations have the highest levels of emissions, they must therefore bear the greatest responsibility for global warming.

1.5 Climate Change

The increase in average global temperatures is the product of climate change. The main contributors to this negative growth are natural disasters, along with human activities that are projected to result in higher average global temperatures (Figure 1.2). The Intergovernmental Panel on Climate Change (IPCC 2007) concluded that climate change has arisen as a result of human activities that have enabled global warming (Myles et al. 2009; Lacis 2012) and the most plausible reason for this is the widespread use of fossil fuels that produce significant greenhouse gas emissions, including methane (CH_4) and carbon dioxide (CO_2). In addition, climate change would do substantial harm to the productive system. The individual discussion of the energy and environmental crisis is also not feasible since it is directly related to it.

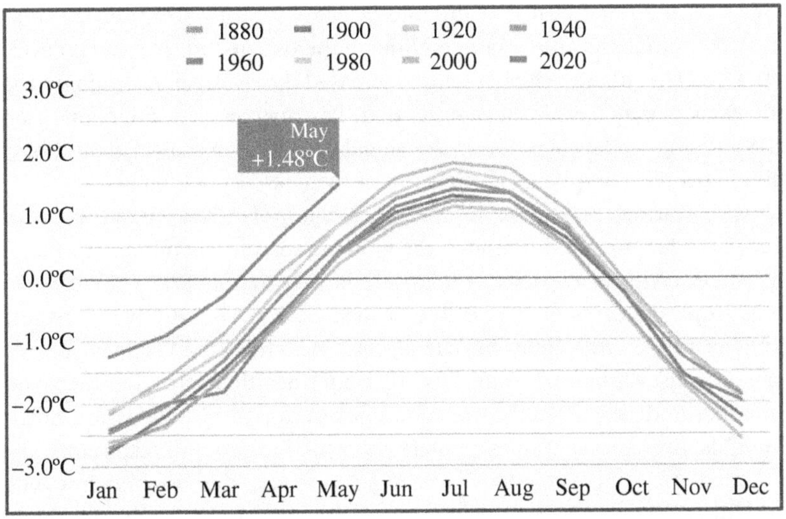

Figure 1.2 The earth is heating up. Monthly divergence from average temperature calculated for 1980–2015 in selected years. *Source:* NASA (2015). Licensed under CC BY ND 3.0.v.

1.5.1 Environmental and Social Consequences of Climate Change

There are a variety of significant consequences of climate change that are environmental, social and economic. These findings would typically be negative, although they may also be beneficial in a few remote circumstances (increase in crop yield). The primary cause of climate change is global warming, with substantial implications for human health and, most importantly, biodiversity. As a result of this, there are some worrying events such as the melting of ice at poles causing ocean levels to increase, the rise in hurricane intensity, shifts in rainfall patterns and ocean acidification (Nathan et al. 2008; Siddall et al. 2009).

Over the last decade, multiple long-term climate shifts have been observed in regional, continental and ocean basin scales, including major changes in Arctic and ice temperatures, heavy precipitation, increased salinity of the seas, altered wind patterns severe events such as drought, and heat waves. Over the last 100 years, normal temperatures have risen in the Arctic region to about twice the global average. In addition, after 1978, satellite data show that the average Arctic sea ice expansion decreased by 2.7%, from 2.1 to 3.3% per decade, with even greater decreases of 7.4%, from 5.0 to 9.8% per decade during the summer (Gregory et al. 2002).

Released in April 2007 (IPCC 2007; Climate Change 2007: Impacts, Adaptation and Vulnerability), the IPCC (Climate Change 2007: Impacts, Adaptation and Vulnerability) report discusses the imminent effects of global warming on human society and ecosystems and outlines scenarios up to 2100. Many effects on the natural environment have already started to emerge as a result of human activities, including improvements in the management of agriculture and forestry in higher latitude areas of the Northern Hemisphere, the timing of planting and growing spring crops, and shifts in pattern disturbance due to fires and pests. Human settlements in mountainous areas, in particular, are at greater risk of floods due to the disruption of glacial lakes caused by glaciers melting. The new IPCC report in September 2019 emphasized that the rate of ocean heating has risen twice as much since 1993 as the initial and consequent melting of the two historic ice sheets covering West Antarctica and Greenland, which has also increased the sea level. Alarmingly, the western Antarctic glaciers may have been so weak that they may have hit the point of no return (IPCC 2019). Rising sea levels pose a major threat to low-lying coastal areas, home to almost 680 million people, comprising around 10% of the world's population.

In conjunction with warmer conditions in the Sahel region of Africa, the eruption of drought has reduced the length and duration of the growing season. Therefore, this dense population area, where adaptive capacity is relatively low, is particularly vulnerable to challenges such as tropical storms or the decline of local coasts. The number of people affected will be higher in the large deltas in Asia and Africa, while small islands are particularly vulnerable because of their isolation. Overall, from a health perspective, there will be a rise in the number of deaths from illnesses and injuries caused by heat waves, hurricanes, floods, droughts, fires, etc. as well as an increase in the incidence of respiratory diseases and the spread of some vectors of infectious diseases due to higher ozone concentrations usually associated with climate change. The extent of the adverse effects will be proportional to the mean global increase in temperature. Even though relative to almost pre-industrial levels, the phenomenon of global warming is limited to 2 °C, harmful effects will be evident, and the planet will be forced to take drastic measures to adjust to current climate conditions. Suppose the increase in temperature crosses the 2 °C mark amid global efforts, in that case, it will lead to an unprecedented situation in which the consequences would certainly have been extreme, widespread and catastrophic.

Renewable water supply is threatened by declines in some areas and expansions in others, both of which are equally significant. In regions where gains are expected, temporary water deficiencies are still possible due to increased flux variability and seasonal shortages (due to reduced accumulation of snow and ice). Clean and fresh water supplies can also decline due to lower water quality caused by warm weather, such as algae-producing toxins, which could degrade the quality of critical sources such as lakes. Such a decline in renewable water supply will exacerbate competition between agriculture, towns, industry and water energy production, impacting local water, food and energy security. In addition, rising sea levels would have major consequences alongside coasts, including flooding, coastal erosion and submergence of low-lying areas, posing severe risks to residents, infrastructure, habitats and near-shore vegetation. Low-lying regions (e.g. Bangladesh) and islands as a whole, such as the Maldives and Kiribati, are at risk of near-term destruction due to various factors such as rising ocean levels, floods and extreme storms.

Over the past millions of years, climate change has occurred progressively and slower, allowing ecosystems to adapt. However, since the early twentieth century, species' extinction rate has risen to more than 100 times the normal rate, i.e. without anthropogenic interference. As a result, we are in the midst of a major biodiversity crisis and may even head towards another mass extinction (Mendenhall et al. 2014). The current rapid changes are suggested to impact both land and ocean ecosystems by 2050. Ecosystem changes, however, include much more than climate change, and a combination of many factors, including urbanization, increased world population and others, causes significant extinctions. However, climate change has shown its impact and will only intensify with time.

Over the past millions of years, climate change has occurred steadily and slowly, allowing ecosystems to adapt. However, since the beginning of the twentieth century, organism extinction rates have risen to more than 100 times the normal rate, i.e. without anthropogenic interference. As a result, we are in the midst of a major biodiversity crisis and maybe even head towards another mass extinction (Mendenhall et al. 2014). It is proposed that by 2050, rapid changes are likely to impact both land and ocean ecosystems.

1.5.2 Process and Causes of Global Warming

The earth gets enough space from radiation coming from the sun. To hold Earth's temperature at an optimal level, greenhouse gases play a crucial role in trapping the solar heat needed to sustain life. This phenomenon is natural, known as the greenhouse effect, and therefore important for sustaining various life forms on Earth. Without the greenhouse effect, Earth's temperature would be around 33 °C lower than it is today (Morice et al. 2012). Human activities have led to substantial increases in atmospheric GHGs due to deforestation and high fossil-fuel combustion rates in recent decades. Over the last century, GHG production is the primary cause of global warming. Earth warming ranged from +0.8 to +1.0 °C after 1900, according to published literature (Figure 1.3; den Elzen and Meinshausen 2006). Since 1950, land-only observations have shown warming trends between +1.1 and +1.3 °C, as land temperatures usually react rapidly in the climate change phase compared to oceans. Various factors affect the Earth's climate, including solar (warming effect), volcanic eruptions (and their cooling effect) and atmospheric GHG levels (warming effect). Methane-led carbon dioxide (CO_2) has been a significant contributor to global warming since the Industrial Revolution of 1750, with CO_2 concentrations rising from 278 parts per million (ppm) in 1960 to nearly double at 401 ppm in 2015 (Levitus et al. 2005). Since 1951, almost 100% of the warming is due to anthropogenic activities. Human activities are now responsible for increasing global temperatures by 1.1 °C, and

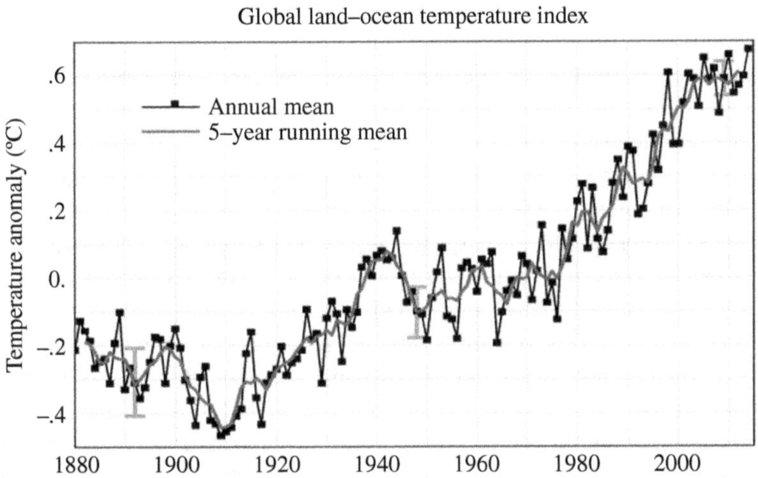

Figure 1.3 Instrumental temperature data 1880–2014. *Source:* NASA Goddard Institute for Space Studies (GISS).

according to reports, global emissions are already heading towards 1.5 °C or near-mid-century targets.

Because of an increase in GHG concentrations, water vapour has a significant indirect impact on temperature changes. High global temperatures increase the atmosphere's ability to absorb water vapour due to GHGs, which increases temperature, as water vapour also contributes to the greenhouse effect. Thus, an increase of 1 °C in global temperature results in an increase in atmospheric water vapour by around 7%. Therefore, it is clear that while CO_2 is the primary candidate for anthropogenic climate, water vapour amplifies the effect and is, therefore, a central agent of climate change (Gillett and Matthews 2010). Limiting global warming to below 2 °C worldwide is widely seen as an effective aim to reduce dangerous warming. Still, it is unlikely to be achieved without major reductions in GHG emissions (Canadell 2007). More than 100 countries have agreed on a global warming limit of 2 °C or less as a benchmark for mitigation steps to reduce the risks, impacts and damage caused by climate change (relative to pre-industrial levels).

1.6 Cleaner Alternatives to Coal to Alleviate Climate Change

1.6.1 Carbon Sequestering and Clean Coal

In order to achieve carbon sequestration, the focus should be on planting trees, reducing deforestation and managing land on a global scale, keeping in mind the human rights and livelihoods of poor people living in forest environments, so as not to threaten their lives through such plantations. Furthermore, micro-algae have been produced to absorb more than 80% of CO_2 emissions from power plants and other greenhouse sources and can therefore be used to generate liquid fuel at an annual rate of 10,000 gal per acre (Makhijani 2008).

According to common knowledge, 'Green coal' has not yet existed, so there should be no dependence on it. It will, however, become available only after 2020 since the climate crisis is already overwhelming. Even if the goal of clean coal is reached, it will be costly and cost about 25% more, and it would also be very difficult to track.

The extreme constraints and attention paid by organizations such as the World Bank to the availability of clean coal and coal efficiency are major obstacles to developing countries trapping them in the dirty energy development loop, a colossal mistake that developed countries pay dearly for. China has already surpassed the United States and, after 2010, will become the world's biggest consumer of oil. For more than 80% of its carbon emissions, China relies on coal. Every week, it opens more than two new 600 MW coal-fired power plants (Martinot and Junfeng 2007), and neither of them can be easily equipped with modern technology for carbon sequestration. This is the essential issue that about 15 000 metric tonnes of CO_2 are generated every day by each coal plant. In order to achieve carbon sequestration, the focus should be on planting trees, reducing deforestation and managing land on a global scale, keeping in mind the human rights and livelihoods of poor people living in forest environments, so as not to threaten their lives through such plantations.

Furthermore, micro-algae have been produced to absorb more than 80% of CO_2 emissions from power plants and other greenhouse sources and can therefore be used to generate liquid fuel at an annual rate of 10,000 gal per acre (Makhijani 2008).

Contrary to common knowledge, 'clean coal' is not yet in operation, so there should be no dependency on it. However, it will only become available after 2020 if the climate crisis is already overwhelming. In any case, it would be costly and cost about 25% more, even though the clean coal goal was met, and it would also be very difficult to track.

1.6.2 Natural Gas and Nuclear Energy

As it emits 70% less carbon per unit of energy than coal, natural gas is considered a cleaner alternative to coal. In combination with a few other hydrocarbons, including ethane, propane, pentane and butane, natural gas consists primarily of methane, with some other trace elements present. Hydrocarbons are responsible for the high combustion properties of natural gas and for the clean-burning properties of methane. Hydrocarbons and impurities are removed, and methane is effectively burned in the process of converting natural gas into a gaseous form. On the other hand, hydrocarbons are needed for good combustion when natural gas is used as a liquid fuel in engines. As both ways can be used, natural gas can also become an efficient fuel of choice. The key benefit of natural gas is its chemical composition, which is mostly methane (CH_4). Since methane has a single carbon atom in its structure, it creates far less carbon emissions. Furthermore, due to their chemical composition, the blue flame formed by the burning of natural gas originates from molecules that help to complete combustion.

In the meantime, natural gas will serve effectively as a bridge fuel, as the transition to renewable energy will be a slow and long-drawn process. However, it has disadvantages that need to be addressed, such as inevitable gas leakage, large amounts of GHG produced by transport and regasification, so that the benefits could be minimal and temporary.

Therefore, nuclear energy is not a healthy option. Strong action is needed to take into account the overall cost of environmental and social aspects, including the possibility of terrorist attacks and accidents and the diversion of radioactive weapons materials. The industry must also be allowed to pay for the permanent disposal of radioactive nuclear waste, and all accident insurance must be issued with immediate effect by the nuclear industry. In addition, all subsidies to the nuclear industry must cease and be diverted to alternatives to renewable energy sources.

1.6.3 Hydrogen

Hydrogen is believed to be a colourless and odourless gas, which makes up 75% of the entire universe's mass. Hydrogen is produced on Earth in combination with other materials such as carbon, nitrogen and oxygen. Hydrogen must first be removed from the other elements to be used for various purposes. The most inspiring and exciting idea and, at the same time, the most challenging challenge is to accept it as a fuel, considering that many businesses will be afraid to do so because it is a huge step forward to transform fully from gas to hydrogen and create a hydrogen-powered car. While this transition in the sea will be a rough trip, some of the benefits will be great. Hydrogen-based cars, for example, can be much more environmentally friendly than those that burn fuel. This will decrease noise and reduce the health problems that asthmatics face as a result of living near contaminated areas. The resource's simple availability would be added, and the great advantage of converting to hydrogen, such as hydrogen, is about 70% of the Earth's mass. All of it is dissolved in water, making it easy to collect. After the collection of water, the energy can be conveniently separated by flowing water. Another benefit is that considering that hydrogen is easier to refine than gasoline, it would be cheaper for the consumer and serve the purpose if cheap fuel were in high demand.

1.7 Climate Change and Energy Demand

The most inspiring and exciting idea and, at the same time, the most challenging task is to accept hydrogen as a fuel, considering that many businesses will be afraid to do so because it is a huge step forward to transform fully from gas to hydrogen and create a hydrogen-powered vehicle. Although this shift in the sea will be a rough trip, some of the benefits will be great. Hydrogen-based cars, for example, can be much more environmentally friendly than those that burn petrol. This will decrease noise and reduce the health problems that asthmatic people face as a result of living near polluted areas. The simple availability of the resource would be added, and the great advantage of converting to hydrogen, such as hydrogen, is around 70% of the Earth's mass. All of it is dissolved in water, making it easy to collect. When water is collected, it can be easily extracted by running electricity via water. Another benefit is that, since hydrogen is easier to refine than gasoline, it would be cheaper for the consumer and serve the purpose if cheap fuel were in high demand.

1.8 Mitigation Measures for the Energy Crisis and Global Warming: Reduce Emissions of Greenhouse Gases (IPCC)

Major measures to decrease GHG emissions from the oil industry and reduce the risk of a global energy crisis and eventual climate change should be a top priority for policymakers. The IPCC reports clearly describe the uninhibited effects of global warming. Sustained warming and far-reaching adverse effects on all climate system components will result in the continued release of greenhouse gases, increasing the occurrence of extreme, imminent and irreversible impacts on the world's population and ecosystems. Significant and sustained reductions in greenhouse gas emissions would be needed to tackle climate change, which, in combination with adaptation, would reduce the risks associated with climate change (IPCC 2018). Technology advancement in conjunction with reduced energy use, decarbonized energy supply, reduced emissions and improved carbon sinks in land-based industries is needed. At the same time, the following steps will save energy for a long time and can help to slow down the warming:

They are developing ways to allow effective use of energy-intensive materials and to improve methods of consumption.

Adoption of low-carbon fuels, particularly in the case of modern refineries.

Developing new sources of energy (e.g. biomass, solar, wind and hydro-electrical power).

In particular, the formulation of effective and strict environmental standards, rules, policies and regulations relating to the oil industry.

Mobilizing and encouraging environmental and emission control activities to cope effectively with the emerging oil industry.

Construction of offshore wind turbines and other related marine-based renewable energy technologies to further transition away from fossil fuel dependency.

Restoration of coastal habitats containing mangroves and salt marshes that are used to store carbon and have multiple benefits, such as providing buffers against extreme tropical storms, filtering contaminants and providing an ideal habitat for fish and other types of wildlife.

These essential and powerful steps are capable of reducing significantly 21% of pollution by 2050. They will minimize warming to 1.5 °C or just above pre-industrial levels by the end of the current century.

1.9 Conclusion

Today, the world is caught up in an energy crisis that has adversely affected substantial measures to decrease GHG emissions from the oil industry and to reduce the risk of a global energy crisis.

Invariably, fossil fuels are a lifeline for human civilization and millions of others around the world. The availability of fuel is not infinite, which is why people are willing to fight for it and to align themselves with their adversaries. Exploration and development are currently underway for other renewable energy sources, most of which have only passed their early stages. The evolution of these technologies can be accelerated by growing government funding and public support, helping free society from the reckless use of fossil fuels. During this massive transition, oil companies will remain resilient because, with the depletion and exhaustion of fossil resources in the future, they would make big profits. To prevent this, it is important to encourage the oil sector and other energy resource companies to look beyond their commercial profits and concentrate on developing alternative strategies when all fossil fuel resources are being depleted.

As a result of the increased concentration of carbon dioxide and other greenhouse gases in the atmosphere resulting from the burning of fossil fuels, we have seen the convincing scientific evidence that the Earth's surface is continuously heating up and its average temperature is increasing. The resulting global warming would eventually manifest itself in the form of significant changes in the Earth's atmosphere, which would have a major impact on human life and the world as a whole. Strong efforts are needed to promote renewable energy and minimize the use of fossil energy to avoid this. The aim must be to reduce energy consumption, which can be done by combining multiple factors, including the minimization of energy demand, the equal use of energy and the abundant use of green energy. The primary objective of this analysis was to take a step forward in achieving this objective. Creating green or sustainable solutions for managing society can be seen as an important response to the energy crisis. The implementation of renewable energy generation approaches and the discovery of ways to use these alternatives as green energy sources are crucial factors in reducing CO_2, which contribute significantly to global warming. While land availability is relatively modest, comprehensive and sustainable agriculture programmes provide significant energy and economic and environmental benefits that could be disseminated in rural areas where they are necessary and can link to further economic growth in rural areas. In this way, all countries will benefit from foreign exchange savings, increased energy security and socio-economic developments. The country's resource base can be dramatically strengthened by tree planting and forest expansion. The community at the international level would benefit greatly from carbon reduction, climate mitigation and enhanced trade opportunities that would lead to new income sources. In addition, environmental and environmental factors, including carbon sequestration and reforestation, renewables as a replacement for CO_2 for fossil fuels, are non-technical approaches that can make a significant contribution to climate restoration. More attention should be paid to the importance of renewable energy and the difficulties of gathering good and accurate renewable energy data at the policy and planning level.

1.10 Future Considerations

We face a terrible global energy crisis, triggered by continued growth in the world's population, continued reliance on fossil fuels, and increased demand for energy for various purposes. It is well understood that unless the rise in greenhouse gas emissions is prevented or reduced, there will be major turbulence in the global atmosphere, with detrimental effects on human civilization and the economy. Such a stubborn problem needs to be addressed by facilitating cooperation between policymakers and designers in different fields of science, engineering and technology.

Healthy creation, evolution and use of technology require major steps in the direction of progress. Therefore, to achieve goals of the optimum scale and magnitude within a limited timeframe, policymakers need to be technologically adept, and aware of policymakers' deeper social and political problems. Similarly, it is also necessary for policymakers to unite with the academic community to achieve technical objectives. The hour's need is to bring these two main groups together closely, with a primary focus on educating socially relevant and competent engineers and technologists to take care of the decades and centuries ahead.

Policies have been adapted to fossil fuels in the previous era and have earned ample preferential treatment, while renewable resources, on the other hand, are an aspect of choice. In economists' opinion, market tax pricing of carbon would allow for more efficient choices between competing technologies to reduce greenhouse gas emissions (Burtraw 2008; Parry and Williams 2013).

The global community is now committed to a consensus on reducing the unhealthful increase in Earth's temperature to 2 °C and to initiate improvements to achieve this goal. This would make it possible for the planet to abolish the burning of fossil fuels in order to achieve an effective zero-carbon emission status. Therefore, the need for the hour is for a radical change in the lives of modern humans on the crusade towards a brighter future.

References

Baumert, K., T. Herzog, J. Pershing. (2005). *Navigating the Numbers: Greenhouse Gas Data and International Climate Policy*, World Resources Institute.

Fawzy, S., Osman, A., Doran, J., Rooney, D. W. (2020). Strategies for mitigation of climate change: a review. *Environmental Chemistry Letters*, pp. 1–26.

Burtraw, Dallas. (2008). Regulating CO2 in electricity markets: sources or consumers? *Climate Policy (Earthscan)* 8 (6), pp. 588–606.

Canadell, J. (2007). Contributions to accelerating atmospheric CO2 growth from economic activity, carbon intensity, and efficiency of natural sinks. *Proceedings of the National Academy of Sciences of the United States of America* 104, pp. 18866–18870.

Dias, G.F. (2006). *Human Ecological Footprint and Sustainability*, Gaia, São Paulo.

den Elzen, M. G. J. & Meinshausen, M. (2006). Meeting the EU 2°C climate target: global and regional emission implications. *Climate Policy* 6, pp. 545–564.

Field, Christopher B., Vicente, R. Barros, David Jon Dokken, Katherine, J. Mach, Michael D. Mastrandrea, (2014). "Technical summary", in Christopher B. Field, Vicente R. Barros, David Jon Dokken, Katherine J. Mach, Michael D. Mastrandrea, et al. (eds.) *Climate Change 2014: Impacts, Adaptation, and Vulnerability. Part A: Global and Sectoral Aspects.* Contribution of Working Group II to the Fifth Assessment Report of the Intergovernmental Panel on Climate Change, New York: Cambridge University Press, pp. 35–94.

Friedman, T. (2005). As Toyota goes . . .. *The New York Times*. http://www.nytimes.com/2005/06/17/opinion/17friedman.html (accessed 15 December 2020).

Giddens, A., (2011). *The Politics of Climate Change*. Cambridge: Polity Press.

Gillett, N., and H. Matthews (2010). Accounting for carbon cycle feedbacks in a comparison of the global warming effects of greenhouse gases. *Environmental Research Letters*, 5, pp. 034011.

Gregory, J. M., Stouffer, R. J., Raper, S. C. B., Stott, P. A. & Rayner, N. A. (2002). An observationally based estimate of the climate sensitivity. *Journal of Climate*, 15, pp. 3117–3121.

Held, D. Hervey, A. and Theros, M., eds., (2011). *The Governance of Climate Change: Science, Economics, Politics and Ethics*. Cambridge: Polity Press.

Helm, D. and Hepburn, C, (2009). *The Economics and Politics of Climate Change*. Oxford University Press.

Helweg-Larsen T. and J. Bull. (2007). *Zero Carbon Britain*. Centre for Alternative Technology, Machynlleth, Powys, Wales.

IPCC. (2007). *Impacts; Adaptation and Vulnerability*. Report of Working Group II, Brussell.

IPCC. (2011). Summary for policymakers. O. Edenhofer, R. Pichs-Madruga, Y. Sokona, K. Seyboth, P. Matschoss, S. Kadner, et al. (Eds.), *IPCC Special Report on Renewable Energy Sources and Climate Change Mitigation*, Cambridge University Press, Cambridge, UK and New York, NY.

IPCC. (2018). Global warming of 1.5 °C. In: Masson-Delmotte, V., Zhai, P., Pörtner, H. O., Roberts, D., Skea, J., Shukla, P.R., Pirani, A., et al. (eds) *An IPCC special report on the impacts of global warming of 1.5 °C above pre-industrial levels and related global greenhouse gas emission pathways, in the context of strengthening the global response to the threat of climate change, sustainable development, and eforts to eradicate poverty*. https://www.ipcc.ch/site/assets/uploads/sites/2/2019/06/SR15_Full_Report_High_Res.pdf (accessed 22 December 2020).

IPCC. (2019). Special report on the ocean and cryosphere in a changing climate. *51st Session of the IPCC*, Monaco, 25 September 2019.

Lacis, A.A. (2012). Greenhouse effect. In: Liu, G., Ed., *Greenhouse Gases-Emissions, Measurement and Management*. Intech, Crocia 5–110.

Levitus, S., Antonov, J. & Boyer, T. (2005). *Geophysical Research Letters* 32, L02604, doi:https://doi.org/10.1029/2004GL021592.

Makhijani, Arjun (2008). *Carbon-Free and Nuclear-Free: A Roadmap for US Energy Policy*. Rdr Books, Berkeley.

Martinot, E. and L. Junfeng. (2007). *Powering China's Development: The Role of Renewable Energy*, Worldwatch Institute, Washington, DC.

Mendenhall, C. D., Karp, D. S., Meyer, C. F., Hadly, E. A. and Daily, G. C. (2014). Predicting biodiversity change and averting collapse in agricultural landscapes. *Nature* 509, pp. 213–217.

Morice, C. P., J. J. Kennedy, N. A. Rayner, and P. D. Jones (2012). Quantifying uncertainties in global and regional temperature change using an ensemble of observational estimates: the HadCRUT4 dataset, *Journal of Geophysical Reearch*, 117 D08101.

Myles, et al. (2009). Warming caused by cumulative carbon emissions towards the trillionth tonne, *Nature*, 458, pp. 30.

Nathan, P., Daithi, A., Peter, A. S. et al. (2008). Attribution of polar warming to human influence. *Nature Geoscience*, 1, pp. 750–754.

Newton, D. E. (2012). *World Energy Crisis: A Reference Handbook*, ABC-CLIO, Incorporated, pp. 334.

OECD/IEA. (2004). *Renewables for Power Generation: Status and Prospect*. OECD/IEA.

Parry, Ian W.H. and Roberton C. Williams. (2013). Efficiency and distributional trade-offs in recycling carbon cap-and-trade revenues. B. E. J. Econ. Anal. Policy.

Rees W. E (1999). The built environment and the ecosphere: a global perspective. *Building Research and Information* 27(4), pp. 206–20.

REN21. (2017). *Renewables 2017 Global Status Report*. Paris: REN21.

Siddall, M., Stocker, T.F. and Clark, P.U. (2009). Constrains on future sea-level from past sea-level change. *Nature Geoscience*, 2, pp. 571–575.

Tsai, W.T. and Chou, Y.H. (2006). An overview of renewable energy utilization from municipal solid waste (MSW) incineration in Taiwan. *Renewable & Sustainable Energy Reviews*, 10, pp. 491–502

United Nations. (2001). *World Urbanization Project: The 1999 Revision*. New York: The United Nations Population Division.

Valentine, S.V. (2011). Emerging symbiosis: renewable energy and energy security. *Renewble & Sustainable Energy Reviews*, 15, pp. 4572–4578.

Varun, R. Prakash and I.K. Bhat. (2009). Energy, economics and environmental impacts of renewable energy systems. *Renewable & Sustainable Energy Reviews*, 13, pp. 2716–2721.

Veritas. Press Release, 2004.

Wang, B., Liang, X.J., Zhang, H., Wang, L., Wei, Y.M. (2014a). Vulnerability of hydropower generation to climate change in China: results based on Grey forecasting model. *Energy Policy*, 65, pp. 701–707.

Wang, B., Nistor, I., Murty, T. and Wei, Y.M. (2014b). Efficiency assessment of hydroelectric power plants in Canada: a multi criteria decision making approach. *Energy Economics*, 46, pp. 112–121.

Wei, Y.M., G. Wu, Liang, Q.M. and Liao H. (2012). *China Energy Report (2012): Energy Security Research*. Science Press, Beijing.

Wei, Y.M., Liao H., Wang, K. and Hao Y. (2014). *China Energy Report (2014): Energy Poverty Research*. Science Press, Beijing.

WWF (2006). Climate Change Impacts in the Amazon – Review of scientific literature, World Wide Fund for Nature. Presentation at the 8th UN Conference of the Parties to the Biodiversity Convention.

2

Advances in Alternative Sources of Energy

Opening New Doors for Energy Sustainability

Jyoti Tyagi

Department of Chemistry, Zakir Husain Delhi College, University of Delhi, New Delhi, India

2.1 Introduction

Energy is regarded as an essential and crucial commodity for the economic growth of a country and more importantly for developing nations. Energy resources whether renewable or non-renewable are consumed to generate electricity to meet the energy demand of a nation (Sorrell 2015). At present, main sources of energy are fossil fuels, namely coal, natural gas and petroleum-based fuels (Balat 2007; Abas et al. 2015). Identification of non-renewable energy sources and past, present and future of these energy sources are exhaustively studied. Industrial revolution has exploited these resources at a rapid rate. Extensive use of these energy resources leads to increase in CO_2 amount on earth. Non-renewable energy sources dependency poses a threat on the earth's environment which ultimately leads to negative impact on global climate (Martins et al. 2019). Further, current consumption of fossil fuels and their continuously increasing demand also endanger biodiversity on the earth (Harfoot et al. 2018). There is a huge pressure on existing energy resources due to exponential increase in the world population. Looking at the reserves of the non-renewable energy resources (Zou et al. 2016) and to achieve the goal of energy sustainability in future, use of renewable energy sources seems to be a viable solution (Güney 2019). So, it becomes important to develop new methods to harness energy from the renewable sources of energy. Further, due to enormous exploitation of conventional resources, a worldwide competition has started for energy production from the renewable energy sources (Ellabban et al. 2014). However, in the past many questions were raised on utility and effectiveness of the renewable energy sources. Can these energy resources really supersede the non-renewable energy sources (York 2012; Stuermer and Schwerhoff 2015)? Looking at energy conversion efficiency of the alternative sources of energy (Yaramasu et al. 2015; Ma et al. 2018; Sahli et al. 2018), it can be realized that these energy sources have a long way to go before replacing fossil fuels as source of energy.

In recent times, worldwide potential research is going in the field of alternative sources of energy. New technologies are evolving to increase the energy conversion efficiency,

Energy: Crises, Challenges and Solutions, First Edition. Edited by Pardeep Singh, Suruchi Singh, Gaurav Kumar, and Pooja Baweja.

reduce the cost of power production and feasibility for large-scale production in nearly all the renewable energy sources sectors i.e. solar (Kannan and Vakeesan 2016; Pandey et al. 2016), wind (IRENA 2019a), hydro (Kougias et al. 2019), geothermal (Olasolo et al. 2016), bioenergy (IRENA 2019b) and ocean energy (Wilberforce et al. 2019). Technologies in the solar and wind energy sector are growing at a faster rate compared with other alternative sources of energy. One of the paradigm shifts in solar energy usage is conversion of CO_2 into fuels and highly valued chemicals (He and Janáky 2020). Such advances will have dual benefits i.e., generating fuels using CO_2 which is a threat to environment and utilizing the renewable solar energy. Further, solar energy is also exploited to resolve the freshwater crises via advanced evaporation systems (Xu et al. 2020). Harnessing onshore and offshore wind energy can become competitive to power generation from the non-renewable energy sources due to evolution of new and innovative technologies in the wind energy sector (IRENA 2019a).

2.2 Need of Novel Research in Alternative Sources of Energy

The world population is expected to be around 9.8 billion by 2050 as projected by the UN's World Population Prospects 2019. This increase in population despite a continuous decline in birth rate globally demands a huge volume of energy resources to accomplish their daily requirements. If we look at the industrial needs of energy resources then also, we can understand the ever-increasing demand due to intense industrialization in the past and will continue to rise in future as well. Reserves of coal, natural gas and crude oil which are available worldwide are predicted to be exhausted by 2300 if consumed at the same rate (Kougias et al. 2019). Geographical location of fossil fuels is not uniformly spread across the world. This uneven distribution of fossil fuels leads to high and unstable prices of petroleum-based fuels. To accomplish the energy demand of the world population, use of renewable resources should also be increased consistently and in an exponential mode so that by 2300 the world can rely only on renewable resources. Further, usage of non-renewable energy resources also increases the carbon emission. Total carbon emission from the consumption of energy resources is around 33,000 million metric tons in 2012 as reported by Global Status Report, 2014 (REN21, Paris 2014). This massive emission of CO_2 leads to global warming and change in the earth atmosphere.

To maintain a balance between carbon emission and usage of energy resources, there is a need for such energy resources which have minimum or zero carbon emission. In the planetary emergency of climate change, there is a need for valuable lessons from the past. Remedy of this problem lies in the usage of renewable sources of energy which are in use from the past few decades but in small percentage. In order to decrease the dependency on fossil fuels, renewable energy sources consumption has to be promoted. At present, if one carefully looks at the various alternative sources of energy, namely solar, wind, biomass, hydropower, ocean and geothermal energy, then it can be realized that fossil fuels cannot be completely replaced by these energy sources. Alternative sources of energy are complementary to non-renewable energy resources so this is the need of the hour to harness as much energy as possible via these sources. For the uniform and sustainable economic growth of a nation, use of renewable energy sources should be enhanced by the policymakers of a country (Bhattacharya et al. 2016). Also, some subsidies must be provided to the industry and end users to encourage the usage of alternative sources of energy. Recently, very promising research is performed in the process of harnessing energy through renewable energy sources in an efficient manner, explained in detail in the following sections of the chapter.

2.3 Recent Advances in Renewable Sources of Energy

Till 2017, global production of electricity was 74% from fossil fuels and nuclear sources and 26% from all the renewable energy sources as reported by the International Energy Agency (www.iea.org). If one looks at the contribution from the different renewable sources, then it was majorly from hydro power (16%) followed by wind (4%), bio power (2%), solar (2%) and others (2%). Figure 2.1 shows the contribution of all the renewable energy technologies and their sub-technologies in electricity generation worldwide from 2010 to 2018.

Further, 2019 data shows that the capacity of power generation by renewable energy sources was 2537 gigawatts (GW) worldwide. In the same year, renewable power generation capacity growth was 7.4% with an addition of 176 GW, slightly lower than that in 2018 (7.9% with an addition of 171 GW). Contribution of hydropower was maximum i.e. 1190 GW. Capacity of wind and solar energy was 623 GW and 586 GW, respectively. Bioenergy share was also noticeable i.e. 124 GW. Geothermal energy and marine energy accounted for 14 GW and 500 MW, respectively. Among the various renewable energy sources, solar

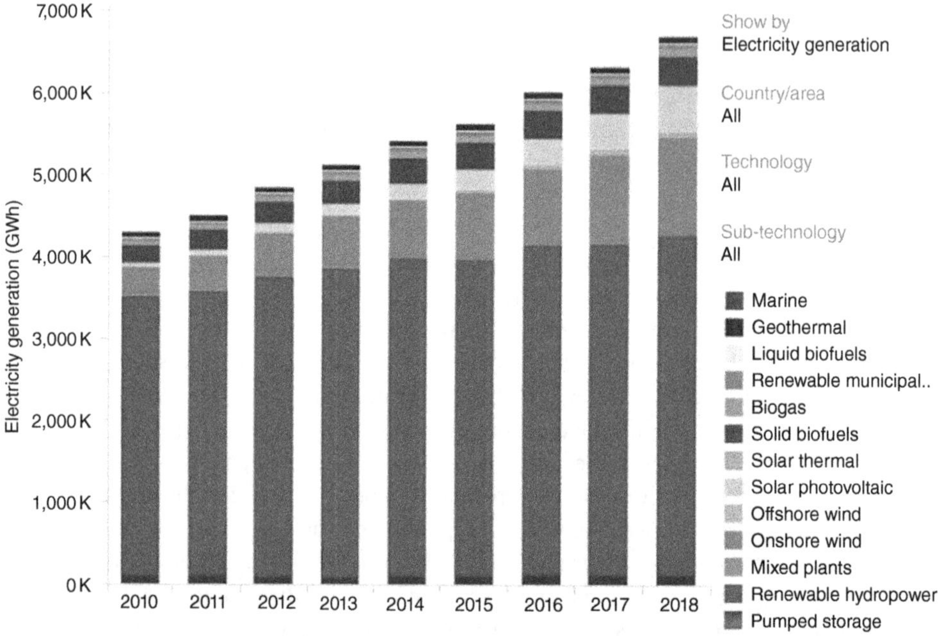

Figure 2.1 Worldwide electricity generation (GWh) between 2010 and 2018 using all the renewable energy technologies. *Source:* www.irena.org © IRENA.

energy growth dominated with a growth rate of 20% followed by wind (10%), hydropower (1%) and bioenergy (5%) as per International Renewable Energy Agency (IRENA) report, 2020 (IRENA 2020b).

Majorly, six countries around the globe are harnessing energy from renewable sources, and these countries are the United States, China, Spain, Germany, Italy and India (REN21, Paris 2014). China leads the overall production of energy from renewable sources. However, the contribution of India is also noticeable among the Asian countries especially in the hydropower sector along with the major contribution from the solar industry. To change the scenario of energy contribution, numerous researches are carried out in the field of utilizing and generating electricity from the renewable energy sources such as solar, wind, biomass, geothermal and others as well. New methods and techniques are being developed to harness energy in an efficient and fruitful manner (Gong et al. 2019). Research based on design, size and new materials formation or invention is more renowned nowadays (Kannan and Vakeesan 2016).

2.3.1 Solar Energy

Solar energy is available in the form of heat and light. Sun emits solar energy at the rate of 3.8×10^{23} kilowatt (kW), from this enormous amount of energy approximately 1.8×10^{14} kW is obstructed by the Earth (Panwar et al. 2011). Worldwide energy requirements can be achieved by utilizing solar energy due to its adequate and free availability. Further, this source of energy is inexhaustible in nature, and output efficiency is also becoming better constantly due to various researches done in the field of photovoltaics. Factors on which harnessing of solar energy depends are its distribution and intensity of radiations, and hence, the efficiency of the solar industry becomes dependent on geographical location of a country (Panwar et al. 2011). Considering the distribution of solar radiation worldwide, it can be clearly understood that Asian countries receive the maximum radiations as compared with the rest of the world. Further, IRENA report 2020 (IRENA 2020b) published that Asian countries continued to dominate the global solar capacity expansion with a 56 GW increase (about 60% of the global expansion in 2019).

Exercising solar energy has dual results as it fulfils the energy demand and does not disturb the ecosystem contrary to the exploitation of fossil fuels. Further, applicability of solar energy is not different for rural and urban due to its easy installation and hence can be easily utilized with the same ease and equal efficiency.

Solar–thermodynamic power plants or concentrating solar thermal power (CSP) and solar photovoltaic (PV) are the two main technologies that can be practically used to transform solar energy into electric power. Nowadays, solar heaters are also becoming very popular which consume heat from the sun to directly increase the temperature of a fluid.

2.3.1.1 Solar Photovoltaic

This is the fastest growing technology with an average increase of 48% since 2002 (Kropp 2009). Six main types of solar PV which are used to transform solar energy directly into electricity are crystalline silicon, thin film solar cells, concentrated solar PV, organic/polymer cells, hybrid solar cells and dye-sensitized solar cells (DSSCs) (Pandey et al. 2016). Apart from these main types, there are some other solar cells based on advanced technologies. Tuning of the band gap of solar cells using nanoscale composites revealed enhanced power conversion efficiency. These are often termed as third-generation PV (tandem cells, impurity-band and intermediate-band devices, hot-electron extraction and carrier multiplication) based on nanostructures. In the field of nanotechnology, carbon nanotubes, quantum dots and 'hot-carrier' flat-plate device based solar PV cells are produced (Razykov et al. 2011).

Under the crystalline silicon solar cells which are one of the categories of solar PV, there are mono-crystalline, poly-crystalline and GaAs-based solar cells. Mono-crystalline is still popular among the manufactures due to high efficiency and easy availability; however, its cost is high for both manufactures and end users. So, other cost-effective options are also evaluated to further decrease the cost, and ploy-crystalline offers a good deal in terms of production cost. Another alternative under the category of crystalline silicon cell is GaAs-based solar cells which provides high efficiency, and these are also low-weight. However, again, its cost is high compared with other types of crystalline solar cells. These are resistant to high heat which makes them suitable for the concentrated PV (used in power generation), hybrid use and space applications (Deb 1998).

Thin-film solar cells are of three types, namely amorphous Si, CdS/CdTe and CIS/CIGS (copper indium gallium selenide). Amorphous Si-based thin-film solar cells are further classified into three types: single junction, double junction and triple junction (El Chaar et al. 2011). Thin-film solar cells require less manufacturing materials which makes them cheaper compared with crystalline Si-based cells. Amorphous Si-based solar cells have higher absorption rate of light (40 times due to non-crystalline and disordered structure) which makes them more popular than CdS/CdTe and CIS/CIGS among the same category owing to the higher efficiency of the former (Pandey et al. 2016). Let us consider a particular example of CdTe solar cell, where an experimental study (Soliman et al. 1996) to enhance the characteristics of CdTe showed that to produce better cells, chemical heat treatment is required. Another example in the same category is CIGS which has been popular because of its laboratory-scale efficiency of about 20.3%. In the area of thin films, there is ongoing research to enhance the efficiency and lifetime of these cells (Pandey et al. 2016).

Concentrated solar PV (CPV) system is gaining popularity nowadays due to its high efficiency which is the major requirement to make it cost-effective technology and also to make it feasible at user end. Different classification of concentrated solar irradiation based on a study (Looser et al. 2014) is shown in Figure 2.2. CPVs are used to generate electricity as well as heating of water to low or medium temperature by extracting heat using active cooling i.e. using heat transfer fluid. For the long-term applications of CPV in different sectors, various studies are conducted worldwide. In a particular example, at the Institute of Nuclear Research in Taiwan, Kuo et al. 2009 worked on the design and development of the 100 kW high-concentration photovoltaic (HCPV) with passive cooling system. This institute receives solar radiation of 850 W/m^2, with this solar radiation system module efficiency reported to be 26.1% with a concentration ratio of 476×.

Organic/polymer solar cells have efficiency between 8 and 10% (Dou et al. 2012). In addition to the low efficiency, these cells are used as an alternative material due to various properties such as low manufacturing cost, low weight and good mechanical flexibility. Globally many laboratories have developed high-performance solar cells using P3HT (poly [3-hexylthiophene]) as the donor and PCBM ([6, 6]-phenyl C60 butyric acid methyl ester)

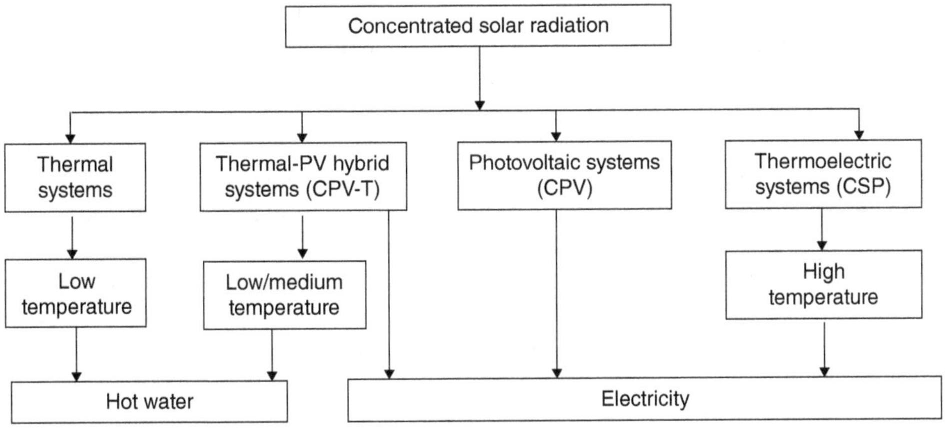

Figure 2.2 Classification of common technologies and system set-up for concentrated solar irradiance conversion. *Source:* Based on ref. Looser et al. (2014).

as the acceptor and/or BHJ (bulk hetero-junction) structures (Bagienski and Gupta 2011; Devi et al. 2011). Further, from an environmental point of view, these types of cells are the most desirable ones.

Hybrid solar cells offer a right blend of inorganic and organic materials. At present, this type of cells are gaining popularity due to cheap processing techniques of organic materials. Choice of organic and inorganic materials opens various options for the chemical synthesis and molecular design of hybrid solar cells (Pandey et al. 2016). Inorganic part of the cell possesses high charge-carrier mobility while the organic part has strong optical absorption which makes them one of good options for energy fulfilment.

DSSCs are simple to manufacture, similar to hybrid solar cells with low cost, low toxicity and ease of production. These cells have the potential in the solar industry in near future. At present, these cells cannot be used in commercialized PV systems owing to their poor efficiency (8–12%), a major concern for the solar cells in this category (Pandey et al. 2016). Recently, a new profitable platinum-free counter electrode for DSSCs has been reported (Ahmad et al. 2014). Graphene nanoplatelets (GNPs) or multi-wall carbon nanotubes (MWCNTs), or various weight % of hybrid GNPs and MWCNTs mixtures were used to make counter electrodes. A marginal increase in conversion efficiency was reported in the study. Using Ru (II) dyes, the efficiency of current DSSCs was reported to reach 12% (Sharma et al. 2018), which is still less as compared with the efficiency of the first- and second-generation solar cells (20–30%).

Several researches are conducted in the techniques for improving the efficiency of PV panels apart from the advancement of cell material itself. Because of fluctuating solar flux, PV systems are not efficient to capture all available energy. So, to capture all the available solar energy, solar tracking (one axis and two axis) is performed. To increase the solar radiation collection, a tracker keeps PV photo thermal panels in a particular position perpendicular to sun rays during the day (Roth et al. 2005). One of the studies concluded that use of two-axis tracking surfaces increases the total daily collection of solar radiation by approximately 41.34% compared with fixed one (Mamlook et al. 2006). A CSP system consisting of parabolic, trough-shaped mirrors focuses the sunlight on the tubes which contain a heat transfer fluid. Temperature is raised to 734 °F by repeated exchange of heat. Heated fluid is used to generate the super-heated steam which powers turbine generators to produce electricity (Devabhaktuni et al. 2013).

A different technique in which solar energy is not concentrated is becoming popular in a short span of time. In this technique, flat plates and evacuated tubes are used as solar energy collectors for heating and cooling purposes. This technique is cost-effective with good efficiency and can be used in low-intensity solar areas. Insulated copper tubes consisting of water or air are used to absorb solar energy. Water or air present in the tubes is heated up before returning to the storage system (Kannan and Vakeesan 2016). In a modification evacuated tube collector is used where heating pipes are shielded by vacuum. This modification is 20–45% more efficient than flat-plate collectors (Mangal et al. 2010).

2.3.1.2 Solar Power Generation

For electrical and mechanical connections, a solar power generation plant has many parts, namely arrays and modules of solar cells and means of controlling systems. Immense research is done in the area of power generation using solar energy at the practical level to evaluate the efficiency, lifespan, cost and durability of such power plants or even small-scale grid systems. For the fulfilment of energy demand of people, use of hybrid power systems (Parida et al. 2011) is suggested where PV panels cannot generate regular electricity for consumption. In hybrid systems, PV systems are combined with hydro or wind turbines and sometimes with diesel or petrol-driven generators for uninterrupted power supply. This hybrid system reduces the usage of fossil fuels.

For power generation using solar energy, different combinations are evaluated from time to time such as hybrid wind/PV or fuel cell power generation systems, wind/PV/battery system and wind/PV/fuel cell electrolyser system. In one of the studies researchers have developed a system by combining PV, wind and fuel cells to transfer maximum power to a fixed direct current voltage bus (El-Shatter et al. 2006). In this study, fuzzy logic control was used to obtain maximum power tracking of PV and wind energy. Recently, this fuzzy scheme was used by different researchers to track the wind and PV energy so that maximum available solar and wind energy can be extracted.

2.3.1.3 Photovoltaic/Thermal (PV/T) Collectors

It is a hybrid system of PV cells and thermal collectors which are flat and tube-shaped material for absorbing heat from sun. PV/T has dual benefits i.e. while generating electricity using PV cells, thermal collectors utilize heat energy from solar radiations which are not consumed by PV cells as well as waste energy from PV cell (Kannan and Vakeesan 2016). Recently, PV/T systems are becoming popular owing to higher efficiency. Numerous researches are conducted from the different aspects to evaluate the performance and efficiency of PV/T collectors (Huang et al. 2001; Tripanagnostopoulos et al. 2002; Zondag et al. 2003). Different materials for PV cell are chosen for research ranging from mono-crystalline, poly-crystalline, amorphous Si and thin-film cells. Many options for collectors are also evaluated which are based on different size and shape of tube (square, rectangular, spiral, round hollow and flat) (Sandnes and Rekstad 2002). For the thermal collector part, different materials such as copper, aluminium, polymer, water and air are being evaluated by many researchers (Sandnes and Rekstad 2002; Tripanagnostopoulos et al. 2002).

In solar industry developments are also taking place to upgrade the solar heaters, improvisation in design and size of solar cells. Further, invention of new materials which can efficiently absorb the light has also been reported (Kannan and Vakeesan 2016).

2.3.2 Wind Energy

Harnessing energy from wind is one of the oldest technologies. Wind power has been the second dominated technology in the domain of renewable energy technologies in recent decades. Due to reduction in cost, its usage is increasing worldwide. Since 2000, wind power has increased at an average compound annual growth rate of >21% (IRENA 2019c). Using wind energy, electricity can be generated by two types of technologies, namely onshore and offshore. As per (IRENA 2020b), worldwide wind-generation capacity including onshore and offshore has increased to 623 GW by 2019 as compared with 7.5 GW in 1997 which is a factor of >80 in the last 22 years. In future, the combination of wind and solar energy can transform the global energy sector. Around 35% of the total electricity requirement can be generated by wind power (onshore and offshore), and it can become one of the major sources of electricity generation by 2050.

Advances in the wind industry can be realized from the milestone achievements in the last four decades. This industry has seen developments with respect to installations, advancement in technologies along with cost reduction. A milestone in the wind industry took place in 2018, with global installed wind capacity of 564 GW and requirement of 1.2 million man power was generated in the sector. In 2019, commercially available offshore wind turbines have reached 10 MW capacity (IRENA 2019a).

Power generation from wind is also geographical location dependent as was solar power generation. Globally, speed of the wind varies in different locations and best locations are the remote ones. Apart from location, size of turbine and length of the blades determine the amount of electricity that can be produced from the wind energy. Further, output of the wind turbine is also proportional to rotor dimensions and cube of wind speed.

2.3.2.1 Onshore Wind Energy Technology

Innovative developments in design and size of the rotor have increased the capacity of wind turbine. At present, capacities of wind turbine have reached approximately 2 MW for the onshore and even more for the offshore (IRENA 2019a). To enhance the growth in onshore wind energy sector, various innovations, advancement in techniques and several practices are adopted in this area as discussed in the following sub-sections.

2.3.2.1.1 Turbine Size and Ratings Important factors for the developments in wind turbine technologies are size of rotor diameter and hub height. Recently, various developments in technology have taken place to manufacture larger-capacity turbines. These advances have increased the efficiency along with reduction in capital and operation costs. Large rotors increase the capacity of wind turbines even in low wind areas. By 2018, rotor diameter has increased to 110.4 m with wind turbine ratings of 2.6 MW from rotor diameter of 50.17 m with 1.0 MW capacity of wind turbine in 2000 (Pérez-Collazo et al. 2015; IRENA 2019d; www.windpowermonthly.com). Expected capacity of the wind turbine by 2022–2025 is 5.8 MW with rotor diameter equal to 170 m. In a particular example, GE has come up with improved onshore turbine technologies rated at 4.8 MW and 5.3 MW, respectively (www.windpowermonthly.com).

2.3.2.1.2 Design and Materials of Rotor Blade Two types of wind turbine designs are horizontal axis wind turbine and vertical axis wind turbine, and the former type dominates in the wind turbine industry. Nowadays, focussed research is carried out to improve the aerodynamic profile of blades as well as enhancement in the quality of materials so that energy production can be increased along with cost reduction of operation and maintenance. Advances in the manufacturing of composite materials leads to better performances particularly in rough and corrosive environments of sea and deserts (Windtrust 2016). Blades made up of composite materials can spin faster and capture winds even at lower speed.

2.3.2.1.3 Power Electronics Optimization To cut the cost of wind turbine installation and operation, optimized design of the various parts of power electronics has been used by different manufactures via concentrating on the several features such as humidity protection, scalability and decreasing the number of components. These features lead to minimizing the failure of power modules, creation of smart and innovative power modules having power density of nearly 30% more than their predecessor and better performance of power electronics by reducing the number of active elements in power modules (Windtrust 2016).

2.3.2.1.4 Smart Wind Turbines These turbines are equipped with novel technologies which control and monitor the turbine. These turbines possess the advanced mechanism to forecast using big data and artificial intelligence along with the automatic regulations of turbines which results in higher energy output (Windtrust 2016). In addition, these digitized turbines also reduce the maintenance costs (www.woodmac.com). Recently, GE Japan has exemplified the manufacturing of such a smart turbine with the help of artificial intelligence which has higher efficiency, low maintenance costs (20% lesser) and higher output (5% more) (IRENA 2019e).

2.3.2.1.5 Recycling of Materials In recent times, recycling of various materials in the wind energy sector has become popular due to reduction in cost for power generation. By increasing the three R's i.e. reduce, reuse and recycle of raw materials, residues, metals and other resources in the wind sector will cut down the overall cost of electricity generation. At present, about 2.5 million tonnes of composite materials are used in the wind energy sector. In the next five years, around 12,000 wind turbines will be out of service in Europe (www.recycling-magazine.com) which will produce large amounts of materials that need to be recycled. For the recycling of materials, mechanical process i.e. cutting the turbine blades into smaller slices for easy transport or thermal processes such as combustion or pyrolysis provide viable options (WindEurope 2017). For the circular economy along with the production of new blades, reuse of decommissioned blades after some processing should be considered (WindEurope 2017). For example, the Dreamwind project (Designing REcyclable Advanced Materials for WIND energy) is working towards the development of a chemical substance which can separate the glass (expensive component) from the plastic fibres by heating the composite materials at 600 °C. After the separation and cleaning, the glass component produced from the recycling process can be used in making new blades for wind turbines (www.dreamwind.dk).

2.3.2.2 Offshore Wind Energy Technology

Offshore wind power generation is an emerging giant technology and has seen more potential owing to advances in technology. Due to limitation of space availability onshore, growth of offshore wind farms is gaining more popularity. In many European countries, offshore wind power projects are in trend. Offshore wind technology is leading in the wind energy sector because it is exploring more resources further offshore. In contrast to onshore wind turbines, installation of offshore wind turbines has many advantages, namely availability of more space, less complaints of noise and visual interference, and winds are stronger and even more regular in the offshore region (IRENA 2019a). In addition to the advantages, offshore wind also comes with some disadvantages i.e. their cost is high as compared with onshore wind turbines and they are difficult to install and maintain due to harsh and changing weather conditions of coastal regions.

The major offshore wind technologies based on the maximum overall potential are future-generation turbines; floating foundations; repowering of sites; integrated turbine and foundation installation; high-voltage direct current (HVDC) infrastructure; direct current (DC) power take-off and array cables; and site layout optimization (IRENA 2019a). These technologies have beneficial impacts ranging from high, medium and low on five different aspects which are decreasing the cost of energy, increasing grid integration, opening up new markets, reducing environmental effect, and improving health and safety levels.

2.3.2.2.1 Future-Generation Turbines Technology It is one of the most relevant technologies in the offshore wind energy sector based on the development of turbines. With the important developments in size of blade, drivetrain and control technologies, wind turbines are becoming more reliable and larger along with the increase in capacity ratings. Rotor diameters of offshore wind turbines have increased to 148 m with an average rated capacity of 5.5 MW (in 2018) from 43.73 m with an average rated capacity of 1.6 MW (in 2000) (IRENA 2019a). Thus, in the last two decades, the average size of offshore wind turbines has grown by a factor of 3.4. Size of turbines is expected to grow (rotor diameters >230 m) with turbine ratings between 15 and 20 MW by 2030. Due to the increased size of turbines, there will be a smaller number of turbines in a wind farm for a particular rated capacity so it will reduce the cost, and impact on the environment will also be less along with some other advantages.

2.3.2.2.2 Floating Foundations It is another major technology in the offshore wind energy sector. In this technology, turbines are rooted in the seabed by monopile or jacket foundations, and these are restricted to waters which are <60 m deep. The spar-buoys, spar-submersible and tension-leg platforms are three main designs for turbines which are under development and have been tested (IRENA 2019a). Under this technology, turbines are installed in water depths of up to 40 m and these are about 80 km away from the shore. If one looks at the largest potential markets for offshore wind sector i.e. Japan and the United States (US), then they have less shallow water sites, which makes current floating foundations as the limiting technology. On the basis of recent advances which are taking place across various regions, floating wind farms can contribute 5–15% in total offshore wind capacity installed worldwide (~1000 GW) by 2050 (IRENA 2019a). Along with the ease of turbine set-up, floating foundations also come with environmental benefits such as less invasive activity on the seabed during installations as compared with fixed bottom designs.

2.3.2.2.3 Repowering of Sites In this technique, turbines and their foundation and array cables are replaced by larger units which are spaced at larger distances. This is an alternative technique to continue operation with the same configuration. This technology faces a major challenge due to the rough and corrosive offshore environment. Further, to reduce the cost of repowering wind farms, transmission assets can be retained after some renovation. Retainment of transmission assets lowers the cost of farm by decreasing the energy cost for the repowered phase. This will also reduce the levelized cost of electricity during the entire lifetime of the farm site (IRENA 2016).

2.3.2.2.4 Integrated Turbine and Foundation Installation Using this technique, wind turbines are assembled and pre-commissioned in nearby place and after that a complete and integrated turbine which includes a rotor, tower and foundation is installed in a single offshore operation (IRENA 2016). Thus, using this technique most of the offshore installation operations can be put to an end. With the help of customized vessels or tugboats, the combined turbine foundation is taken to the wind farm site. Use of this technology in the floating foundation technique will further make the usage of the latter more promising. As far as fixed foundations are concerned, use of gravity-based foundations that can be floated out and sunk at the site is more useful. Installation cost as well as the exposure to health and safety risks decreases by these innovative techniques. Projected commercialization of this technique is by 2025 with more advances in technology to fulfil the requirement of large turbines.

2.3.2.2.5 HVDC Infrastructure This technique is based on the electrical interconnections. For the offshore wind farms which are located between 80 and 150 km, HVDC transmission is preferred over HVAC (high-voltage alternating current). At present, this is an expensive technique; however, with the help of invention in the offshore installations and component techniques, its cost can be reduced in future. This technique comes with an advantage i.e. installation of a wind farm in a further offshore region which has rich wind resources. In future, this technique can lead to higher annual production of energy with less planning. Further this technology opens new doors for the regions where near-shore developments are not possible (IRENA 2016). With the invention of direct current nodes in upcoming time, a multi-nodal network can be built which is required for integration of this technique. With point-to-point grid connections used on few projects in European waters, commercialization of this technique is ongoing.

2.3.2.2.6 DC Power Take-off and Array Cables Same as HVDC infrastructure, this technique is also based on the electrical interconnections. It is in its infancy; however, using this technique, the industry can get rid of the other half of the conventional turbine power conversion system which converts back to grid-frequency alternating current (AC). This technique saves the capital cost and enhances reliability. By shifting towards DC collection, the array of cable cores decreases to two from three and the amount of material also reduces by 20–30% (IRENA 2016). Progression of this technique is happening at a slower rate due to non-acceptance of higher-voltage AC array cables by the industry.

2.3.2.2.7 Site Layout Optimization For energy production from wind farms, different options of site layout are analysed with the help of software tools commonly described as wind farm design tools. One of the tools is WAsP model which calculates wind flow of the site (IRENA 2019a). With the help of these tools, various predictions are done by calculations, namely variation in wind speed with height, variation in wind speed over the site area and the wake interaction between wind turbines. In this technology, usage of software tools enables more informed and complete layout of wind farm which enhances the characterization of wind resources, aerodynamic wake effects, meteorology oceanic climate and seabed conditions (IRENA 2016). If installation process methods and foundation technique are further combined with layout, then energy production from the wind farm will be more effective.

2.3.2.2.8 Other Techniques Along with major techniques as discussed above, there are some techniques which are attaining a commercial stage. One of the techniques is 'Airborne Wind' in which flying turbine devices which have 90% less mass than traditional wind turbines are used (IRENA 2016). At 300 m above altitudes, these turbines can generate power from the low-speed winds. This technique covers a range of concepts such as rigid wings and airborne rotors. This technique can generate electricity at a lower cost due to less material usage, low capital expenditure and higher energy production compared with conventional turbines (IRENA 2016).

Apart from innovations in technologies, developments are also taking place in operation, maintenance and services (IRENA 2016). Due to higher growth rate in offshore wind turbines, there is a need of advances in the technology systems to keep track of turbines and to avoid failures. One such operation and maintenance tool is 'Condition Monitoring Systems' which help operators to forecast any problem with the working of wind turbines or failure of any components so that replacement can be done in advance to avoid any accident (Froese 2017).

2.3.3 Hydropower

Energy resulting from the running water is termed as hydropower. Harnessing energy from water is one of the ancient methods used by the Greeks to run wheels for grinding grain. It is an inexhaustible and clean source of energy without consumption, pollution, emission of waste gas and other wastes. Utilizing hydropower is one of the efficient ways to achieve low-carbon emission and meet the Paris climate goals. Its contribution in power generation is maximum among all renewables, World's largest hydropower plant in China – 'Three Gorges Dam' with 22.5 GW capacity is a great example of this renewable source. First time in the history, contribution of hydropower reached just 50% in the year 2018 due to more growth in solar and wind energy sector. In recent times, its growth rate (2% in 2018) continued to slow, with only China adding a substantial amount of 8.5

GW in 2018, as per IRENA report (IRENA 2019f). Hydropower plants are classified into two types, namely with dams and reservoirs or without. Hydropower plants with dams and reservoirs produce electricity at a large scale while the other type produces at a smaller scale.

Hydropower technology has reached a mature level as compared with other renewable sources, namely solar and wind. Thus, advances in identification and implementation of radical design which can change the operative method of hydropower have less potential. However, development of new methods in design, planning and operation of a hydropower station still has many opportunities (Kougias et al. 2019). Recent advances in the hydropower sector show increment in efficiency, flexibility of operations, durability and cost reduction of installation, operation and maintenance. Developments in technologies are needed for this sector to respond to changing climate conditions, expanding markets and variabilities of electrical power systems. Further, novel developments for upgradation and refurbishment of current facilities in hydropower are required in accordance with the environmental standards.

2.3.3.1 Flow Control Technologies

The flexibility in electricity generation from the wind and solar renewable source poses some challenges on power generation from hydro. To fulfil the demand of variable energy production along with limited capability for energy storage, hydraulic turbines need to operate at a wide range and changing conditions (Valero et al. 2017). Recently, some developments have taken place in control technology to reduce the flow instabilities which occur due to self-induced instability in the operation of hydro turbines. Several techniques have emerged which decrease the flow instabilities in hydro turbine operation. These comes under passive and active control techniques (Kougias et al. 2019). Number of techniques are reported under these two types of control technologies. Some other techniques have also been reported by a recent research known as 'Magneto-rheological control techniques' (Trivedi et al. 2015) which uses magneto-rheological brake to mitigate flow instability along with the decreasing speed of the runner.

2.3.3.2 Digitalization of Hydropower Plants

Hydropower plants were established and designed decades ago, so to meet today's requirement of variable energy demand and face the present environment conditions, they need digitalization of operative methods used in existing plants. Present hydropower plants also require developments with respect to flexibility which can be done by providing storage capacity and advanced system services. However, practically there are some hydrodynamic phenomena which are associated with the working of hydropower plants which leads to limit their flexibility. Digitalization in the hydro sector will change the methodology for design, development, operation and maintenance of new hydropower plant projects. Thus, renovation and digitalization of various equipment of existing plants offer the opportunities of enhanced operations. These developments will increase the lifespan of power plants along with increased efficiency and power production (Kougias et al. 2019). The effect of digitalization can be best understood by the projected increase in capacity of the world's 1225 GW installed hydropower by 42 terawatt-hour (TWh) annual production. This increment will help in achieving the low-carbon emission and thus maintaining a balance in the ecosystem. On the basis of recent advances reported by various research groups for digitalization of hydropower plants, one of the studies (Kougias et al. 2019) concluded that a 'Digital Avatar' of hydropower plant dynamics can be developed by following a multidisciplinary approach which covers the hydraulic machinery, electrical power systems and its associated control and components fatigue modelling.

2.3.3.3 Evolution in Hydroelectric Energy Storage

Continuous increase of variable renewable energy (solar and wind) in electrical power system (EPS) leads to advances in the field of energy storage. For energy storage, pumped-storage power plant (PSPP) is a well-developed technology with continuous improvements taking place in PSPP (Kougias et al. 2019) in order to reduce response time for transition mode i.e. from pump to turbine and vice versa. Recently, two technologies are evolving under PSPP which can play an important role in fast energy storage systems, namely flywheels and supercapacitors (Sarasúa et al. 2016; Gevorgian et al. 2017). Synergy between fast energy storage systems and hydropower operation permits an improved frequency control in EPS. Integration of these technologies into the operational hydro plants can be easily done in a span of a few months which can provide advantages in voltage control in the proximity of the hydro plant. Coordinated operation of fast energy storage systems and hydropower plants is also possible for one or set of hydro plants which are connected to the transmission power system in other geographical locations. As an example, globally, by the end of 2017, total installed hydropower capacity including pumped storage was 1267 GW (IHA 2018) as evaluated by International Hydropower Association. Now, assume that every operational hydro plant and PSPP are connected by fast energy storage systems with 5% (assumption) of the installed capacity, then new 65 GW fast energy storage systems are needed to be manufactured and installed across the world.

A new concept in pumped energy storage is emerging, known as underwater pumped hydro energy storage (UPHES) owing to advance research in the energy storage area. Its technical feasibility was investigated between 2008 and 2011 (Kougias et al. 2019). In contrast to the PSPP, it is not restricted to specific geographical locations. In UPHES, sea acts as the upper reservoir while a hollow deposit situated at the seabed is the lower reservoir. Seawater enters the deposit, which in turn drives the turbine and leads to electricity production.

2.3.3.4 Technology Evolution: Small-Scale Hydropower Plants

Construction of large-scale hydropower plants poses a great threat on the environment, balance of ecosystem and the life of aquatic species. Small-scale hydropower plants offer a solution to these persisting problems with large-scale hydropower plants. Further, small hydro plants do not disturb the flow of the river to generate electricity which makes them environment-friendly. At present, small hydro plants are not profitable; however, it can become cost-effective with the usage of digitalized operation and control technologies (Kougias et al. 2016a) along with combination of other variable renewable energy technologies (Kougias et al. 2016b).

2.3.3.5 Gravity Hydropower Converters

Gravity hydropower converters include hydrodynamic screws and gravity water wheels (Quaranta and Revelli 2018). Gravity water wheels are classified into three types (based on the head differences and the maximum flow rate per metre width), namely overshot, breastshot and undershot. Efficiency of gravity water wheels ranges between 75 and 85% (Quaranta and Revelli 2018), and their cost is lower than that of Kaplan turbines and hydrodynamic screws. With improved designs owing to recent developments, these hydropower converters are becoming cost-effective and the effect on fish migration is decreasing along with better efficiency. Installation of these water wheels at old mill sites and in canals leads to electricity production which makes them more profitable (Quaranta and Revelli 2018). The hydrostatic pressure machine (HPM) and the turbine water wheel (TWW) are the examples of the novel water wheels which are in their primary stage. Hydraulic efficiency of HPM is 60–65% with an advantage of using running water without any canal drop (Kougias et al. 2019). The TWW finds its usage in head differences (up to 6 m) of an overshot water wheel and flow rates (few m^3s^{-1}) of an undershot water wheel (Helmizar 2016).

2.3.3.6 Pump as Turbines (PAT)

When a hydraulic pump operates in reverse mode i.e. as a turbine, it is termed as PAT. With the help of a connected induction motor which works as a generator, PAT generates electricity (Agarwal 2012) opposite to what a pump does. Worldwide, pumps are produced at a massive scale, so if they are used as turbines, then it comes with many benefits such as reduced dimensions, simple maintenance along with easy availability of spare parts. PAT is 5–10 times cheaper than the conventional turbines along with low installation cost (Novara et al. 2017). Using PAT in small hydro plants having installed capacity <100 kW makes them cost-effective as the cost of a turbo-generator unit is normally 35% of the total cost of the plant (Ogayar and Vidal 2009). Along with many advantages over conventional turbine, PATs have some disadvantages also, such as lower/lesser peak efficiency, design and operation uncertainties and lack of in-built regulation devices. At the ground level, contribution of PATs in the market is very small partially due to lack of information or awareness from its manufactures and hydropower consulting firms (Kougias et al. 2019). PATs find its significant application in rural and remote off-grid power generation projects where conventional turbines cannot be used due to high cost and low availability (Agarwal 2012).

In addition to the evolution in technology in small-scale hydropower plants to harness more energy, there is an opportunity of development in existing small dams. Some small dams, established in rural areas, are non-powered dams as the aim was to fulfil other needs of rural areas such as irrigation and drinking water supply (Kougias et al. 2014). Thus, the transformation of small dams to small hydropower plants provides a good opportunity for power generation along with the reduction in cost and saving time.

2.3.3.7 Developments in Fish-Friendly Hydropower

Hydropower plants have some adverse impacts on the environment and preservation of ecosystems. Various studies have been conducted to understand the effect of hydro plants on aquatic species populations, especially fishes. In preservation of ecosystems, downstream and upstream migration of fishes is very significant. Recent advances in the development of fish-friendly turbines along with progress in water-lubricated bearings (used in turbines) have been reported by several research projects (Hogan et al. 2014; Quaranta et al. 2017; Fjeldstad et al. 2018). Gravity water wheels and hydrodynamic screws which come under low head gravity turbines are fish-friendly as opposed to conventional turbines. However, these turbines can be used in small hydro plants (very low head sites). For the large-scale hydropower plant, two different approaches have emerged to allow fish migration, namely fishways facilities and usage of fish-friendly turbines (Kougias et al. 2019).

Conventional turbines hinder the migration of fishes (upstream and downstream) so to allow the migration, hydraulic structures are used, termed as fishways (Quaranta et al. 2017). For making eco-efficient fishways having optimal design, more studies are needed to understand their swimming ability and at what position inlet or outlet should be installed so that fishes can use them efficiently (Fjeldstad et al. 2018).

In the high head hydropower plants, fishways find some restrictions, so in order to overcome these restrictions, some advances are taking place in the turbine designs. For safe passage of fishes through hydraulic turbines, a technology termed as fish-friendly turbine has emerged, owing to the developments in the area of advances in turbines. Recently, two different turbines with better designs have been introduced, the Alden turbine and the Minimum Gap Runner turbine. Usage of these turbines leads to more power generation, reduction in fish mortality rate and injury along with better downstream water quality (Hogan et al. 2014).

2.3.4 Geothermal Energy

Geothermal energy is generated within the sub-surface of earth in the form of heat. Direct use of this energy is for heating purposes (Lund and Boyd 2016), and/or it can also be harnessed for electricity generation using a geothermal power plant (Bertani 2016). Compared with the intermittent sources of energy (solar and wind), geothermal energy is constant around the year and it is available worldwide. Electricity generation from the geothermal energy comes with many benefits as compared with some other renewable energy sources such as higher capacity factors (>90%), flexibility in power production, ability of efficiently providing baseload electricity, low cost of electricity generation due to lower cost of operation, lower impact on environment along with less CO_2 emissions and land usage is also small (Geirdal et al. 2015). Electricity production from geothermal has lower life-cycle greenhouse gas (GHG) emissions as compared with fossil fuels (https://archive.ipcc.ch/pdf/special-reports).

Despite many advantages of geothermal energy, its share in worldwide energy production is very small owing to many challenges such as substantial initial capital cost, longer time period before a geothermal plant becomes operational and high risk at early stage (pre-survey, exploration and test drilling) (IRENA 2017). As mentioned earlier in Section 3, contribution of geothermal energy was 14 GW by the end of 2019 which is a relatively small percentage (<0.6%) of the total generation capacity of all the renewable sources of energy.

Geothermal energy has tremendous potential for development in near future owing to the huge amount of energy present within 10 km of the earth's surface which contains

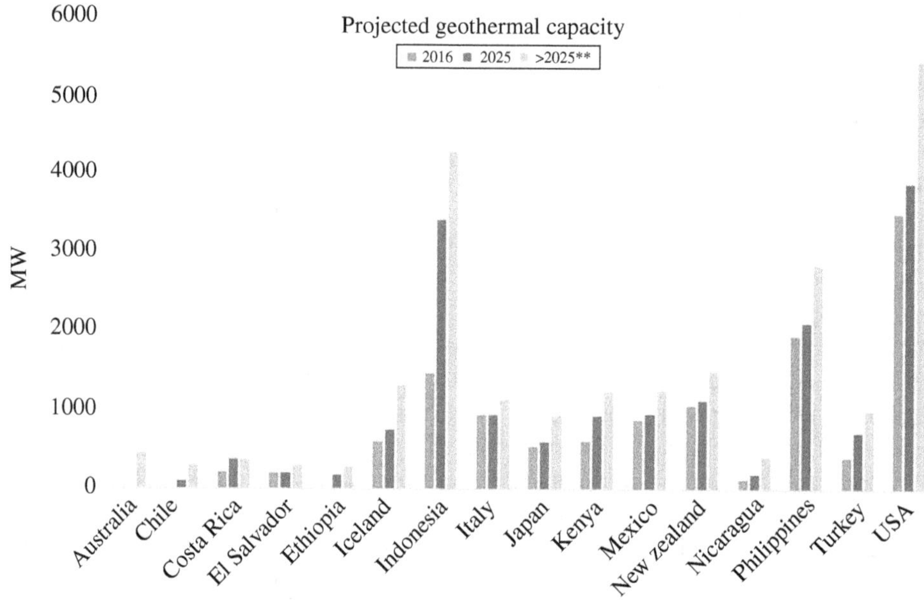

Figure 2.3 Comparison of geothermal power capacity (MW) by different countries in 2016 and projected values in 2025 and beyond 2025. *Source:* Based on data in Ref. (IRENA 2017).

50000 times more energy than all the fossil fuel reserves existing globally (Shere 2013, pp. 201). Depending upon the amount of heat present in a geothermal reservoir, different technologies are used for power generation. Medium or high-temperature resources are required for power production which are generally located in the vicinity of tectonically active regions. To transfer the heat present within the earth's surface, water or steam is used. At present, four different technologies are used for power production from geothermal energy, namely dry steam plants, flash plants (single, double and triple), binary plants, and combined-cycle or hybrid plants (Long et al. 2003). These technologies are based on exploitation of conventional geothermal resources.

Recently, due to more developments in enhanced geothermal systems (EGS), which are the deep geothermal resources, power production from these types of plants will add a new dimension in the geothermal energy sector. Power generation will increase by many folds and can reach to 70 GWe (installed capacity) by 2050 (Lu 2018). Figure 2.3 shows the geothermal power capacity of some countries in 2016 along with projected values in 2025 and ahead of 2025 (IRENA 2017).

2.3.4.1 Direct Dry Steam Plants

Direct dry steam geothermal power plant uses steam which is at 150 °C or higher, produced from the production well, to generate electricity with the help of geothermal steam turbines. Owing to design, these turbines effectively use the low-pressure and high-volume fluid produced in the steam field for conversion. Usually, these plants make use of condensing turbines, and thus produced condensate is re-injected (closed cycle) or evaporated in wet cooling towers (https://iea-etsap.org). The power generation capacity of these plants lies between 8 and 140 MW (www.platts.com).

2.3.4.2 Flash Power Plants

At present, most of the operational geothermal power plants are flash power plants. These plants work similar to dry steam plants; the only difference is steam production step. Steam is produced from flashing, a separation process for the two-phase fluid, and then passed into the turbine. Flash plants are of three types, single, double and triple depending upon the capacity of the plant. In a single flash plant, condensate resulting from the turbine is re-injected into the well while in a double-flash plant, the condensate is directed for further separation at lower pressure to generate more steam. In a triple-flash plant, the process of flashing is repeated one more time. Flash power plants are suitable for those reservoirs which possess well temperature >180 °C (IRENA 2017). Triple-flash plants have the maximum power capacity (60–150 MW) followed by double (2–110 MW) and then single having the least capacity between 0.2–80 MW (www.platts.com).

2.3.4.3 Binary Plants

These plants are established for the reservoirs which possess well temperature between 100 and 170 °C. These plants use a process fluid which obtains heat from the geothermal fluid through heat exchangers in a closed loop. Depending upon the well-matched boiling and condensation points of the geothermal fluid, different process fluids can be used such as ammonia/water mixtures used in Kalina cycles or hydrocarbons in organic Rankine cycles (https://iea-etsap.org). The capacity of these plants varies between <1 MW and 50 MW (www.platts.com).

2.3.4.4 Combined-Cycle or Hybrid Plants

In these plants, an additional Rankine cycle is employed to harness more electricity from the heat generated from the binary cycle, thus increasing the efficiency of the plant. These plants have power capacity up to 10 MWe (DiPippo 2015). On the other hand, hybrid plants make use of an additional source of heat such as CSP to enhance the efficiency of the plant by increasing the temperature of geothermal fluid. For example, a hybrid plant in Italy uses biomass energy to increase the temperature of brine and the Stillwater project in the United States which has combined CSP and solar photovoltaics with a binary plant (IRENA 2017).

2.3.4.5 Enhanced Geothermal Systems (EGS)

Conventional geothermal plants use the heat reservoirs which are present at intermediate (0.1–4 km) or shallow (up to 0.1 km) depths (Olasolo et al. 2016). Heat content of these reservoirs is limited so to access more heat from deep geothermal reservoirs (4–5 km or more), it has been proposed that reservoirs possessing huge amount of heat can be created artificially. This technique of creating large heat reservoirs using advanced drilling technology to make artificial fractures in the deep rocks having low permeability is known as EGS (Johnston et al. 2011). Figure 2.4 depicts the basic outline of a geothermal power plant using EGS technology. EGS decreases the dependency on conventional hydrothermal reservoir along with the generation of large quantities of

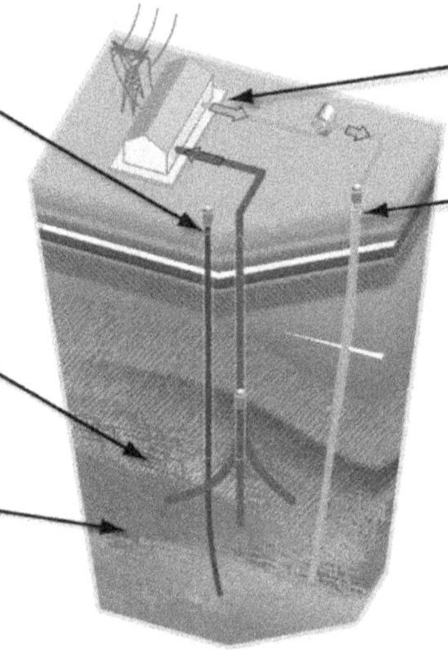

Figure 2.4 Outline of an EGS power plant. *Source:* IRENA (2017).

electricity. Hydraulic or chemical stimulation is used to create artificial fractures to make the flow of water easier through the rocks (less permeable) which are at greater depths. In the process of stimulation, water and a small amount of chemicals are injected to create or reopen fissures in the rocks that exist in greater depths. When the injection pressure decreases, there is a probability that artificial fractures in the rocks are blocked so proppants are added to avoid the closing of fractures. Binary power plants are used in the EGS technology to generate electricity using the hot brine. Since there is no natural flow of water in the deep hard rock so to exploit the heat from these hot rocks, water is used as the working fluid. To maintain the constant pressure and production, brine solution has to be re-injected into the well, thus avoiding any release of air during the lifespan of the power plant. By the end of 2018, there were 18 important EGS sites around the globe which are harnessing electricity using the advanced drilling techniques (Lu 2018). Recently, due to advances in the area of EGS systems, some techniques are used in EGS plants to make their performance better such as use of dense fluids for hydraulic stimulation and CO_2 as the working fluid.

By making use of hydraulic stimulation, artificial fractures are created in deep and hot rock. These fractures increase permeability of the rock via increasing the natural pathways (Olasolo et al. 2016). While creating fractures in a rock, there is a high probability that this hydraulic overpressure may cause micro-seism in the vicinity. So, use of dense fluid for hydraulic stimulation is one of the options to control overpressure at the surface. Further, use of dense fluid has many advantages such as increased and improved underground circulation pathways; less permeability of these pathways; and better mass flow ratio and heat extraction ratio (Olasolo et al. 2016).

Use of CO_2 as a working fluid is more efficient at medium and high pressures while at low pressure, water is a better working fluid. Due to its high heat extraction ratio

(50% higher) than water along with its better performance at low temperatures, its use as working fluid in the EGS has increased (Olasolo et al. 2016).

Many developments are taking place with respect to software packages which are used for estimating and simulating costs of the EGS plants. GEOPHIRES and EURONAT are the most suitable software packages (Olasolo et al. 2016). EURONAT is a European and GEOPHIRES is the US software package. These packages have their own pros and cons; however, GEOPHIRES has the unique feature of being capable to simulate cost for electricity generation as well as for direct use heating and combined heat and power.

Worldwide, installed capacity of geothermal can be enhanced by using some advances which increase the efficiency of operational plants. Low-temperature bottoming cycle is one such way which increases the efficiency of power generation by making use of a binary cycle in a conventional flash power plant. Another way is co-generation, this technique uses the heat of condensate to raise the temperature of different water source before re-injecting into the well. To cut down cost of electricity production from the geothermal plants, use of co-produced resources i.e. by-product of some industrial methods as a geothermal fluid offers a viable option (IRENA 2017).

Supercritical geothermal systems are the geothermal reservoirs where fluid is present in its supercritical state i.e. at very high temperature and pressure. Use of these supercritical geothermal systems as the wells for injection can increase the efficiency of current power plant (Friðleifsson et al. 2015). Another option is establishing a plant which directly uses the supercritical systems as the heat reservoirs. These plants will have better economic performance than the traditional plant due to high-temperature wells (IRENA 2017).

2.3.5 Bioenergy

Bioenergy is one of the alternative forms of energy derived from biological sources and/or residues i.e. biomass. Thus, source of energy is biomass. Use of bioenergy is classified mainly into two types, traditional/conventional and modern (Gupta et al. 2014, pp. 1–17). Conventional use is the direct utilization of source i.e. biomass (e.g. wood, agricultural waste, traditional charcoal etc.) combustion to obtain energy. Modern use is the indirect use i.e. conversion of biomass using different technologies into biofuels. Apart from biofuels, other recent bioenergy technologies includes bio-refineries (a range of bio-products are produced consuming biomass), biogas (obtained via anaerobic digestion of bio-residues) and wood pellet heating systems. Bioenergy can be stored for a long period as opposed to other renewable energy sources. By 2015, its share in total final energy consumption was nearly 10%, and 1.4% in worldwide electricity production. Approximately, 80% of bioenergy is consumed traditionally in the developing countries i.e. for cooking, space heating, and lighting (Lund et al. 2016, pp. 36–38). In 2019, bioenergy power generation capacity was 124 GW with a 5% growth in the year. Its contribution is roughly 5% in the worldwide renewable generation capacity, and China contributed more than half (3.3 GW) of new capacity added (6 GW) in the same year (IRENA 2020b).

Large-scale production of biofuels can be achieved when these are produced using agricultural crops, crop residues, agro-industries wastes, forests residue etc. At present, production cost of biofuels is high, conversion efficiency is low and its cost-effectiveness is also low compared with other energy sources. To mitigate these issues, scientific community is doing lot of research worldwide. Advances are taking place in technologies for pre-treatment of lignocelluloses and cost-effective production method for biofuels with no or minimum by-products (Lund et al. 2016, pp. 36–38). Bioenergy can significantly contribute in energy sector in the twenty-first century. It will be a key component for the development of rural areas, agricultural sector and forestry.

2.3.5.1 Biopellets and Biogas

Biopellets are the simplest form of bioenergy yet competitive to crude oil or natural gas for the household uses in Western countries. Research in pelletization process of leaves in order to use them as additives or substitute to wood provides an alternative for biopellets production (Verma et al. 2012). Transforming biomass to biogas is a modern use of bioenergy, anaerobic digestion of biomass using a digestor leads to formation of biogas. Different types of digestors used for the anaerobic digestion are fixed dome, floating drum and plug flow (Gupta et al. 2014, pp. 1–17). Chemical composition of biogas is a function of the type of feedstock, design of digestor used and processing conditions (Lund et al. 2016, pp. 36–38). Replacing wood and coal with biogas in rural areas will decrease the emission of GHGs (Pathak et al. 2009). Moreover, biogas and biomethane produced from medium and large-scale plants can substitute the natural gas usage as fuel in transportation, leading to decarbonization (IRENA 2019b).

2.3.5.2 Bioethanol and Biodiesel

Most frequently used biofuel is bioethanol. It can be formed from three types of feed stocks, namely sugar-based (most common), starchy and lignocellulosic (require pre-treatment). Production of bioethanol from sugar-based crops was the most common method till 2003 and contributed nearly 60% in the worldwide production (Gupta et al. 2014, pp. 1–17). After sugar-based crops (sugarcane, sorghum etc.), starch feedstock (corn, wheat, barley etc.) is used for bioethanol production followed by lignocellulosic feedstock (wood, straw, corncob etc.). Chemically lignocelluloses are mainly composed of carbohydrates (cellulose and hemi-cellulose) and lignin along with some other quantities of biomass. Owing to the varying composition of lignocelluloses, several pre-treatment processes were explored to obtain fermentable sugars. Advance researches are conducted for the various pre-treatment processes because usage of lignocellulosic feedstock can lead to cost-effective biofuel (Lund et al. 2016, pp. 36–38). Pre-treatment process can be biological, physical and chemical depending upon the source and means of conversion procedure. Among the different methods for pre-treatment, biological method which is a microbial process consumes less energy compared with the other two processes and moreover, a safe and environment-friendly process (Potumarthi et al. 2013). Physical pre-treatment can be done via mechanical comminution, steam explosion, ultrasonic radiation and extrusion process (Gupta et al. 2014, pp. 1–17). Under the chemical pre-treatment methods, acid hydrolysis, alkaline hydrolysis, ammonia hydrolysis and ozonolysis are different available methods. Use of bioethanol-blended gasoline in automobiles offers a viable option to cut down the use of petrol. This will also reduce the GHG emission along with contribution in achieving goals of Paris agreement.

Biodiesel is another biofuel that can be produced from oil-based feedstock, animal fat wastes and waste cooking oils. Animal fat wastes are commonly obtained from chicken, cow, pork lard, fish and other animals. Oil-based feedstock can be edible (soybean, corn etc.) or non-edible. But, according to food first principle, use of edible feedstock is not promoted worldwide. However, continuous evolving technologies for converting biomass to fuels have enabled the bioenergy industry to utilize more of lignocellulosic biomass. Technological developments open doors for new types of feedstock for bioenergy production e.g. forest biomass, agricultural residue, perennial grasses and trees and even municipal waste (Lund et al. 2016, pp. 36–38).

2.3.5.3 Advanced or 2G Biofuels

Currently, biofuels production is based on edible feedstock termed as conventional biofuels or first-generation (1G) biofuels. In the last 20 years, conventional biofuels were promoted by the policymakers in view of changing climate conditions. However, after 2007, various discussions have started worldwide on the sustainability issues of these biofuels (IRENA 2019b). Biofuel usage was raising concern on food security, price of food and feed, and land use change. Production of 1G biofuels has negative impact on food resources, and there is a high probability of land use change and indirect land use change because of high demand of edible feedstock. Recent advances in production technologies lead to advanced biofuels or second-generation (2G) biofuels, future of sustainable bioenergy (IRENA 2019b). Advanced biofuels are derived from non-edible feedstock and other biomass such as agricultural waste, animal fat wastes and municipal wastes etc. On the basis of production method adopted by advanced biofuel industry, they can be classified into four categories which are tabulated in Table 2.1.

Potential research is going on in search of algae (3G) biofuels which are based on algae consumption for biofuel production. Recent study (www.energy.gov) has found that algae could be richer in biofuel production compared with conventional feedstocks. There are some limitations associated with algae usage such as its mass production can increase the total cost by 40% (Oh et al. 2018). Thus, more inputs are needed from scientific community in future to pave the path for 3G biofuels.

Table 2.1 Categorization of advanced biofuel industry based on production process.

Method	Biomass type	Biofuel
Microbial conversion	Lignocellulosic e.g. stalks, corn stover etc.	Bioethanol or biobutanol
Transesterification	Waste and/or non-edible vegetable oils or animal fats	Fatty acids and methyl esters (FAME) i.e. biodiesel
Hydro-treatment followed by alkane isomerization and cracking	Waste and/or non-edible vegetable oils or animal fats	Drop-in fuels (HVO/HEFA)*
Thermochemical/ gasification	Waste and/or non-edible vegetable oils or animal fats	Biocrude/ syngas (converted to renewable gasoline)

Note1: HVO – hydro-treated vegetable oils and HEFA – hydro-processed esters and fatty acids.
Source: Based on Ref. IRENA (2019b).

2.3.6 Ocean Energy

As we know, 70% of the earth's surface is covered by water in different forms. Ocean energy also known as blue energy is present in the form of tidal, current, wave, temperature gradient and salinity gradient energy. Electricity could be generated using energy possessed by all the different types of ocean energy. IRENA 2020 report (IRENA 2020b) states that electricity generation capacity of ocean energy till 2019 was 500 MW, very small contribution compared with capacity of other alternative sources of energy. Approximately 1.7 GW is under development (Wilberforce et al. 2019). Compared with other alternative sources of energy, ocean energy is steady and predictable (especially tidal range technology) making it a right choice for power production in future (Crus 2008, pp. 70–75). Looking at the promising potential of ocean energy, various researches are going worldwide to see the feasibility of how to harness this energy (Wilberforce et al. 2019). Most of researches were aimed to pinpoint best possible locations for ocean energy exploitation, energy conversion efficiency of various technologies used in ocean energy sector and consequences of several technologies on the marine ecosystem as well as on the environment, as this is a raising concern about the usage of ocean energy technologies. This is a clean form of energy, GHGs emissions are negligible and assumed to be environment-friendly. Despite all these factors, its share in energy sector is very small at present.

Around the globe, mainly four types of ocean energy are harnessed to generate electricity, namely wave, ocean thermal energy conversion (OTEC), tidal and salinity gradient energy. Out of all these forms, tidal range share with respect to generation capacity is approximately 99%. Currently, contribution of other forms is not significant but many projects are under development based on OTEC, tidal stream and wave energy (Kerr et al. 2015). On completion of various current projects share of OTEC, tidal stream and wave energy will become noticeable. Theoretical maximum capacity of all projects which are under construction will be 15 GW (Wilberforce et al. 2019). Major contributors in the commercialized marine technology are the United States and United Kingdom (UK) along with small share of South Korea, Ireland, The Netherlands and China.

2.3.6.1 Wave Energy

Wave energy convertor devices capture the energy possessed by the waves (kinetic and potential) and transform it to electric power. Wave energy convertors are classified into different types of categories based on the deployed location. Onshore devices are of two types 'Oscillating water column' (trap air to run a turbine) and 'Overtopping devices' which uses height differences of waves to generate electricity. 'Oscillating wave surge converters', 'Point Absorber' and 'Submerged Pressure Differential devices' come under the near-shore location category. Offshore devices are 'Attenuator', 'Bulge wave devices' and 'Rotating mass converters'. Each of these devices has its own characteristics (Wilberforce et al. 2019). Worldwide many companies are involved in research and development (R&D) to exploit wave energy so that this technology can be commercialized (Lorente et al. 2011; www.aquaret.com). Some studies have revealed that theoretical global capacity of wave energy was 32 PWh (petawatt-hour) per annum (Mørk et al. 2010; Lagoun et al. 2014) which was double of the global electricity provided in the year 2008 (Leonard and Michaelides 2018). These studies have suggested potential of this energy in future.

2.3.6.2 Tidal Energy

This form of energy can be generated by using tidal range technologies, tidal current (often called tidal stream) technologies and hybrid technologies. Tidal range technologies produce electricity by making use of a barrage which can be a dam similar to that in hydropower or some other barrier. Devices used in tidal current technologies work similarly to a turbine deployed in wind energy to harness energy of tides, but these turbines use water to supply the captured energy. Technologies involved in tidal current are matured than wave energy owing to similarity of devices to wind turbine of the former. Tidal current energy converters are classified into three, which are further divided into six types (Wilberforce et al. 2019). These devices transform the kinetic energy of water into electricity. Tidal current energy could be mainly exploited along the coastal areas due to maximum availability of tides in these regions. Tidal stream energy is location-dependent and at present, 106 locations are known in Europe which possess great potential for electricity production and 48 TWh/yr power can be generated using all these locations (Wilberforce et al. 2019).

2.3.6.3 Ocean Thermal Energy Conversion (OTEC)

As the name signifies, this form of energy is based on heat possessed by the ocean. This energy makes use of temperature difference existing between the water layers at 800–1000 m depth to drive the turbine and thereby produce electricity using OTEC technologies. For effective operation of technologies, temperature difference of approximately 20 °C is needed. Water vapors produced from hot sea layers act as working fluid which runs the turbines. Key techniques used in the OTEC plant are open cycle (function via water from the sea), closed cycle (operates using NH_3 as working fluid) and hybrid cycle (blend of both the cycles) (Wilberforce et al. 2019).

2.3.6.4 Salinity Gradient Energy

This energy originates because of difference in salt concentrations at the junction of river and ocean or sea. Energy generation from salinity gradient is a two-stage process. The 'pressure retarded osmosis' is used where freshwater flows through a semi-permeable membrane to increase the pressure in a reservoir of seawater and 'reverse electro dialysis' with ions of salt passing through alternating reservoirs of sea and river water. Recently, a research estimated the potential of salinity gradient energy as approximately 1650 TWh/yr. (Cornett 2008, p. 9). At present, the major challenge faced by utilization of this energy is its high cost of power generation. In 2009, this energy was harnessed for the first time at Tofte site (Nihous 2010).

Currently, harnessing ocean energy using different technologies is in its infancy due to costly technologies, high capital cost, socio-economic impact on shipping and tourism and environmental effects. There are several technical issues which are associated with this energy such as device manufacturing, installation, maintenance and grid and power transmission. Further, due to lack of studies which establish the impact of ocean energy exploitation on marine environment, aquatic ecosystem and water pollution due to corrosion of various device obstructs its development.

2.4 Future Fuel: Hydrogen

Use of fossil fuels to meet the energy demands of world population is consistently increasing the amount of GHGs which ultimately changes the earth's environment. Temperature of the earth is increasing at a faster rate due to GHGs emission. Replacing fossil fuels with such an energy source which accomplishes energy requirements along with reducing the percentage of GHGs in earth's environment is the need of present era. So, hydrogen seems to be the future fuel as it meets both the requirements (Pareek et al. 2020). Due to its high calorific value and no emission of GHGs, it is considered as a clean, green and sustainable fuel as long as it is produced from renewable sources of energy. It also possesses high energy and low volumetric density which further enhances its prospects as future fuel. Moreover, it decreases the dependency on oil which is mainly imported and thus provides energy security. Although hydrogen is present in abundant amount in the universe, it only exists in amalgamation with other elements such as oxygen, carbon and nitrogen. Splitting hydrogen from other elements by a process using some other source of energy (renewable and non-renewable) provides an alternative for this high energy-carrier fuel. Producing hydrogen in pure form using renewable sources of energy from a cost-effective method along with possessing high energy conversion efficiency is the main challenge in exploiting it as a fuel (Pareek et al. 2020). Currently, it is majorly produced from non-renewable sources due to easy and cost-effective methods. Steam reforming and

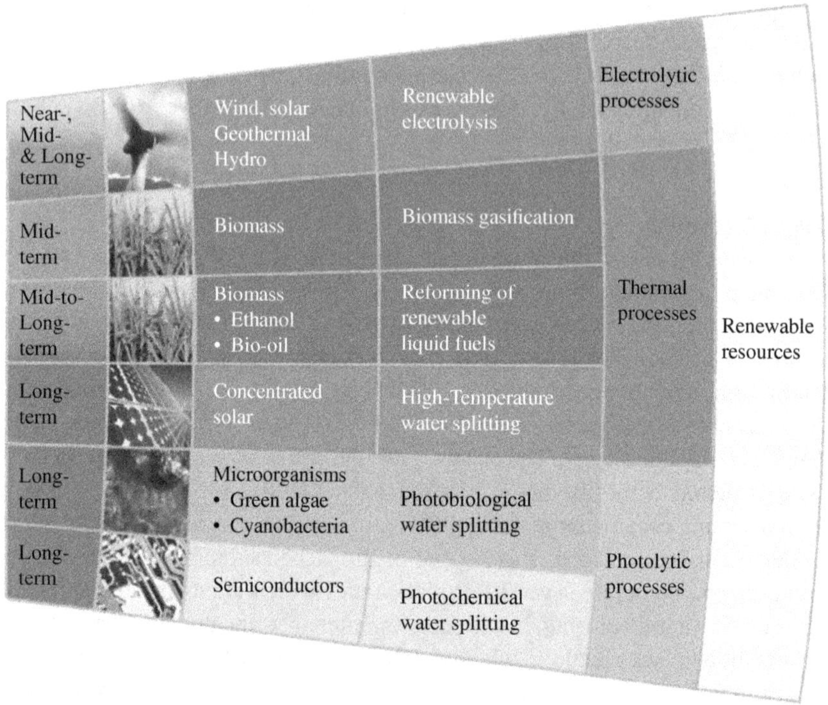

Figure 2.5 Hydrogen production methods using renewable energy sources. *Source:* Reprinted from Ref. Pareek et al. 2020 with Licence under http://creativecommons.org.

gasification are two such methods which are used to produce hydrogen using natural gas and coal, respectively. Using a mature process i.e. steam reforming, 96% of hydrogen is generated industrially (Balat 2008). Other two processes which utilize nuclear energy for hydrogen production are thermochemical water splitting and high-temperature electrolysis (Acar and Dincer 2014). Electrolysis method can either use solar photovoltaic or wind as energy source for H_2 production. Biomass gasification and bio-hydrogen are two green methods which utilize biomass remains, agricultural, human and animal wastes and other biodegradable waste materials. Figure 2.5 shows how hydrogen can be produced from various processes by consuming different renewable energy sources with respect to near-, medium- and long-term objective. Producing hydrogen from renewable sources is at developing stages, and status of technology used in many processes is at R&D or early R&D level (Pareek et al. 2020).

2.4.1 Hydrogen Production Methods Using Renewable Sources

2.4.1.1 Renewable Electrolysis

This technique uses electric current, generated via solar PV or wind energy to split water into its constituent elements (hydrogen and oxygen). Utilization of solar PV for hydrogen production is the costliest method using current technologies. However, decreasing cost of solar PV due to advances in technology will decrease the cost of this process in future (Karlsson and Oparaocha 2009). Exploiting wind energy for electricity production needed for electrolysis is also a sustainable method; however, wind turbine and electrolyser cost is again a matter of concern. This is again a costly method compared with methods which are fossil fuels based.

2.4.1.2 Biomass Gasification

Biomass gasification process, carried out in a gasifier, uses steam and oxygen at high temperature and pressure to transform different types of biomass residue into a mixture of gases such as hydrogen, carbon monoxide and carbon dioxide. This method has an advantage of consuming biomass remains and wastes from animal, municipal and agricultural along with clean form of energy generation. However, large-scale hydrogen production using this process at affordable cost is not viable, at present. By this process, overall emission of CO_2 decreases compared with methods which utilize fossil fuels. Main challenge faced by this technique is availability of massive amounts of natural resources and land, needed for growing feedstock required for the process (Acar and Dincer 2014).

2.4.1.3 Thermochemical Water Splitting

Hydrogen production from this method needs very high temperature heat at 500–2000 °C which can be produced from CSP. This is a closed-loop reaction in which only water is utilized and splits into hydrogen and oxygen and rest of the chemicals are recycled (Acar and Dincer 2014). It is a cost-effective technology with high efficiency and apt for mass-scale production (Wang et al. 2019). However, high cost of solar concentrators and heat transfer medium renders this technique to reach at commercial level, currently. With advances in the CSP, this process can be commercialized in future.

2.4.1.4 Bio-Hydrogen Production

This is a biological way of producing hydrogen via microorganisms which are normally present in aqueous atmosphere at normal temperature and pressure conditions (Kotay and Das 2008). Last two decades have seen tremendous input at R&D level to advance various technologies needed in this process to make these methods economically viable. This process consumes bio-wastes and offers an interesting path for their minimization along with energy generation. Energy from this process can be harnessed locally due to availability of primary resources on which microorganisms will act upon, thus making possible production of distributed hydrogen. Bio-hydrogen generation processes have further categories, namely direct bio-photolysis, indirect bio-photolysis, photo fermentation, dark fermentation and bio-catalysed electrolysis (Pareek et al. 2020). All these methods need further advances due to their poor hydrogen conversion efficiency and high cost, at present.

2.5 Challenges

2.5.1 Efficiency

Energy conversion efficiency of various technologies which utilize renewable energy sources is still small except hydropower (Sahli et al. 2018). Advances are taking place worldwide to increase the efficiency. Since, solar and wind sectors are expanding at much larger scale compared with other energy sources. These two are further discussed with respect to efficiency increment.

In solar industry, scientific community is evaluating various new materials to produce electricity more efficiently. Thin-film solar cells are cheaper in cost and flexible in nature; however, efficiency of these cells is still a matter of concern for researchers so experimental studies are carried out on the range of materials such as amorphous silicon, CdS/CdTe and CIS to overcome this barrier (El Chaar et al. 2011). To improve the efficiency of organic and polymer solar cells, use of a controlled layer of multi-wall carbon nanotubes is also reported (Capasso et al. 2014) which improves the efficiency of these cells from 8 to 10%. DSSCs also fall in the same category of low efficiency which is again between 8 and 12%. To increase the efficiency of PV system, generated current is transferred to grid systems. This grid is either mounted on ground or built on the roof of a building so that PV system provides better efficiency as it mainly depends on the intensity of solar radiation. On an average, capacity of PV panel ranges from 10 to 60 MW. Lifetime of a well-developed PV panel is approximately 10 years at 90% capacity, and it increases to 25 years with a 10% decrease in capacity (Devabhaktuni et al. 2013).

A major challenge in CPV system is reduction in efficiency of PV due to increase in temperature while concentrating light as conversion efficiency of CPV is still small. Remaining solar radiation gets converted into thermal energy which raises the temperature of junction in the cell, thereby decreasing efficiency. Lifetime of PV also decreases due to increase in temperature. This leads to overall degradation in performance of PV system (Pandey et al. 2016). So, more studies are required to explore ways to decrease temperature of the CPV system.

In wind energy sector, replacement of components of wind turbine system with latest components having cost-effective technology will increase the efficiency of the existing wind farm. In addition to this, repowering of onshore wind assets can enhance operating lifetime of the current wind project. Repowering can be either full or it can be partial.

2.5.2 Large-Scale Production

Large-scale production of solar PV systems is at infant stage due to low efficiency and high cost. Government is supporting these installations to promote renewable energy sources. Technically solar PV systems are viable but not economical. So, this area needs more R&D and innovative methods to increase the production at large scale.

One of the major challenges in deployment of offshore wind energy at a large scale is its high cost of installations and maintenance. With the latest technology developments in the field of manufacturing, installations and operations, its cost has significantly decreased. By 2050, cost of offshore wind energy is expected to decrease by a large factor, and it can directly compete with electricity generation using fossil fuels (IRENA 2019d). So, this challenge could be tackled by the wind industry based on the advanced technologies developed by scientific research community.

Other than solar and wind sector, all alternative sources of energy technologies are not economically viable compared with conventional sources of energy. Thus, large-scale production in these sectors needs more advances in technology from cost and production viewpoint.

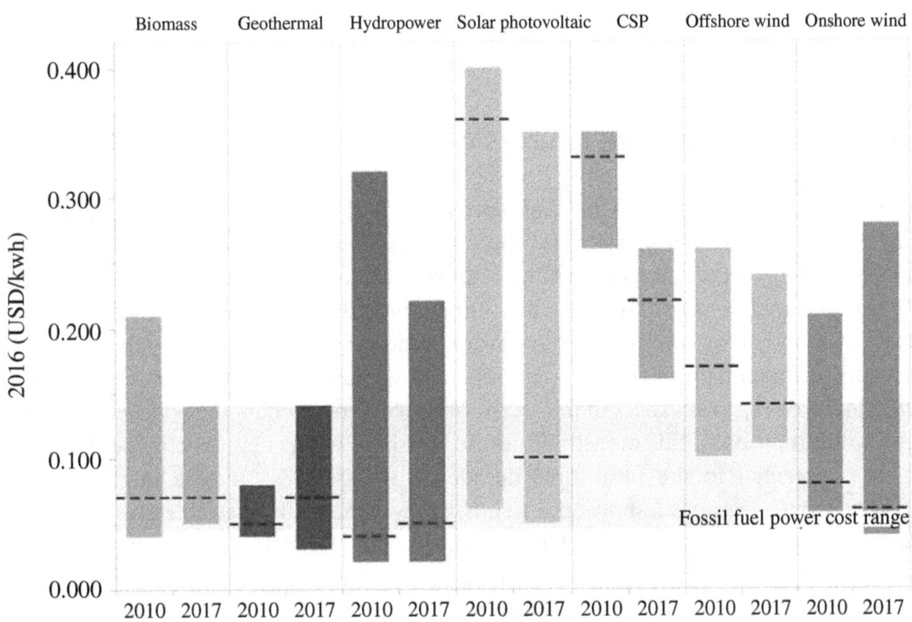

Figure 2.6 Worldwide, LCOE from the alternative energy sources for the period 2010–2017. *Source:* https://www.irena.org/Statistics © IRENA. Note2: Dashed line represents the global weighted average LCOE and band shows the fossil-fuel-fired electricity generation cost range.

2.5.3 Cost-Effective Production

Globally, levelized cost of electricity (LCOE) from solar and wind energy is decreased by a large factor from 2010 to 2017 as depicted in Figure 2.6. Electricity production from other energy sources has not realized any major change. In fact, global weighted average LCOE as represented by the dashed line in Figure 2.6 has increased for geothermal and hydropower while for biomass it remains the same for the period from 2010 to 2017.

Onshore and offshore wind technologies are competitive to fossil fuels and in some regions, LCOE from wind energy is less than that from fossil fuels. Electricity production cost from PV is also approaching the cost from fossil fuels. Cost of crystalline PV has reduced by 99.5% in the last 30 years due to tremendous research in PV industry with the development of novel materials and techniques (www.economist.com). As reported by (IRENA 2019a), the global weighted average total installation cost for offshore wind projects will decline in the near future, and it will be USD 1700–3200/kW by 2030 and USD 1400–2800/kW by 2050.

Electricity generation using hydro technology is the most cost-effective and preferred over the other renewable energy sources method where available. Using gravity water wheel instead of Kaplan turbines in small-scale hydro plants reduces the cost as the former cost is 33–60% of the latter (Quaranta and Revelli 2018). Further, usage of PATs also decreases the installation cost which leads to overall reduction in electricity generation cost from the small hydro plants as well.

Cost of power generation using geothermal energy largely depends on the site. Normally, cost of geothermal plant lies between USD 1870 and 5050 per kW. Compared with direct dry steam and flash plants, the cost of binary plant is high. The LCOE from a geothermal plant is between USD 0.04 and 0.14 per kWh, assuming maintenance costs of USD 110 per kW per year and a life span of 25 years (IRENA 2017).

2.6 Future: Alternative Sources of Energy

Future of solar energy is very promising especially for solar PV systems such as PV/T, CPV and PV-powered desalination systems. These systems can be path breaking in solving the hurdles of energy requirement in coming centuries. More R&D is needed in the PV/T and CPV systems to make economically viable at the user end.

Onshore wind energy sector can drive overall renewables growth worldwide due to the availability of abundant wind resource, large and competitive market and cost-effective technologies owing to advances in the recent times. Moreover, changing policies across the globe will further boost this industry in future. Offshore wind energy sector can have tremendous potential in the next three decades with 10 times increase in its capacity (IRENA 2019g). Figure 2.7 shows the trends in investment in renewable energy, worldwide. Maximum investments in solar and wind sectors indicate a promising future for these two sources. Bioenergy sector is also ahead of other energy sources with respect to investments which displays its prospects in future.

Despite small investments in geothermal, the future of EGS has tremendous potential to expand and contribute more in power generation. By 2050, the installed capacity of EGS electricity production will be 70GWe assuming 85% probability. Due to advances in the area

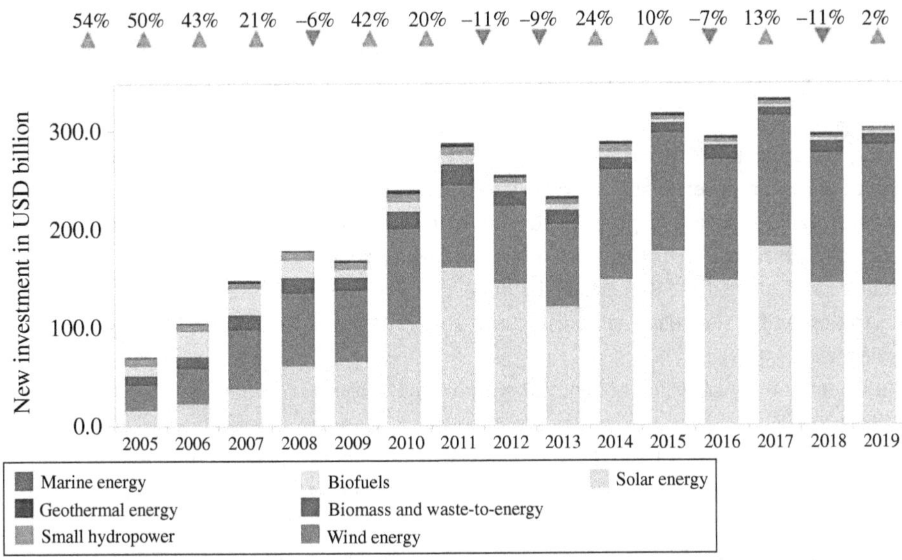

Figure 2.7 Global investment in renewable energy. *Source:* Modified from Frankfurt School-UNEP Centre/BNEF (2020). Global Trends in Renewable Energy Investment 2020. https://www.irena.org..

of EGS technology, more EGS power plants will be added, thus making it as one of the emerging technologies in the geothermal power sector. Investments in small hydropower are consistent in the last decade and are relatively small compared with other renewable sources.

2.7 Conclusions

Recent advances in technology, cost-effectiveness of power production and ecological effects of all alternative sources of energy are thoroughly deliberated in the chapter. Currently, exploitation of alternative energy sources is small compared with the non-renewable energy sources. Despite the negligible emission of GHGs, green nature and least impact on environment, renewable energy sources usage are still small. Cost of electricity generation is the main obstacle for harnessing renewable sources which could be fixed through further research. Out of all the alternative sources of energy, the solar and wind sectors are experiencing higher growth due to advances in technologies, reduction in cost and better energy conversion efficiency. Hydropower is a mature technology. New investments are taking place in small hydro plants owing to their environment-friendly nature. EGS is an emerging technology in the geothermal sector because of developments in drilling technology. Bioenergy sector is also increasing at a consistent pace, however, growth is not rapid as this technology needs more developments with respect to cost-effectiveness, land usage, economical pre-treatment methods for lignocellulosic, efficiency increment and policy amendments. In the future, ocean energy can possibly diversify the renewable energy arena by addition in power generation and will become a support to the economy. Hydrogen could be a promising fuel in future as it possesses significant potential. However, its production methods from renewable energy sources are still at primary stage. Commercialization of these processes has a long journey ahead.

References

Abas, N., Kalair, A., and Khan, N. (2015). Review of fossil fuels and future energy technologies. *Futures* 69: 31–49.

Acar, C., and Dincer, I. (2014). Comparative assessment of hydrogen production methods from renewable and non-renewable sources. *International Journal of Hydrogen Energy* 39: 1–12.

Agarwal, T. (2012). Review of pump as turbine (pat) for microhydropower. *International Journal of Emerging Technology and Advanced Engineering* 2 (11): 163–169.

Ahmad, I., McCarthy, J. E., Bari, M. et al. (2014). Carbon nanomaterial based counter electrodes for dye sensitized solar cells. *Solar Energy* 102: 152–161. https://doi.org/10.1016/j.solener.2014.01.012

Aquatic energy renewable technology. (2008). Aqua RET. www.aquaret.com (accessed 30 July 2020)

Bagienski, W., and Gupta, M. (2011). Temperature dependence of polymer/fullerene organic solar cells. *Solar Energy Materials & Solar Cells* 95 (3): 933–941. https://doi.org/10.1016/j.solmat.2010.11.026

Balat, M. (2007). Status of fossil energy resources: a global perspective. *Energy Sources, Part B: Economics, Planning, and Policy* 2: 31–47.

Balat, M. (2008). Possible methods for hydrogen production, energy sources. Part a recover. *Util. Environ. Eff.* 31: 39–50.

Bertani, R. (2016). Geothermal power generation in the world 2010–2014 update report. *Geothermics* 60: 31–43.

Bhattacharya, M., Paramati, S.R., Ozturk, I. et al. (2016). The effect of renewable energy consumption on economic growth: evidence from top 38 countries. *Applied Energy* 162: 733–741.

Capasso, A., Salamandra, L., Chou, A., et al. (2014). Multi-wall carbon nanotube coating of fluorine-doped tin oxide as an electrode surface modifier for polymer solar cells. *Solar Energy Materials & Solar Cells* 122: 297–302.

Cornett, A.M. (2008). A global wave energy resource assessment. *Proc. ISOPE 8*(March):9.

Crus, J. ed. (2008). *Ocean Wave Energy: Current Status and Future Perspectives*. Berlin: Springer

Deb, S. K. (1998). Recent developments in high efficiency photovoltaic cells. *Renewable Energy* 15: 467–472.

Devabhaktuni, V., Alam, M., Depuru, S. S. S. R. et al. (2013). Solar energy: trends and enabling technologies. *Renewable and Sustainable Energy Reviews* 19: 555–564. https://doi.org/10.1016/j.rser.2012.11.024

Devi, B. P., Wu, K-C., and Pei, Z. (2011). Gold nanomesh induced surface plasmon for photocurrent enhancement in a polymer solar cell. *Solar Energy Materials & Solar Cells* 95 (8): 2102–2106. https://doi.org/10.1016/j.solmat.2011.02.031

DiPippo, R. (2015). *Geothermal Power Plants: Principles, Applications, Case Studies and Environmental Impact*. Amsterdam: Elsevier. https://doi.org/10.1016/C2014-0-02885-7

DOE. (2018). *Algal biofuels*, Office of Energy Efficiency & Renewable Energy, Department of Energy. www.energy.gov (accessed 30 July 2020)

Dou, L., You, J., Yang, J. et al. (2012). Tandem polymer solar cells featuring a spectrally matched low-band gap polymer. *Nature Photonics* 6:180–185. https://doi.org/10.1038/nphoton.2011.356

Dreamwind. (2020). Increased sustainability of wind energy through the development of new materials. www.dreamwind.dk (accessed 29 July 2020).

Economist. (2012). Price of crystalline silicon photovoltaic cells (2012). www.economist.com (accessed 30 July 2020)

El Chaar, L., Lamont, L.A., and El Zein, N. (2011). Review of photovoltaic technologies. *Renewable and Sustainable Energy Reviews* 15 (5): 2165–2175. https://doi.org/10.1016/j.rser.2011.01.004

Ellabban, O., Abu-Rub, H., and Blaabjerg, F. (2014). Renewable energy resources: current status, future prospects and their enabling technology. *Renewable and Sustainable Energy Reviews* 39: 748–764.

El-Shatter, T. F., Eskander, M. N., and El-Hagry, M. T. (2006). Energy flow and management of a hybrid wind/PV/fuel cell generation system. *Energy Conversion and Management* 47 (9): 1264–1280. https://doi.org/10.1016/j.enconman.2005.06.022

Fjeldstad, H.P., Pulg, U., and Forseth, T. (2018). Safe two-way migration for salmonids and eel past hydropower structures in Europe: a review and recommendations for best-practice solutions. *Marine and Freshwater Research* 69: 1834–1847. https://www.publish.csiro.au/mf/MF18120

Frankfurt School-UNEP Centre/BNEF. (2020). Global Trends in Renewable Energy Investment 2020. www.irena.org

Friðleifsson, G.O., Pálsson, B., Albertsson, A.L. et al. (2015). IDDP-1 drilled into Magma -world's first Magma-EGS system created. *Proceedings World Geothermal Congress* Melbourne, Australia. http://iddp.is/wp-content (accessed 30 July 2020)

Froese, M. (2017). What are the new ideas in condition monitoring for wind turbines. *Wind Power Engineering and Development*. www.windpowerengineering.com (accessed 30 July 2020).

Geirdal, C. A. C., Gudjonsdottir, M. S., and Jensson, P. (2015). Economic comparison of a well-head geothermal power plant and a traditional one. *Geothermics* 53:1–13

Gevorgian, V., Muljadi, E., Luo, Y. et al. (2017). Supercapacitor to provide ancillary services. *IEEE Energy Conv Congress Expo (ECCE)*, pp: 1030–1036.

Gong, J., Li, C., and Wasielewski, M. R. (2019). Advances in solar energy conversion. *Chemical Society Reviews* 48: 1862–1864.

Güney, T. (2019). Renewable energy, non-renewable energy and sustainable development. *International Journal of Sustainable Development and World Ecology* 26(5): 389–397.

Gupta, V. K., Kubicek, C. P., Saddler, J. et al. ed. (2014). *Bioenergy Research: Advances and Applications*. Amsterdam: Elsevier.

Harfoot, M. B. J., Tittensor, D. P., Knight, S., et al. (2018). Present and future biodiversity risks from fossil fuel exploitation. *Conservation Letters* 11: e12448.

He, J., and Janáky, C. (2020). Recent advances in solar-driven carbon dioxide conversion: expectations versus reality. *ACS Energy Letters* 5(6): 1996–2014.

Helmizar, H. (2016). Turbine wheel-a hydropower converter for head differences between 2.5 and 5 m. PhD Thesis University of Southampton.

Hogan, T. W., Cada, G. F., and Amaral, S. V. (2014). The status of environmentally enhanced hydropower turbines. *Fisheries* 39 (4): 164–172. https://doi.org/10.1080/03632415.2014.897195

Huang, B. J., Lin, T. H., Hung, W. C. et al. (2001). Performance evaluation of solar photovoltaic/thermal systems. *Solar Energy* 70 (5): 443–448. https://doi.org/10.1016/S0038-092X(00)00153-5

IEA. (2020). Webpage. www.iea.org (accessed 28 July 2020)

IEA-ETSAP (International Energy Agency – Energy Technology Systems Analysis Programme) (2010). Technology Brief E07 Geothermal Heat and Power. https://iea-etsap.org (accessed 30 July 2020)

IHA. (2018) *Hydropower Status Report Technical Report London*, UK: International Hydropower Association.

IPCC. (2011). Renewable energy sources and climate change mitigation. https://archive.ipcc.ch/pdf/special-reports (accessed 30 July 2020)

IRENA. (2016). *Innovation Outlook for Offshore Wind Technology*, International Renewable Energy Agency, Abu Dhabi.

IRENA. (2017). *Geothermal Power: Technology Brief*, International Renewable Energy Agency, Abu Dhabi.

IRENA. (2019a). *Future of Wind: Deployment, Investment, Technology, Grid Integration and Socio-Economic Aspects (a Global Energy Transformation Paper)*, International Renewable Energy Agency, Abu Dhabi.

IRENA. (2019b). Advanced biofuels. *What Holds them Back?* International Renewable Energy Agency, Abu Dhabi.

IRENA. (2019c). *Renewable Capacity Statistics 2019*, International Renewable Energy Agency, Abu Dhabi.

IRENA. (2019d). *Renewable Power Generation Costs in 2018*, International Renewable Energy Agency, Abu Dhabi.

IRENA. (2019e). *Innovation Landscape Brief: Artificial Intelligence and Big Data*, International Renewable Energy Agency, Abu Dhabi.

IRENA. (2019f). *Renewable Energy Statistics 2019*, The International Renewable Energy Agency, Abu Dhabi.

IRENA. (2019g). *Global Energy Transformation: A Roadmap to 2050 (2019 Edition)*, International Renewable Energy Agency, Abu Dhabi.

IRENA. (2020a). Data & Statistics. Webpage. www.irena.org (accessed 28 July 2020)

IRENA. (2020b), *Renewable Capacity Statistics 2020 International Renewable Energy Agency (IRENA)*, Abu Dhabi.

IRENA. (2020c). Data & Statistics. https://www.irena.org/Statistics (accessed 30 July 2020).

Johnston, I. W., Narsillo, G. A., and Colls, S. (2011). Emerging geothermal energy technologies. *KSCEJ Civil Engineering* 15 (4): 643–653. http://dx.doi.org/10.1007

Kannan, N., and Vakeesan, D. (2016). Solar energy for future world: a review. *Renewable and Sustainable Energy Reviews* 62: 1092–1105.

Karlsson, G., and Oparaocha, S. (2009). The road to Johannesburg and beyond: networking for gender and energy. *Energy for Sustainable Development* 7 (3): 62–67. https://doi.org/10.1016/S0973-0826(08)60366-X

Kerr, S., Colton, J., Kate, J. et al. (2015). Wright Glen. Rights and ownership in sea country: implications of marine renewable energy for indigenous and local communities. *Marine Policy* 52: 108–115.

Kotay, S. M., and Das, D. (2008). Biohydrogen as a renewable energy resource-prospects and potentials. *International Journal of Hydrogen Energy* 33: 258–263.

Kougias, I., Patsialis, T., Zafirakou, A. et al. (2014). Exploring the potential of energy recovery using micro hydropower systems in water supply systems. *Water Utility Journal* 7: 25–33.

Kougias, I., Karakatsanis, D., Malatras, A. et al. (2016a). Renewable energy production management with a new harmony search optimization toolkit. *Clean Technologies and Environmental Policy* 18(8): 2603–2612. https://doi.org/10.1007/s10098-016-1173-4

Kougias, I., Szabó, S., Monforti-Ferrario, F. et al. (2016b). A methodology for optimization of the complementarity between small-hydropower plants and solar PV systems. *Renewable Energy* 87: 1023–1030. https://doi.org/10.1016/j.renene.2015.09.073

Kougias, I., Aggidis, G., Avellan, F. et al. (2019). Analysis of emerging technologies in the hydropower sector. *Renewable and Sustainable Energy Reviews* 113:109257.

Kropp̌, R. (2009) *Solar Expected to Maintain its Status as the world's Fastest-Growing Energy Technology*. U.S.A: Sustainability Investment News.

Kuo, C-T., Shin, H-Y., Hong H-F. et al. (2009). Development of the high concentration III–V photovoltaic system at INER, Taiwan. *Renewable Energy* 34 (8): 1931–1933. https://doi.org/10.1016/j.renene.2008.12.014

Lagoun, M.S., Benalia, A., and Benbouzid, M.E.H. (2014). A predictive power control of doubly fed induction generator for wave energy Converter in irregular waves. *Proceedings of the IEEE ICGE*, p. 26–31 Sfax, Tunisia. hal.archives-ouvertes.fr/hal-01023509/document.

Leonard, M. D., and Michaelides, E. E. (2018). Grid-independent residential buildings with renewable energy sources. *Energy* 148 (1): 448–460.

Long, M.P.E. and Lewis, B. et al. (2003). Geothermal power production: steam for free. *POWER Engineers*, Idaho, US. Available at: www.powereng.com (accessed 30 July 2020)

Looser, R., Vivar, M., and Everett, V. (2014). Spectral characterisation and long-term performance analysis of various commercial heat transfer fluids (HTF) as direct-absorption filters for CPV-T beam-splitting applications. *Applied Energy* 113:1496–1511.

Lorente, I. L., lvarez, A., and Blanco, D. J. L. (2011). Performance model for parabolic trough solar thermal power with thermal storage: comparison to operating plant data. *Solar Energy* 85: 2443–2460.

Lu, S-M. (2018). A global review of enhanced geothermal system (EGS). *Renewable and Sustainable Energy Reviews* 81:2902–2921.

Lund, J. W., and Boyd, T. L. (2016). Direct utilization of geothermal energy 2015 worldwide review. *Geothermics* 60: 66–93.

Lund, P. D., Byrne, J., Berndes, G. et al. ed. (2016). *Advances in Bioenergy the Sustainability Challenge*. Chichester, UK: Wiley.

Ma, C., Liu, J., Ye, M., et al. (2018). Towards utmost bioenergy conversion efficiency of food waste: pretreatment, co-digestion, and reactor type. *Renewable and Sustainable Energy Reviews* 90: 700–709.

Mamlook, R., Nijmeh, S., and Abdallah, S. M. (2006). A programmable logic controller to control two axis sun tracking system. *Information Technology Journal* 5 (6): 1083–1087. https://doi.org/10.3923/itj.2006.1083.1087

Mangal, D., Lamba, D., Gupta, T. et al. (2010). Acknowledgement of evacuated tube solar water heater over flat plate solar water heater. *International Journal of Engineering* 4 (4): 279–284.

Martins, F., Felgueiras, C., Smitkova, M., et al. (2019). Analysis of fossil fuel energy consumption and environmental impacts in European countries. *Energies* 12: 964.

Mørk G, Steve Barstow, Alina K. Kabuth & M. Teresa Pontes et al. (2010). Assessing the global wave energy potential. *Proceedings of OMAE2010 (ASME), 29th international conference on ocean, offshore mechanics and arctic engineering*. Shanghai, China.

Nihous, G. C. (2010). Mapping available ocean thermal energy conversion resources around the main Hawaiian islands with state-of-the-art tools. *Journal of Renewable and Sustainable Energy* 2 (4): 043104. https://doi.org/10.1063/1.3463051

Novara, D., Derakhshan, S., McNabola, A. et al. (2017). Estimation of unit cost and maximum efficiency for pumps as turbines. *9th Eastern European Iwa Young Water Professionals*, Budapest, Hungary (24–27 May 2017). http://iwa-ywp.eu/wp-content/uploads/2017/05/Final-Conference-Program.pdf.

Ogayar, B., and Vidal, P. G. (2009). Cost determination of the electro-mechanical equipment of a small hydro-power plant. *Renewable Energy* 34 (1): 6–13. https://doi.org/10.1016/j.renene.2008.04.039

Oh, Y., Hwang, K., Kim, C. et al. (2018). Recent developments and key barriers to advanced biofuels: a short review. *Bioresource Technology* 257: 320–333.

Olasolo, P., Juárez, M. C., Morales, M. P. et al. (2016). Enhanced geothermal systems (EGS): a review. *Renewable and Sustainable Energy Reviews* 56: 133–144.

Pandey, A. K., Tyagi, V.V., Selvaraj, J. A/L., et al. (2016). Recent advances in solar photovoltaic systems for emerging trends and advanced applications. *Renewable and Sustainable Energy Reviews* 53: 859–884.

Panwar, N., Kaushik, S., and Kothari, S. (2011). Role of renewable energy sources in environmental protection: a review. *Renewable and Sustainable Energy Reviews* 15(3):1513–1524.

Pareek, A., Dom, R., Gupta. J. et al. (2020). Insights into renewable hydrogen energy: recent advances and prospects. *Materials Science for Energy Technologies* 3:319–327.

Parida, B., Iniyan, S., and Goic, R. (2011). A review of solar photovoltaic technologies. *Renewable and Sustainable Energy Reviews* 15 (3): 1625–1636. https://doi.org/10.1016/j.rser.2010.11.032

Pathak, H., Jain, N., Bhatia, A. et al. (2009). Global warming mitigation potential of biogas plants in India. *Environmental Monitoring and Assessment* 157: 407–418. https://doi.org/10.1007/s10661-008-0545-6

Pérez-Collazo, C., Greaves, D., and Iglesias, G. (2015). A review of combined wave and offshore wind energy. *Renewable and Sustainable Energy Reviews* 42: 141–153.

Potumarthi, R., Baadhe, R. R., Nayak, P. et al. (2013). Simultaneous pretreatment and sacchariffication of rice husk by Phanerochete chrysosporium for improved production of reducing sugars. *Bioresource Technology* 128: 113–117. https://doi.org/10.1016/j.biortech.2012.10.030

Quaranta, E., and Revelli, R. (2018). Gravity water wheels as a micro hydropower energy source: a review based on historic data, design methods, efficiencies and modern optimizations. *Renewable and Sustainable Energy Reviews* 97: 414–427. https://doi.org/10.1016/j.rser.2018.08.033

Quaranta, E., Katopodis, C., Revelli, R. et al. (2017). Turbulent flow field comparison and related suitability for fish passage of a standard and a simplified low-gradient vertical slot fishway. *River Research and Applications* 33 (8): 1295–1305. https://doi.org/10.1002/rra.3193

Razykov, T., Ferekides, C., Morel, D. et al. (2011). Solar photovoltaic electricity: current status and future prospects. *Solar Energy* 85:1580–1608.

RECYCLING. (2019). Joint project to advance wind turbine recycling. *RECYCLING magazine*. www.recycling-magazine.com (accessed 29 July 2020).

REN21, P.S. (2014). Renewables 2014: Global Status Report. *2014: Secretariat Renewable Energy Policy Network for the 21st Century (REN21)*, Paris.

Roth, P., Georgiev, A., and Boudinov, H. (2005). Cheap two axis sun following device. *Energy Conversion and Management* 46 (7): 1179–1192. https://doi.org/10.1016/j.enconman.2004.06.015

S&P Global Platts. (2016). UDI world electric power plants data base. www.platts.com (accessed 30 July 2020)

Sahli, F., Werner, J., Kamino, B.A., et al. (2018). Fully textured monolithic perovskite/silicon tandem solar cells with 25.2% power conversion efficiency. *Nature Materials* 17: 820–826. https://doi.org/10.1038/s41563-018-0115-4

Sandnes, B., and Rekstad, J. (2002). A photovoltaic/thermal (PV/T) collector with a polymer absorber plate. Experimental study and analytical model. *Solar Energy* 72 (1): 63–73. https://doi.org/10.1016/S0038-092X(01)00091-3

Sarasúa, I., Torres, B., Pérez-Díaz, J.I. et al. (2016). Control strategy and sizing of a flywheel energy storage plant for the frequency control of an isolated power system. *15th wind integration workshop*.

Sharma, K., Sharma, V., and Sharma, S. S. (2018). Dye-sensitized solar cells: fundamentals and current status. *Nanoscale Research Letters* 13:381.

Shere, J. (2013). *Renewable: The World-Changing Power of Alternative Energy*. St Martin's Press: New York, pp. 201.

Soliman, M. M., Shabana, M. M., and Abulfotuh, F. (1996). CdS/CdTe solar cell using sputtering technique. *Renewable Energy* 8 (1–4): 386–389. https://doi.org/10.1016/0960-1481(96)88883-5

Sorrell, S. (2015). Reducing energy demand: a review of issues, challenges and approaches. *Renewable and Sustainable Energy Reviews* 47: 74–82.

Stuermer, M. and Schwerhoff, G. (2015). Non-renewable resources, extraction technology, and endogenous growth. FRB of Dallas Working Paper No. 1506. http://dx.doi.org/10.24149/wp1506r1

Tripanagnostopoulos, Y., Nousia, Th., Souliotis, M. et al. (2002). Hybrid photovoltaic/thermal solar systems. *Solar Energy* 72 (3): 217–234. https://doi.org/10.1016/S0038-092X(01)00096-2

Trivedi, C., Cervantes, M. J., Dahlhaug, O. G. et al. (2015). Experimental investigation of a high head Francis turbine during spin-no-load operation. *Journal of Fluids Engineering* 137: 061106. https://doi.org/10.1115/1.4029729

Valero, C., Egusquiza, M., Egusquiza, E., et al. (2017). Extension of operating range in pump-turbines influence of head and load. *Energies* 10 (12): 2178. https://doi.org/10.3390/en10122178

Verma, V. K., Bram, S., Delattin, F. et al. (2012). Agro-pellets for domestic heating boilers: standard laboratory and real life performance. *Applied Energy* 90 (1): 17–23. https://doi.org/10.1016/j.apenergy.2010.12.079

Wang, Z., Li, C., and Domen, K. (2019). Recent developments in heterogeneous photocatalysts for solar-driven overall water splitting. *Chemical Society Reviews* 48: 2109–2125.

Wilberforce, T., Hassan, Z. E., and Durrant, A. et al. (2019). Overview of ocean power technology. *Energy* 175:165–181.

Wind Power Monthly. (2018). GE launches 5.3MW onshore turbine. www.windpowermonthly.com (accessed 29 July 2020).

WindEurope. (2017). *Discussion Paper on Managing Composite Blade Waste*, WindEurope, Brussels.

Windtrust. (2016), *Designing the Wind Turbine of the Future*, Windtrust.

Wood Mackenzie. (2019). Unplanned wind turbine repairs to cost industry $8 billion+ in 2019. www.woodmac.com (accessed 29 July 2020).

Xu, Z., Li, Z., Jiang, Y., et al. (2020). Recent advances in solar-driven evaporation systems. *Journal of Materials Chemistry A* https://doi.org/10.1039/D0TA08869B

Yaramasu, V., Wu, B., Sen, P. C., et al. (2015). High-power wind energy conversion systems: state-of-the-art and emerging technologies. *Proceedings of the IEEE* 103: 740–788. doi:https://doi.org/10.1109/JPROC.2014.2378692

York, R. Do alternative energy sources displace fossil fuels?. (2012). *Nature Climate Change* 2: 441–443. https://doi.org/10.1038/nclimate1451

Zondag, H. A., de Vries, D. W., van Helden, W. G. J. et al. (2003). The yield of different combined PV-thermal collector designs. *Solar Energy* 74 (3): 253–269. https://doi.org/10.1016/S0038-092X(03)00121-X

Zou, C., Zhao, Q., Zhang, G., et al. (2016). Energy revolution: from a fossil energy era to a new energy era. *Natural Gas Industry B* 3: 1–11.

3

Recent Advances in Alternative Sources of Energy

Pradeep Pratap Singh[1], *Ambika*[2], *and Maya Verma*[3]

[1] *Department of Chemistry, Swami Shraddhanand College, University of Delhi, New Delhi, India*
[2] *Department of Chemistry, Hansraj College, University of Delhi, New Delhi, India*
[3] *Department of Physics, Hansraj College, University of Delhi, New Delhi, India*

3.1 Introduction

Energy has occupied a predominant space in the existence and development of mankind. However, humans seek more and more sources of energy from the time immemorial. The demand of energy has drastically increased with the growth in population. To meet the energy requirements, there is a constant search for the alternate sources of energy. In the quest of meeting energy requirements, one should also be considerate towards its effects on the environment. At present, most of the energy requirements are fulfilled by non-renewable resources. However, these resources suffer from several disadvantages like limited availability, long time duration for their formation and non-sustainable, greenhouse gas emissions. The emission of carbon dioxide, methane and other harmful gases that are the main cause of global warming may result change in the weather patterns which in turns affect the ecosystems (Wang et al. 2020a). Apart from the air pollution, excessive usage of fossil fuels results in water pollution, land pollution, climate change etc. All these pollutants severely affect the human health and environment. Burning of coal emits mercury, which on inhalation by the human beings and other animals may lead to neurological disorders (Munawer 2018). Many pollutants emitted due to burning of fossil fuels may enter the water cycle and cause acidic rain. This acidic rain when mixes with the water bodies can severely harm the aquatic animals and plants. Acidic rain may also cause corrosion to machineries and monuments (Munawer 2018).

Thus, there is an urgent need of technology which can replace these resources with sustainable energy sources. Renewable energy resources are the potential candidates that could produce energy much greater than the actual energy requirements. These are the sources which can be continuously replenished by nature such as sunlight, wind, water etc. Our ancestors had realized the importance of these resources very long back and described them as powerful energy sources in Vedas. In the twenty-first century, huge investment and

Energy: Crises, Challenges and Solutions, First Edition. Edited by Pardeep Singh, Suruchi Singh, Gaurav Kumar, and Pooja Baweja.

growth have been proposed in alternative energy sources all over the world (IEA 2019). IEA also expects the growth of renewable energy share to reach 30% by 2024. Recent research is based on either to make use of renewable resources or to develop new technologies which could replace non-renewable sources and meet the future energy demands (Chel and Kaushik 2018; Gielen et al. 2019; Peter and Mbohwa 2019). Recently, the generation of energy from waste products has drawn the attention of scientists due to their reduced costs, increased profits and reduced environmental impacts (Shariar and Bustam 2012). In this chapter, recent developments in the alternate energy sources and their contribution in sustainable development have been discussed.

3.2 Different Innovations Employed in Major Types of Alternative Sources of Energy

Renewable energy sources are the clean and sustainable energy supplies. Various types of renewable energy resources have been used to generate energy (Scheme 3.1). The driving force behind the development of renewable resources and innovative technologies is energy security. The security of energy is the prime requirement of a country's development policies. Apart from ensuring energy security, these sources have resulted in improving the economy of the countries due to more industrial set-ups and air quality due to less pollution. Importing fuels can affect country's economy. Instead of importing non-renewable fuels, if a country set up its own industries making use of renewable resources, it will open the scope of investment in other sectors. In this manner, country's economic growth will be improved. As these resources do not cause pollution and contribute towards sustainable development, these are also called clean energy sources. Recently, various alternative sources of energy utilize innovative technique for their improvement in working and energy generation (Table 3.1). Various types of renewable energy sources are generating energy by using different techniques which are discussed below.

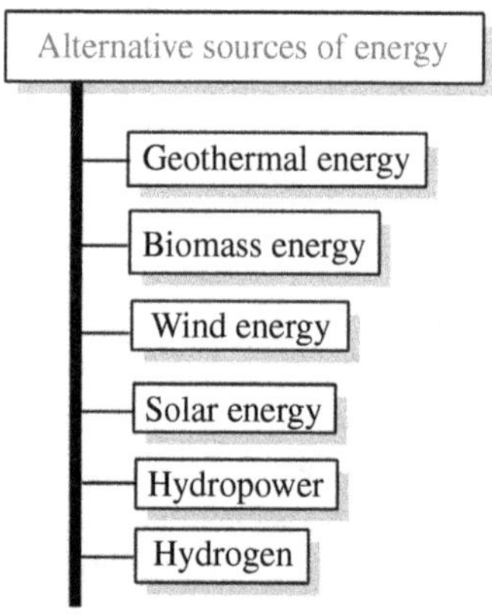

Scheme 3.1 Different alternative sources of energy.

3.2.1 Solar Energy (Semiconductor Technology to Harness Solar Power)

The energy from sun is one of the predominant and sustainable renewable energy resources for generating energy. It is obtained from the sun and utilized for heating, lighting and generating electricity. Passive solar and active solar are the two types of solar energy. Passive solar energy utilizes the direct and indirect thermal energies from the sun to heat water, whereas active solar energy can utilize the electromagnetic radiation from the sun for the generation of electrical energy (Alrikabi 2014). However, the topographical region and seasons are the major factors which affect the solar energy harnessing. In addition, it involves

Table 3.1 Various technologies employed in the development of renewable resources and their applications.

Renewable energy sources	Technologies involved	Applications	Challenges
Solar	Semiconductor technology for solar panels, photovoltaic cells, solar cooker, dye-sensitized solar cell, quantum dots	Electricity generation, heating, cooking, satellite, remote industries	Costly due to semiconductor technology involved, weather dependent
Wind	Conventional wind mills, onshore and offshore wind turbines	Power generation	Depends on direction and speed of wind, maintenance cost of wind tower and turbine blades
Geothermal	Geothermal heat pumps	heating, bathing, cooking, power generation	Location specific, high cost involved in geothermal heating/ cooling systems
Hydropower	Building dams, control techniques for its operation	Power generation	Cost involved in building dams, environmental concerns, relocating residents
Biomass	Combustion, gasification, hydrolysis and fermentation, pyrolysis, anaerobic digestion	Electricity generation, heating, fuel for cooking and transport	Undesirable high moisture content, low density causes handling of large amount of biomass
Hydrogen	Nanomaterials for storage, hybrid fuel cell turbines, e-fuel	Fuel cell electric vehicle, electricity generation	Difficult to store and transport

high cost for implanting solar panels as compared to conventional resources based on plants to generate electricity.

Solar energy technologies (i.e. photovoltaics [PV], solar thermal) offer a clean, renewable and domestic energy source for sustainable energy future in comparison to the traditional energy sources. PV can be used to generate electricity through semiconductor technology in the form of solar panels (Scheme 3.2). Solar PV power can be used for a number of applications such as in satellites, remote industries, solar home systems etc.

The photons received from the sun excite electrons and produce electricity. PV cell is an example of sustainable energy harvester which can transform solar energy into electricity. It makes use of semiconductor technology. There are three different generations of PV cell, namely first generation, second generation and third generation (Scheme 3.3).

Crystalline silicon cells are the first generation and the most common types of PV cells due to the matured technology and high efficiency. They can absorb photons with energies lying in the visible light as well as some photons from the infrared range of the spectrum. However, due to the high cost of monocrystalline cells, different alternative materials such as polycrystalline, GaAs materials are used to manufacture solar cells (Deb 1998). Recently,

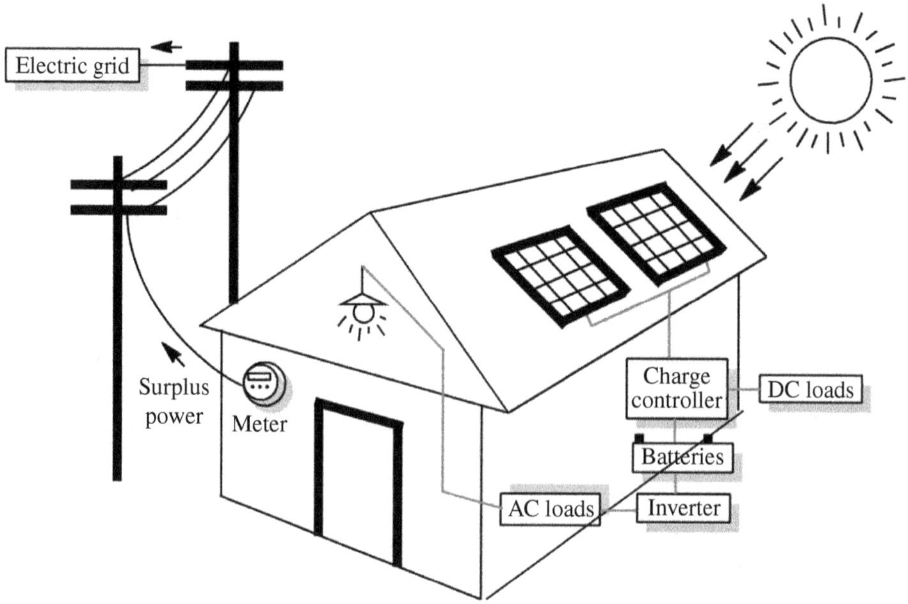

Scheme 3.2 Schematic representation of photovoltaic cell.

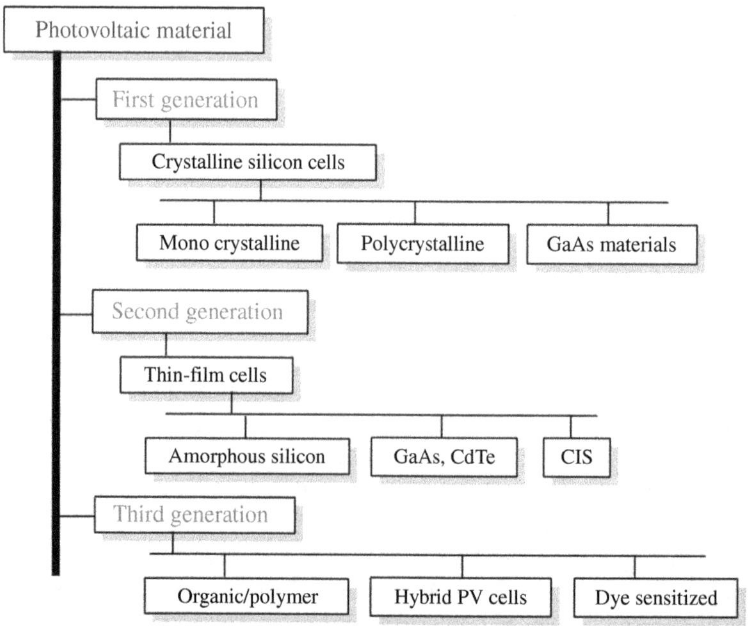

Scheme 3.3 Schematic representation of different types of photovoltaic cells.

low-cost, highly flexible and lightweight thin-film-based solar panels have been utilized in the power generation; it is also known as second-generation cell (McCann et al. 2001).

Thin-film solar panels with enhanced efficiency have been developed using various materials such as amorphous silicon, cadmium tellurides, copper indium gallium diselenide etc. (El Chaar and El Zein 2011). Third-generation solar cells constitute a variety of PV cells such as the concentrated solar PV (CPV) system, hybrid cells, organic PV cells, dye-sensitized solar cells (DSSCs) etc. (Scheme 3.3). CPV technique has been used to generate electricity by making use of mirrors to focus sunlight on the receiver using concentrating solar power technique (Looser et al. 2014). In comparison to the other solar cells, these cells involve the utilization of lesser materials and offer viable and cost-effective technology with higher efficiency. Organic solar cells are another PV cell which is in demand nowadays as an alternative for solar cells due to their mechanical flexibility, low cost, lightweight etc. However, these cells possess lower efficiency in contrast to other types of solar cells (Dou et al. 2012). Recently, the high-performance solar cells have been developed using different donor, acceptor and/or bulk hetero junction (BHJ) structures (Bagienski and Gupta 2011; Parvathy Devi et al. 2011). Recently, the hybrid cells have been fabricated by using a combination of both inorganic (e.g. Si, nanoparticles) and organic (e.g. conjugated polymers or small organic molecules) materials (Moule et al. 2012). Also, DSSCs have drawn the attention of researchers due to their less expensive and simpler methods of manufacturing as compared to other methods (Figure 3.1) (Shalini et al. 2015). This method makes use of dyes to convert solar energy to electricity. The dyes used may consist of organic, metal–organic and inorganic molecules or may be extracted naturally from different parts of plants (Xing et al. 2016; Narayan 2012; Hao et al. 2006).

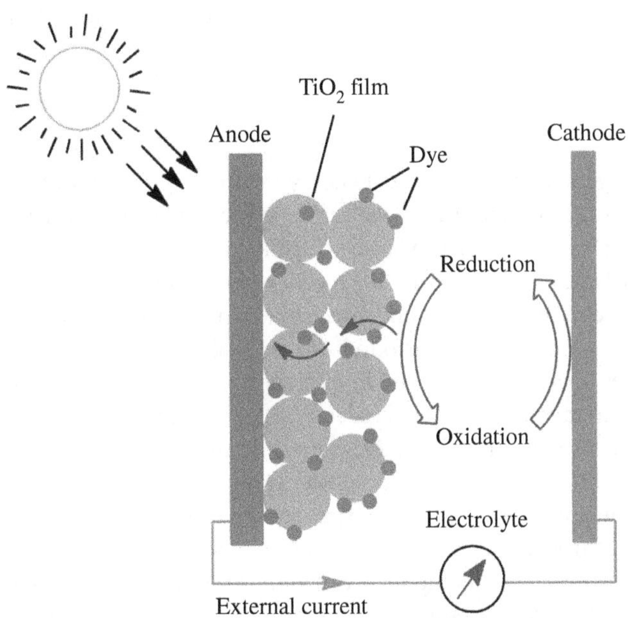

Figure 3.1 Schematic representation of dye-sensitized solar cells. *Source:* Modified from Shalini et al. (2015).

Nowadays, researchers are attracted towards development of smart windows which act as energy regulators and energy harvesters (Wang et al. 2016). It has been reported that using a coating of semi-transparent perovskite solar cell and multilayer nanophotonic on smart window device can be used as an energy harvester (Wang et al. 2020b). Various PV technologies have been employed in different areas with a cost-effective and enhanced technology. Similarly, Quantum dots (QD)-based LEDs have been employed in high-power devices with enhanced efficiency. They posses several advantages such as colour purity, broad wavelength range, stability and low production cost. Recently, QDs have been employed in solar PVs for the development of QD-based solar cells (Ambika and Singh 2018). Solar energy can also be used for heating systems with different types of solar collectors. Flate plate solar collector (FPSC), integrated collector storage systems and evacuated tube collectors are some of the examples of solar collectors (Kabeel et al. 2016; Mallah et al. 2019). FPSC transfers solar energy radiations to circulating fluid in collector pipe to increase its temperature. Nowadays, researchers are using nanofluids instead of regular fluid due to its more effective thermal properties of heat absorption and transfer (Zayed et al. 2019). Solar cookers with compound parabolic concentrators have been developed which is very beneficial in rural communities, where wood is still used as a fuel and its burning causes respiratory problems (Lopez Sosa et al. 2014). It also provides the most important source of electricity for carrying out space programmes. According to IEA report 2019, there is a prediction that contribution of solar energy will be 60% of the renewable growth and expansion due to solar PV in renewable power capacity by 50% till 2024 (IEA 2019). Though it is advantageous due to low maintenance cost and forever availability, this energy can be utilized in favourable seasons and at suitable geographical regions. The main concern to utilize this energy source is its dependence on weather conditions, resulting in reduced solar power generation.

3.2.2 Hydropower

This methodology involves the generation of electrical energy by utilizing the renewable source of water (Kaunda et al. 2012). Generally, hydroelectric power is generated by constructing the dams near moving water bodies. The basic principle involves the utilization of the kinetic energy of moving water which operates the mechanical dynamos to generate electric power. However, this technique suffers from some disadvantages like ecological imbalance due to geological changes in the area surrounding the dams.

Various innovative techniques have been employed in hydropower for the reduction of dam construction and power generation costs (Kougiasa et al. 2019). There are some control techniques such as passive control techniques, active control techniques and magneto-rheological control techniques for the operation of hydropower plants effectively. Recently, fish-friendly turbine such as Alden turbine and Minimum gap runner turbine was designed for the smooth passage of fishes, through the turbine with minimum pressure within water passage. Alden turbine operates at a slower rotational speed with three blades to decrease fish death due to blade strike. In case of minimum gap runner turbine, the gaps between the adjustable runner blade and the hub have been reduced as compared to conventional turbines (Hogan et al. 2014).

3.2.3 Wind Energy

A wind mill is a machine which utilizes the kinetic energy of wind for the production of electricity using a generator. Generally, wind turbine can work in the horizontal axis and vertical axis. Wind turbines location has huge impact on the performance of the machine. Wind energy is eco-friendly, non-polluting and produces no hazardous emissions. Also, while utilization of wind, the land can be used for farming purpose and to collect solar energy (Alrikabi 2014; Zhang et al. 2020). Recently, it is reported that the use of wind energy can save twice of the fossil fuel which in turn has a positive impact on socio economy and environment in Europe (Izquierdo and Rio 2020). As the wind farms are set up on agricultural or fallow lands, they fetch extra income to the farmers or other persons involved in these activities, hence improving the economic condition. The challenge to harness wind energy is the cost involved in repairing wind towers and turbine blades which may get damaged due to high winds.

Various innovative techniques have been used in the development of generation of electricity using wind energy. The power generation using wind energy is reported to be significant in European countries (Monforti et al. 2016; Zhang et al. 2020). Recently, an advanced concept airborne wind energy (AWE) which can transform wind energy into electricity by the use of aerodynamic or aerostatic lift devices has been developed (Ahrens et al. 2013). AWE technology can operate at higher altitude and utilizes less material, lower cost of manufacturing in comparison to conventional turbines. Similarly, offshore floating wind turbine has been developed to generate electricity in water depths (Jonkman and Matha 2011). These types of turbines can significantly increase the sea area available for offshore wind farms. Floating hybrid energy platforms are another innovative technique used to generate electricity. Hybrid energy platforms are a technique which utilizes the combination of at least two different energy types or technology types for power generation (Roy et al. 2018). For example, blackbird system utilizes the combination of wind with wave energy (Golightly 2017). SCDnezzy is another example of hybrid system which exploits both conventional and airborne wind energy converter on the same platform (De Vries 2018). Japan is also in the process of replacing nuclear reactors with offshore wind turbines after Fukushima nuclear disaster. The hybrid power system comprising distributed wind energy source and solar energy results in enhanced power generation (Allan et al. 2015). The modifications in other parts of wind mill have also been developed for the enhanced functioning. For example, the development of smart rotors involves the adjustment of the blade size according to harsh environment wind conditions, etc. (De Vries 1979; Barlas and van Kuik 2010). Nowadays, diffuser-augmented wind turbines have been developed with a funnel-shaped wind that can collect and concentrate the entering wind. Further, the performance of the diffuser can be enhanced by modifying the diffuser by the addition of a broad ring around the exit point and an inlet shroud at the entrance (Ohya et al. 2017). Recently, large single rotor has been replaced with a multiple-rotor system (MRS) to enhance the efficiency of a wind turbine.

3.2.4 Geothermal Energy

Geothermal energy is derived from the earth's sub-surface to produce steam, which then powers turbines to produce electricity (Alrikabi 2014). It has historical significance and is used for heating, bathing and cooking purposes. The advantage of using this source of energy over solar and wind energy is its availability all the time, i.e. it could be utilized at any day of the year.

Geothermal energy is the form of energy which can be harnessed by using geothermal heat pumps (GHPs) for heating and cooling purposes. GHP transfers the heat of a building into the ground during summer and carries heat from the ground into the building during winters (Pareek et al. 2020). In United States, geothermal energy has brought economic growth in both rural and urban areas. Recently, a model for the estimation of economic potentials of geothermal energy in European countries has been developed (Limberger et al. 2014). It is estimated that the contribution of geothermal plants towards electricity generation in European countries will be enhanced significantly by 2050 (Longa et al. 2020; Olabi et al. 2020). These are expected to be useful energy resources in future due to their presence in abundance. Recently, a dual-source heat pump prototype (DSHP) as air-conditioning system has been developed with overall energy saving (Bottarelli 2019). Integrating geothermal energy with other renewable sources results in increased output efficiency (McTigue et al. 2020).

3.2.5 Biomass Energy

Biomass is the organic matter obtained from animal wastes and plants residues. Biomass energy is generated when organic matter is converted to energy. It comprises solid biomass, liquid biofuels, biogas and other wastes. Solid biomass includes organic matter such as wood, other agricultural products and animal dung. Liquid biofuels such as biodiesel, mainly used in transport sector, are produced using plant and animal wastes. Biogas is released in the form of methane and carbon dioxide when organic matter undergoes anaerobic digestion. It is mainly used for heating and power generation applications (Caposciutti et al. 2020).

The non-renewable fossil fuels like petroleum, diesel, natural gas etc. are depleting rapidly with the rise of industrialization. Innovations are required to generate energy from wastes, called bioenergy, for a sustainable approach. In this manner, bioenergy generation solves the current problem of waste disposal and hence resolving the issues related to climatic changes (Dhanya et al. 2020). The energy derived from biomass is used for heating, electricity and fuels purposes. Biomass is the organic matter obtained from animal wastes and plants residues. Recently, a variety of biomass conversion processes have been developed and exploited such as thermochemical processes, bio-refinery, biogas etc. Biomass is converted into gas such as methane by gasification process, which in turn is used to generate electricity. Biofuels in liquid form such as ethanol can also be derived from biomass for power generation purposes. Biomass has been utilized as the potential source of alternative energy in Bangladesh for cooking purposes (Uddina et al. 2019). Recently, researchers have developed sustainable and innovative methods for the production of energy. For example, various food and plant wastes have been utilized for the production of biomethane and bioethanol which may further be used as energy harvesters for various purposes such as transportation, power plants and composting (Shuit et al. 2009; Chiew and Shimada 2013; Ma et al. 2017; Gutierrez et al. 2018). Pomelo peels when treated with phosphoric acid and KOH to prepare carbonaceous structures result in cost-effective, eco-friendly, high capacitance materials (Hong et al. 2014; Liang et al. 2014). Bioelectricity has been generated from orange peel and other food waste-derived microbial fuel cells without chemical treatment

(Miran et al. 2016; Xin et al. 2018). The cellular imaging has been reported using highly luminescent orange juice-derived C-dots (Sahu et al. 2012). *Allium sativum*-derived C-dots have been used as in vitro; biomarker for living cells (Das et al. 2019). Residues of banana crops yielded glucose, ethanol and, a very important component, polyhydroxybutyrate, which reduced the requirement of energy and water by 30.6 and 35%, respectively (Moncada et al. 2013; Naranjo et al. 2014; Quinaya and Alzate 2014).

3.2.6 Hydrogen as a Fuel

It is present in abundance in the form of water and hydrocarbons. Recently, researchers gained much interest in studying hydrogen as an alternative energy source due to its emission-free burning and higher efficiency. Hydrogen has a wide range of applications as a fuel, namely domestic purposes, vehicles and spacecraft. Many techniques such as electrolytic, thermal and photolytic processes have been reported to produce pure hydrogen (Pareek et al. 2020). In spite of all these techniques, there are problems with the generation of hydrogen for commercial purposes. Natural gas has been employed as the least expensive fuel for the generation of hydrogen at commercial level (Pareek et al. 2020). The hydrogen production with the help of renewable energy resources such as sunlight and water involves photo-electrochemical and biohydrogen techniques (Kotay and Das 2008).

Hydrogen gas is a clean fuel which can be employed for a variety of applications. However, hydrogen storage is associated with some problems as compared to traditional fossil fuels due to its low density. Recently, carbon nanomaterials such as carbon nanotubes, fullerenes, graphene, activated carbon, carbon fibre, etc. have been employed in the hydrogen storage. Metal doping on carbon nanostructures may increase the hydrogen storage capacity; however, higher storage capacities can be obtained at cryogenic temperature and higher pressure (Mohan et al. 2019). Single-walled carbon nanotubes displayed enhanced hydrogen uptake due to which they can be used to fabricate tanks for hydrogen storage in FCV (Dillon et al. 1997). Carbon fibre reinforced resin matrix composite materials (CFRC) have been employed in the aerospace industry to reduce the weight of vehicle (Zheng et al. 2018). Hydrogen storage can be efficiently performed below 3 or 4 MPa above room temperature by utilizing metal hydration (Das 2016). Large amounts of hydrogen can be reversibly stored in lithium nitride in the temperature range of 170–210 °C (Chen et al. 2002). Zeolites have also been used in hydrogen storage which may be attributed to their high thermal stability, large surface area, low cost etc. (Langmi et al. 2005). Nowadays, fuel cell has been used to generate electricity by the electrochemical oxidation of hydrogen (Figure 3.2).

Fuel cell electric vehicles (FCEVs), using hydrogen fuel cell, have attracted interest of researchers in electric vehicle industries (Cano et al. 2018). Due to the drawbacks associated with the production and transportation of hydrogen fuel cells, it has been replaced by direct liquid fuel cells (DLFCs) (An et al. 2015). Recently, the hybrid fuel cells have been used in electric hybrid vehicles. A hydrogen fuel cell in combination with batteries and other control system can be used to develop a sustainable hybrid vehicle (Briguglio et al. 2010). For example, hybrid fuel cell gas turbine systems have been developed by the combination of high-temperature fuel cells (HTFCs) into cycles with gas turbines which can operate on natural gas, landfill gas, biomass syngas, etc. The above technology can appreciably enhance fuel-to-electricity conversion efficiency with reduced greenhouse

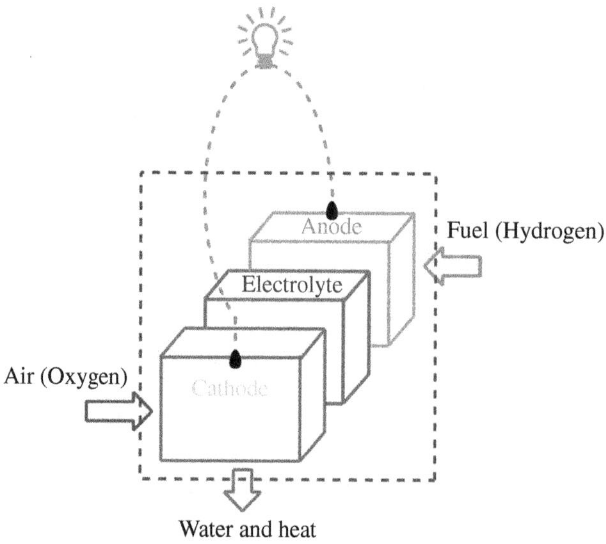

Figure 3.2 Schematic representation of fuel cell and its working.

emissions and other pollutants (Brouwer 2012). Recently, researchers are attracted towards electrically rechargeable liquid fuel, also called e-fuel (Jiang et al. 2019). E-fuel consists of electroactive species such as organic and inorganic materials and suspended particles. Various advantages of e-fuels cells have been reported as compared to conventional liquid fuel cells (Shi et al. 2020). These e-fuel cells are emerging as a powerful technology for power generation systems in an efficient and economical manner. Recently, perovskite materials, as mixed conductors, have been employed for improving the operational efficiency of existing materials with improved cell designs (Skinner 2001).

3.3 Environmental Impacts

Various types of renewable energy sources can have different environmental impacts. For example, solar technologies are very expensive and space consuming. The installation of solar cells may have environmental impact on area and the topography of covered land, ecosystem and biodiversity due to the thermal imbalance of the area under application. Also, solar cell batteries are prepared by the use of harmful and toxic chemicals like heavy metals (indium, tellurium and cadmium). Although solar cells do not produce any emissions during operation, but there is always a potential risk of emissions during manufacturing, dismantling and in case of any accident like fire etc. The harmful substance from solar technologies on release may pose severe environmental threat like soil and water pollution (Turney and Fthenakis 2011). Hydroelectricity is also associated with some disadvantages like the construction of storage hydropower plants involves blocking, diverting or changing the natural course of river systems. The building of dam on a river system results in the flooding of significant amount of land due to which the flora and fauna in the river surroundings are badly affected and may result in loss of biodiversity and ecosystems.

In some cases, the flooding can also force the rehabilitation of the residential population (Anjaneyulu and Valli 2007). Disruption of natural flow of a river may pose negative impact on the natural environment of aquatic lives (Ghosh and Prelas 2011). Various migratory animals like fishes have strong impact of environment on the different stages of their life cycle (Bergengren et al. 2013). Water quality of the stored water in the hydroelectric power plant decreases due to the low concentration of oxygen and lower temperature as compared to normal river water. Also, the release of this water may have negative impacts on downstream plants and animals (Bergkamp et al. 2000). Moreover, the hydropower also results in the emission of greenhouse gases (Kumar et al. 2011). Geothermal power plant leads to the surface instability and earthquakes which may be due to the removal of water and steam from reservoirs within the earth that may result in the sinking of the land above the reservoirs (Bosnjakovic et al. 2019). Wind energy also poses some environmental impacts such as forest clearing and disruption of the natural habitat for some species. The collision of local species populations, like birds and bats, with the blades of wind turbines may result into death which may vary from one wind farm to the other (Jaber 2013). The use of traditional biomass may result in the deforestation, loss of biodiversity, soil degradation, pollution by the emissions of greenhouse gases (GHG) and loss of nutrients due to combustion of cattle dung and crop residues (Trimble et al. 1984). Hydrogen is very difficult to store and transport. Thus, a huge risk is associated with its storage.

3.4 Future Prospects

Renewable energy sources and innovative technologies for energy production will keep on gaining popularity in the coming years. With the more investments global and domestic industries are making to generate energy with renewable resources, the future seems to be very bright in this direction. Huge energy requirements will be fulfilled through renewable energy resources with the use of green technologies. This is the way to achieve sustainable energy system. Though the growth in renewable energy sources is remarkable, still their share in the overall energy consumption is not up to the mark. Various policies regarding that will emerge in near future. Further research should be focused to develop new and improved technologies which can reduce the cost of setting up the power plant. With the advancement towards these resources, the cost of energy sources will decline, industries would be set up, jobs will be created and standard of living will be improved, which will result in better socio-economic condition of a country. So, the alternative energy sources would serve dual purposes, i.e. cost-effective and sustainable.

3.5 Conclusions

Alternative sources of energy offer huge opportunities in the development of energy sector. The adequate energy supply can be maintained by integrating different technologies from diversified fields, such as solar energy, wind energy, hydropower, geothermal energy, biomass, hydrogen fuel etc. Different types of alternative resources can be developed and utilized by different countries depending on their location and availability of resources.

Various innovative techniques have been employed in various classes of renewable energy sources. Solar energy technologies have been employed to develop a variety of innovative techniques. Various PV technologies have been employed in different areas with cost-effective and enhanced technology. In case of hydropower, several modifications have been introduced in the design and operation of turbines which can reduce the mortality of fishes. Similarly, several innovations have been introduced in the generation of power using wind energy which involves the adjustment of blade size according to harsh environment wind conditions, location, etc. The geothermal energy has been integrated with other renewable sources for enhanced output efficiency. Biomass from various food and plant wastes has been utilized for the production of biomethane and bioethanol which may further be used as energy harvesters for various purposes such as transportation, power plants and composting. Recently, hydrogen has been used as an alternative resource for energy generation which can be used for various applications. Although these resources offer several advantageous due to low maintenance cost and sustainability, they are also associated with few drawbacks like dependence on the seasons and at suitable topographical regions, deforestation, loss of biodiversity and ecosystems, soil degradation, surface instability, earthquakes, difficulty in storage and transport etc. These resources can offer sustainable supply of energy with the reduction in local and global atmospheric emissions. Thus, alternative sources of energy are potential substitutes for non-renewable energy resources.

References

Ahrens, U., Diehl, M. and Schmehl, R. ed. (2013) (Eds.), *Airborne Wind Energy*, Berlin and Heidelberg: Springer.

Allan, G., Eromenko, I., Gilmartin, M., Kockar, I., McGregor, P. (2015). The economics of distributed energy generation: a literature review. *Renewable and Sustainable Energy Reviews* 42: 543–556.

Alrikabi, N. Kh. M. A. (2014). Renewable energy types. *Journal of Clean Energy Technologies* 2(1): 61–64.

Ambika Singh P.P. (2018). Nanotechnology: greener approach for sustainable environment. In: *Nanotechnology in Environmental Science* (eds. C. M. Hussain, A. K. Mishra), Germany: Wiley-VCH Verlag 805–824.

An, L., Zhao, T.S., Li, Y.S. (2015). Carbon-neutral sustainable energy technology: direct ethanol fuel cells. *Renewable and Sustainable Energy Reviews* 50: 1462–1468.

Anjaneyulu, Y., Valli, M. (2007) *Environmental Impact Assessment Methodologies* (2.), India: BS Publication.

Bagienski, W., Gupta, M. (2011). Temperature dependence of polymer/fullerene organic solar cells. *Solar Energy Materials and Solar Cells* 95: 933–941.

Barlas, T. K., van Kuik, G. A. M. (2010). Review of state of the art in smart rotor control research for wind turbines. *Progress in Aerospace Sciences* 46: 1–27.

Bergengren, J., Näslund, I., Kling, J. (2013). *Hydroelectric Impact on Aquatic Ecosystems.* Maritime and Water Authority, Gothenburg.

Bergkamp, G., Mccartney, M., Dugan, P., Mcneely, J. (2000). Dams, Ecosystem Functions and Environmental Restoration.

Bosnjakovic, M., Stojkov, M., Jurjevic, M. (2019). Environmental impact of geothermal power plants. *Technical Gazette* 26(5): 1515–1522.

Bottarelli, M. (2019). Testing of a dual-source heat pump coupled with flat-panel ground heat exchangers. *International Journal of Low Carbon Technologies* 14(2): 247–255

Briguglio, N., Andaloro, L., Ferraro, M., Antonucci, V. (2010). Fuel cell hybrid electric vehicles. S. Soylu, (ed.), *Electric Vehicles-the Benefits and Barriers*, DOI: https://doi.org/10.5772/18634.

Brouwer, J. (2012). Hybrid fuel cell gas turbine (FC/GT) combined cycle systems. A.D. Rao (ed.) *Combined Cycle Systems for Near-Zero Emission Power Generation*. Woodhead Publishing Series in Energy, 265–282.

Cano, Z. P., Banham, D., Ye, S. Hintennach, A., Lu, J., Fowler, M., Chen, Z. (2018). Batteries and fuel cells for emerging electric vehicle markets. *Nature Energy* 3: 279–289.

Caposciutti, G., Baccioli, A., Ferrari, L., Desideri, U. (2020). Biogas from anaerobic digestion: power generation or biomethane production? *Energies* 13: 743.

Chel, A., Kaushik, G. (2018). Renewable energy technologies for sustainable development of energy efficient building. *Alexandria Engineering Journal* 57(2): 655–669.

Chen, P., Xiong, Z., Luo, J., Lin, J., Tan, K. L. (2002). Interaction of hydrogen with metal nitrides and imides. *Nature* 420: 302–304.

Chiew, Y. L., Shimada, S. (2013). Current state and environmental impact assessment for utilizing oil palm empty fruit bunches for fuel, fiber and fertilizer – a case study of Malaysia. *Biomass and Bioenergy* 51: 109–124.

Das, L. M. (2016). Hydrogen-fueled internal combustion engines. In: *Compendium of Hydrogen Energy, Volume 3, Hydrogen Energy Conversion*, Woodhead Publishing Series in Energy, 177–217.

Das, P., Ganguly, S., Maity, P.P., Srivastava, H.K., Bose, M., Dhara, S., Bandyopadhyay, S., Das, A.K., Banerjee, S., Das, N.C. (2019). Converting waste *Allium sativum* peel to nitrogen and Sulphur co-doped photoluminescence carbon dots for solar conversion, cell labeling, and photobleaching diligences: a path from discarded waste to value-added products. *Journal of Photochemistry and Photobiology B: Biology* 197: 111545.

De Vries, O. (1979). *Fluid Dynamic Aspects of Wind Energy Conversion*. Natl Aerosp Lab Rep AGARD-AG-243, published by NATO, 156.

De Vries, E. (2018). Aerodyn prepares SCDnezzy for Japan tests. *Wind Power Mon*, 34: 32–34.

Deb, S. K. (1998). Recent developments in high efficiency photovoltaic cells. *Renewable Energy* 15: 467–72.

Dhanya, B. S., Mishra, A., Chandel, A. K., Verma, M.L. (2020). Development of sustainable approaches for converting the organic waste to bioenergy. *Science of the Total Environment* 723: 138109.

Dillon, A. C., Jones, K., Bekkedahl, T. A., Kiang, C. H., Bethune, D. S., Heben, M. (1997). Storage of hydrogen in single-walled. *Nature* 386: 377–379.

Dou, L., You, J., Yang, J., Chen, C. C., He, Y., Murase, S., Moriarty, T., Emery, K., Li, G., Yang, Y. (2012). Tandem polymer solar cells featuring a spectrally matched low-bandgap polymer. *Nature Photonics* 6: 180–185.

El Chaar, L., El Zein, N. (2011). Review of photo voltaic technologies. *Renewable and Sustainable Energy Reviews* 15: 2165–75.

Ghosh, T. K., Prelas, M. (2011). *Energy Resources and Systems*, Volime 2, Netherlands: Springer.

Gielen, D., Boshell, F., Saygin, D., Bazilian, M. D., Wagner, N., Gorini, R. (2019). The role of renewable energy in the global energy transformation. *Energy Strategy Reviews* 24: 38–50.

Golightly C.R. (2017). Blackbird: A hybrid CAES storage anchored mono-TLP VAWT-WEC, Wind Eur. Conf. 28–30 November, Amsterdam.

Gutierrez, E. C., Wall, D. M., Oshea, R., Novelo, R. M., Gomez, M. M., Murphy, J. D. (2018). An economic and carbon analysis of biomethane production from food waste to be used as a transport fuel in Mexico. *Journal of Cleaner Production* 196: 852–862.

Hao, S., Wu, J., Huang, Y., Lin, J. (2006) Natural dyes as photosensitizers for dye-sensitized solar cell. *Solar Energy* 80: 209–214.

Hogan, T. W., Glenn, F. C., Amaral, S. V. (2014). The status of environmentally enhanced hydro*power turbines. Fisheries* 39 *(4):* 164–172.

Hong, K. L., Qie, L., Zeng, R., Yi, Z. Q., Zhang, W., Wang, D., Yin, W., Wu, C., Fan, Q., Zhang, W., Huang, Y. (2014). Biomass derived hard carbon used as a high performance anode material for sodium ion batteries. *Journal of Materials Chemistry A* 2: 12733–12738.

IEA. (2019), *Renewables 2019*, IEA, Paris https://www.iea.org/reports/renewables.

Izquierdo, M. O., del Rio, P. (2020). An analysis of the socioeconomic and environmental benefits of wind energy deployment in Europe. *Renewable Energy* 160: 1067–1080

Jaber, S. (2013). Environmental impacts of wind energy. *Journal of Clean Energy Technologies* 1: 251–254.

Jiang, H., Wei, L., Fan, X., Xu, J., Shyy, W., Zhao, T. (2019). A novel energy storage system incorporating electrically rechargeable liquid fuels as the storage medium. *Science Bulletin* 64(4): 270–280.

Jonkman, J. M., Matha, D. (2011). Dynamics of offshore floating wind turbines-analysis of three concepts. *Wind Energy* 14: 557–569.

Kabeel, A. E., Khalil A., Shalaby, S. M., Zayed, M. E. (2016). Investigation of the thermal performances of flat, finned, and v-corrugated plate solar air heaters. *Journal of Solar Energy Engineering* 138: 51004.

Kaunda, C.S., Kimambo, C.Z., Nielsen, T.K. (2012). Hydropower in the context of sustainable energy supply: a review of technologies and challenges. *ISRN Renewable Energy* 2012: 730631.

Kotay, S. M., Das, D. (2008). Biohydrogen as a renewable energy resource-prospects and potentials. *International Journal of Hydrogen Energy* 33: 258–263.

Kougiasa, I., Aggidisb, G., Avellanc, F., Denizd, S., Lundine, U., Moroa, A., Munteanf, S., Novarag, D., Perez-Diazh, J. I., Quarantai, E., Schildj, P., Heodossiouk, N. (2019). Analysis of emerging technologies in the hydropower sector. *Renewable and Sustainable Energy Reviews* 113: 109257.

Kumar, A., Schei, A., Ahenkorah, A., Caceres Rodriguez, R., Devernay, J. M., M. Freitas, D. Hall, Å. Killingtveit, Z. Liu (2011). Hydropower. In *IPCC Special Report on Renewable Energy Sources and Climate Change Mitigation*. O. Edenhofer, R. Pichs-Madruga, Y. Sokona, K. Seyboth, P. Matschoss, S. Kadner, T. Zwickel, P. Eickemeier, G. Hansen, S. Schlömer, C. von Stechow (eds) Cambridge University Press, Cambridge, United Kingdom and New York, NY, USA 443–496.

Langmi, H. W., Book, D., Walton, A., Johnson, S. R., Al-Mamouri, M. M., Speight, J. D., Edwards, P. P., Harris, I. R., Anderson, P. A. (2005). Hydrogen storage in ion-exchanged zeolites *Journal of Alloys and Compounds* 404–406: 637–642

Liang, Q., Ye, L., Huang, Z. H., Xu, Q., Bai, Y., Kang, F., Yang, Q. H. (2014). A honeycomb-like porous carbon derived from pomelo peel for use in high-performance supercapacitors. *Nanoscale* 6: 13831–13837.

Limberger, J., Calcagno, P., Manzella, A., Trumpy, E., Boxem, T., Pluymaekers, M. P. D., van Wees, J. D. (2014). Assessing the prospective resource base for enhanced geothermal systems in Europe. *Geothermal Energy Science* 2: 55–71.

Longa, F. D., Nogueira, L. P., Limberger, J., Wees, J. D., von der Zwaan, B. (2020). Scenarios for geothermal energy deployment in Europe. *Energy* 206: 118060.

Looser, R., Vivar, M., Everett, V. (2014). Spectral characterisation and long-termper-formance analysis of various commercial heat transfer fluids (HTF) as direct-absorption filters for CPV-T beam-splitting applications. *Applied Energy* 113: 1496–511.

Lopez Sosa, L. B., Gonzalez Avilesa, M., Gonzalez Pereza, D., Gutierrez, Y. S. (2014). Rural solar cookers, an alternative to reduce the timber resource extraction through the use of renewable energy sources: technology transfer and monitoring project. *Energy Procedia* 57: 159–160.

Ma, Y., Cai, W., Liu, Y. (2017). An integrated engineering system for maximizing bioenergy production from food waste. *Applied Energy*, 206, 83–89.

Mallah, A. R., Mohd Zubir, M. N., Alawi, O. A., Salim Newaz, K. M., Mohamad Badry, A. B. (2019). Plasmonic nanofluids for high photothermal conversion efficiency in direct absorption solar collectors: fundamentals and applications. *Solar Energy Materials and Solar Cells* 201: 110084.

McCann, M. J., Catchpole, K. R., Weber, K. J., Blakers, A. W. (2001). A review of thin-film crystalline silicon for solar cell applications. Part 1: native substrates. *Solar Energy Materials & Solar Cells* 68: 135–71.

McTigue, J. D., Wendt, D., Kitz, K., Gunderson, J., Kincaid, N., Zhu, G. (2020). Assessing geothermal/solar hybridization -integrating a solar thermal topping cycle into a geothermal bottoming cycle with energy storage. *Applied Thermal Engineering* 171: 115121.

Miran, W., Nawaz, M., Jang, J., Lee, D.S. (2016). Conversion of orange peel waste biomass to bioelectricity using a mediator-less microbial fuel cell. *Science of the Total Environment* 547: 197–205.

Mohan, M., Sharma, V. K., Kumar, E. A., Gayathri, V. (2019). Hydrogen storage in carbon materials-a review. *Energy Storage* 1: e35.

Moncada, J., Matallana, L. G., Cardona, C. A. (2013). Selection of process pathways for biorefinery design using optimization tools: a Colombian case for conversion of sugarcane bagasse to ethanol, poly-3-hydroxybutyrate (PHB), and energy. *Industrial and Engineering Chemistry Research* 52: 4132–4145.

Monforti, F., Gaetani, M., Vignati E. (2016). How synchronous is wind energy production among European countries? *Renewable and Sustainable Energy Reviews* 59: 1622–1638.

Moule, A., Chang, L., Thambidurai, C., Vidu, R., Stroeve, P. (2012). Hybrid solar cells: basic principles and the role of ligands. *Journal of Materials Chemistry* 22: 2351–2368.

Munawer M. E. (2018). Human health and environmental impacts of coal combustion and post-combustion wastes. *Journal of Sustainable Mining* 17 (2): 87–96.

Naranjo, J. M., Cardona, C. A., Higuita, J. C. (2014). Use of residual banana for polyhydroxybutyrate (PHB) production: case of study in an integrated biorefinery. *Waste Management* 34: 2634–2640.

Narayan, M. R. (2012). Review: dye sensitized solar cells based on natural photo – sensitizers. *Renewable and Sustainable Energy Reviews* 16: 208–15.

Ohya, Y., Miyazaki, J., Goltenbott, U., Watanabe, K. (2017). Power augmentation of shrouded wind turbines in a multirotor system. *Journal of Energy Resourses Technology* 139: 051202.

Olabi, A. G., Mahmoud, M., Soudan, B., Wilberforce, T., Ramadan, M. (2020). Geothermal based hybrid energy systems, toward eco-friendly energy approaches. *Renewable Energy* 147(1): 2003–2012.

Pareek, A., Dom, R., Gupta, J., Chandran, J., Adepu, V., Borse, P. H. (2020). Insights into renewable hydrogen *energy:* Recent advances and prospects. *Materials Science for Energy Technologies* 3: 319–327.

Parvathy Devi, B., Wu, K. C, Pei, Z. (2011). Gold nanomesh induced surface plasmon for photocurrent enhancement in a polymer solar cell. *Solar Energy Materials and Solar Cells* 95: 2102–2106.

Peter, O., Mbohwa, C. (2019). Renewable energy technologies in brief. *International Journal of Scientific and Technology Research* 8(10): 1283–1289.

Quinaya, S. H. D., Alzate, C. A. C. (2014). Plantain and banana fruit as raw material for glucose production. *Journal of Biotechnology* 185: S34.

Roy, A., Auger, F., Dupriez-Robin, F., Bourguet, S., Tran, Q. T. (2018). Electrical power supply of remote maritime areas: a review of hybrid systems based on marine renewable energies. *Energies* 11(7): 1904

Sahu, S., Behera, B., Maiti, T. K., Mohapatra, S. (2012). Simple one-step synthesis of highly luminescent carbon dots from orange juice: application as excellent bioimaging agents. *Chemical Communications* 48: 8835–8837.

Shalini, S., Balasundaraprabhu, R., Prasanna, S., Mallick, T. K., Senthilarasu, S. (2015). Review on natural dye sensitized solar cells: operation, materials and methods. *Renewable and Sustainable Energy Reviews* 51: 1306–1325.

Shariar, K.F., Bustam, H.A. (2012) Waste to energy: a new dimension in generating Electricity in Bangladesh. *International Journal of Engineering and Technology* 4 (4): 480–483.

Shi, X., Huo, X., Ma, Y., Pan, Z., An, L. (2020). Energizing fuel cells with an electrically rechargeable liquid fuel. *Cell Reports Physical Science* 1(7): 100102.

Shuit, S. H., Tan, K. T., Lee, K. T., Kamaruddin, A. H. (2009). Oil palm biomass as a sustainable energy source: a Malaysian case study. *Energy* 34: 1225–1235.

Skinner, S. J. (2001). Recent advances in perovskite-type materials for solid oxide fuel cell cathodes. *International Journal of Inorganic Materials* 3(2): 113–121.

Trimble, J. L., Hook, R. I. V., Folger, A. G. (1984). Biomass for energy: the environmental issues. *Biomass* 6(1–2): 3–13.

Turney, D., Fthenakis, V. (2011). Environmental impacts from the installation and operation of large-scale solar power plants. *Renewable and Sustainable Energy Reviews* 15: 3261–3270.

Uddina, M. N., Taweekunb, J., Techatoa, K. Rahman, M.A., Mofijur, M., Rasul, M.G. (2019). Sustainable biomass as an alternative energy source: Bangladesh perspective. *Energy Procedia* 160: 648–654

Wang, Y., Runnerstrom, E. L., Milliron, D.J. (2016). Switchable materials for smart windows. *Annual Review of Chemical and Biomolecular Engineering* 7: 283–304.

Wang, C., Yu, S., Guo, X., Kearney, T., Guo, P., Chang, R., Chen, J., Chen, W., Sun, C. (2020b). Maximizing solar energy utilization through multicriteria pareto optimization of energy harvesting and regulating smart windows. *Cell Reports Physical Science* 1(7): 100108.

Wang, J., Obradovich, N., Zheng, S. (2020a). A 43-million-person investigation into weather and expressed sentiment in a changing climate. *One Earth* 2 (6): 568–577.

Xin, X., Ma, Y., Liu, Y. (2018). Electric energy production from food waste: microbial fuel cells versus anaerobic digestion. *Bioresource Technology* 255: 281–287.

Xing, Q., Li, S., Li, H., Yan, R., Wang, X., Hou, L. (2016). Indolo[3,2-b] carbazole-based multi-donor-π-acceptor type organic dyes for highly efficient dye-sensitized solar cells. *Journal of Power Sources* 319: 39–47.

Zayed, M. E., Zhao, J., Du, Y., Kabeel, A. E., Shalaby, S. M. (2019). Factors affecting the thermal performance of the flat plate solar collector using nanofluids: a review. *Solar Energy* 182: 382–396.

Zhang, M., Zhang, J., Li, Y., Yu, J., Qin, J., Wei, K., Song, L. (2020). Multi-site measurement for energy application of small distributed wind farm in complex mountainous areas. *Energy Reports* 6: 1043–1056.

Zheng, H., Zeng, X., Zhang, J., Sun, H. (2018). The application of carbon fiber composites in cryotank *Solidification*. doi: https://doi.org/10.5772/intechopen.73127.

4

Energy and Development in the Twenty-First Century – A Road Towards a Sustainable Future

An Indian Perspective

Shikha Menani and Kiran Yadav

Department of Commerce, PGDAV College, University of Delhi, New Delhi, India

4.1 Introduction

The green planet – Our Earth which is the only known planet that could sustain life because of its water, oxygen and other life sustaining features has been deteriorated by the so-called civilization. The condition has been deteriorated to such an extent that even if all the emissions stop still the reversal would require huge efforts in terms of time and resources (European Commission 2020). Destruction of beautiful coral reefs, damage to the rainforests, depleting air quality, there is nothing that has been left untouched because of the unmindful usage of natural resources by the specie called human being (The Guardian 2019). One of the biggest problems that all the countries face is that of a fall in agricultural productivity, increasing temperatures and lack of water availability. All this is because of the unmindful use of energy sources which has not only exhausted the reservoirs of a few of the essential sources but also polluted the environment to such an extent that many countries have come up with various rules and regulations related to chemicals and hazardous substance usage. Scientists have high confidence that global temperatures will continue to rise for decades to come, largely due to greenhouse gases produced by human activities. The Intergovernmental Panel on Climate Change (IPCC), which includes more than 1300 scientists from the United States and other countries, forecasts a temperature rise of 2.5–10° F over the next century (NASA 2020). Various international organizations like the United Nations Environment Programme, Greenpeace, Global environment facility, International Union for Conservation of Nature and many more are trying their level best to minimize the impact of affluent usage of energy sources on the environment. Even at the micro/country level, various non-government organizations are working towards promoting non-conventional and renewable sources of energy like wind, solar among others. The outcome is visible by the latest reports of carbon emissions which provide that in spite of the carbon emission showing a terrible increase of 2.7% in the year 2018–2019 which was among the highest in the last seven years, the year 2019–2020 has in fact flattened the curve by reducing the amount of carbon emissions compared to the

Energy: Crises, Challenges and Solutions, First Edition. Edited by Pardeep Singh, Suruchi Singh, Gaurav Kumar, and Pooja Baweja.

previous year (Carbonbrief.org 2020). The largest decline across the countries has been by few advanced countries like the United States, European Union, Germany, Japan and the United Kingdom, while other emerging countries like China and India could not contribute much in the decline. The situation is nothing to be elated about as there has been a reduction of only half a percent over the last three decades and the advanced countries are in fact still the ones who emit the maximum carbon in the atmosphere. The rate at which the population is growing in the world would put even more pressure on the countries to deduce energy efficient, cost effective and cleaner sources of energy to achieve the target of carbon neutrality by the year 2050. It is expected that the steel sector in India will be the biggest carbon emitter in the years to come because of increased demand in the country. Though the country has been successful in reducing the carbon emission in certain sectors, the same cannot hold for the steel and iron industry. These industries have to start researching decarbonization technologies to reduce carbon footprints and achieve the goal of sustainable development.

4.2 Energy Consumption and Economic Development

Energy consumption and energy demand is considered as the trigger for economic development. The trend has shown that there is a positive relationship between economic development and energy demand as with the growth in a country's production, the energy demand also peaks to keep pace with the demand of the rising population (Rühl et al. n.d.). Because of the high positive correlation found between energy demand and economic development, many economists started using the word 'Energy based Economic Development' as is synonyms with the development in industrialized and developed countries of United States, United Kingdom, Japan and Europe. There have been various studies that provide how crucial is the role of energy in the economic development as provided by Stern, 2012, in his study, wherein it was empirically revealed that there is a strong cointegration between energy consumption and economic growth, and it is the energy usage that granger causes the economic development, however over a period of time, the energy usage per unit of economic growth measured by gross domestic product (GDP) has drastically reduced especially in case of developed countries and that is because of the realization of the fact that it is the sustainable development that is the need of the hour and not just unmindful use of energy led growth. The change in the usage is also an indication of how few countries have shifted from the use of conventional (polluting) sources of energy like coal to better sources of energy (electricity, solar and wind; Stern 2012). Similar results have been obtained in the studies of Carley, Toman and Jemelkova, Bhamra, Damtoft and many more wherein energy consumption is not only found to be an important element but also the causation factor for economic development.

There is no second thought that energy is a crucial element for any development, however for sustainable development what is required is that countries move to alternative sources of energy that are not only efficient but also clean and do not pose any threat to sustainable development.

In case of India too, there has been a high spike in the economic growth which has benefitted all the sectors of the economy but there is also a negative aspect to it and that is the degradation of its environment bringing it to the forefront as being one of the most polluted countries and also facing the scarcity of energy sources. In fact, India has come down on the 'Environmental Performance Index' which shows the adverse impact of the development on the environment and inability of the country to sustain its development in the near future. Moreover, with the target of making India a $5 trillion economy by the year 2024–2025 would also necessitate that the energy supply is spiked up which would further aggravate the environmental issue if the country does not shift to cleaner and sustainable sources of energy. As per an; estimate to achieve the target of economic growth, India would have to scale up the energy demand that would make India the largest user of the energy by the year 2035. The usage of renewable sources of energy would not only help in solving the environmental issues but also help in generating employment opportunities along with bringing social welfare.

'Fossil fuels currently make up about 75% of India's energy mix but as the world moves towards cleaner technologies, the country will play a significant role in shaping the planet's future climate picture. The Organisation of the Petroleum Exporting Countries (OPEC) has projected India's oil demand to rise by 5.8 million barrels per day (bpd) by 2040 and account for about 40% of the overall increase in global demand during the next 20 years (India 2020 – Analysis – IEA 2020)'

To meet the ever-increasing energy demand as well as to meet the development criteria, the country has come up with various Acts, regulations and innovative ideas which are mentioned in the sections ahead. The success however is dependent upon the adaptability and the cost effectiveness of the techniques being used. A table representing the data of GDP and CO2 emission for India from 1960 to 2016 is presented in Table 4.1 which provides how the GDP growth and carbon emission has reached at an all-time high over the

Table 4.1 India Carbon (CO2) Emissions 1960–2020.

	Total	Per Capita	GDP	Per Capita	Growth
Date	Kilotons of Co2	Metric Tons Per Capita	GDP (Billions of US $)	Per Capita (US $)	Annual % Change
31 December 1960	120 581.961	0.2676	37.02988	82.1886	
31 December 1961	130 402.187	0.2837	39.23244	85.3543	3.7227
31 December 1962	143 467.708	0.3059	42.16148	89.8818	2.9311
31 December 1963	154 083.673	0.3218	48.42192	101.1264	5.9944
31 December 1964	150 647.694	0.3082	56.48029	115.5375	7.453
31 December 1965	165 972.087	0.3325	59.55485	119.3189	−2.6358
31 December 1966	171 765.947	0.337	45.86546	89.9973	−0.0553
31 December 1967	172 238.99	0.331	50.13494	96.3391	7.826
31 December 1968	187 336.029	0.3525	53.08546	99.876	3.3879
31 December 1969	190 724.337	0.3512	58.448	107.6223	6.5397
31 December 1970	195 143.072	0.3515	62.42248	112.4345	5.1572
31 December 1971	205 869.047	0.3625	67.35099	118.6032	1.6429
31 December 1972	217 849.136	0.3749	71.46319	122.9819	−0.5533
31 December 1973	224 343.393	0.3772	85.51527	143.7787	3.2955

Table 4.1 (Continued)

	Total	Per Capita	GDP	Per Capita	Growth
Date	**Kilotons of Co2**	**Metric Tons Per Capita**	**GDP (Billions of US $)**	**Per Capita (US $)**	**Annual % Change**
31 December 1974	231 992.755	0.3811	99.5259	163.4781	1.1853
31 December 1975	252 201.592	0.4048	98.4728	158.0362	9.1499
31 December 1976	263 785.645	0.4137	102.7172	161.0921	1.6631
31 December 1977	279 051.366	0.4277	121.4873	186.2135	7.2548
31 December 1978	283 096.067	0.4241	137.3003	205.6934	5.7125
31 December 1979	296 891.321	0.4347	152.9917	224.001	−5.2382
31 December 1980	314 016.211	0.4493	186.3253	266.5778	6.7358
31 December 1981	338 838.134	0.4736	193.4906	270.4706	6.0062
31 December 1982	349 637.449	0.4775	200.7151	274.1113	3.4757
31 December 1983	378 669.088	0.5053	218.2623	291.2381	7.2889
31 December 1984	388 118.947	0.5061	212.1582	276.668	3.8207
31 December 1985	426 673.785	0.544	232.5119	296.4352	5.2543
31 December 1986	457 571.927	0.5706	248.986	310.4659	4.7766
31 December 1987	488 481.07	0.5959	279.0336	340.4168	3.9654
31 December 1988	527 563.956	0.63	296.589	354.1493	9.6278
31 December 1989	579 008.299	0.6769	296.0424	346.1129	5.9473
31 December 1990	619 154.615	0.709	320.979	367.5566	5.5335
31 December 1991	658 189.83	0.7385	270.1053	303.0556	1.0568
31 December 1992	699 087.881	0.7688	288.2084	316.9539	5.4824
31 December 1993	723 697.118	0.7803	279.296	301.159	4.7508
31 December 1994	764 730.848	0.8087	327.2756	346.103	6.6589
31 December 1995	811 562.105	0.8419	360.282	373.7665	7.5745
31 December 1996	882 324.204	0.8982	392.8971	399.9501	7.5495
31 December 1997	917 685.085	0.9169	415.8678	415.4938	4.0498
31 December 1998	936 221.77	0.9183	421.3515	413.2989	6.1844
31 December 1999	995 766.516	0.9593	458.8204	441.9988	8.8458
31 December 2000	1 031 853.463	0.9766	468.3949	443.3142	3.841
31 December 2001	1 041 152.975	0.9685	485.441	451.573	4.824
31 December 2002	1 054 258.833	0.9643	514.9379	470.9868	3.804
31 December 2003	1 099 597.621	0.9893	607.6993	546.7266	7.8604
31 December 2004	1 154 320.262	1.0219	709.1485	627.7742	7.9229
31 December 2005	1 222 563.132	1.0653	820.3816	714.861	7.9234
31 December 2006	1 303 717.509	1.1186	940.2599	806.7533	8.0607
31 December 2007	1 407 607.286	1.1897	1216.735	1028.335	7.6608

(Continued)

Table 4.1 (Continued)

	Total	Per Capita	GDP	Per Capita	Growth
Date	Kilotons of Co2	Metric Tons Per Capita	GDP (Billions of US $)	Per Capita (US $)	Annual % Change
31 December 2008	1 568 379.567	1.3063	1198.896	998.5223	3.0867
31 December 2009	1 738 645.711	1.4278	1341.887	1101.961	7.8619
31 December 2010	1 719 690.988	1.3933	1675.615	1357.564	8.4976
31 December 2011	1 836 561.945	1.4689	1823.05	1458.104	5.2413
31 December 2012	2 013 788.055	1.5909	1827.638	1443.88	5.4564
31 December 2013	2 030 971.617	1.5856	1856.722	1449.606	6.3861
31 December 2014	2 232 729.957	1.7233	2039.127	1573.882	7.4102
31 December 2015	2 337 749.17	1.7843	2103.588	1605.605	7.9963
31 December 2016	2 407 671.526	1.8178	2294.798	1732.564	8.2563

Source: https://www.macrotrends.net/countries/IND/india/carbon-co2-emissions (Modified Data from India Carbon (CO2) Emissions 1960-2021, Macrotrends LLC).

years strengthening the argument that economic development and carbon mission is not mutually exclusive but tend to coexist (Macrotrends.net).

4.3 Environmental Issues – A Corollary of Economic Development

The biggest invention for human civilization was the invention of fire and other sources of energy like coal, oil and gas. A nation that had no access to energy was considered weak, backward and not suitable for development as energy is considered to be the soul of any economic progress. Energy is still considered as one key element for not only enhancing the growth of an economy but also has the power to change the face of the entire nation. Though such an inevitable tool is now under scrutiny because of the harmful impact it has on the environment. In the words of Douglas Fay, 'At the heart of modern society lies an economy driven by energy use. Unfortunately, the same energy that brings us comfort, convenience, and prosperity also bring us pollution, impoverishment, and global warming.'. Energy consumption has been on an increase since the beginning of industrialization and there is usually a positive relationship between population increase and energy usage and hence the onus of the maximum use of energy is upon the heavily and densely populated countries of the world – India and China. Few environmental issues that our planet is facing because of the unmindful use of energy is listed using a mindmap given in Figure 4.1 below:

All the energy sources can be basically divided into renewable and non-renewable where the latter ones are not only exhaustible but also cause a lot of pollution like coal and oil. These traditional sources of energy in fact have been the main contributors to environmental degradation. The renewable and non-polluting sources of energy like solar energy and

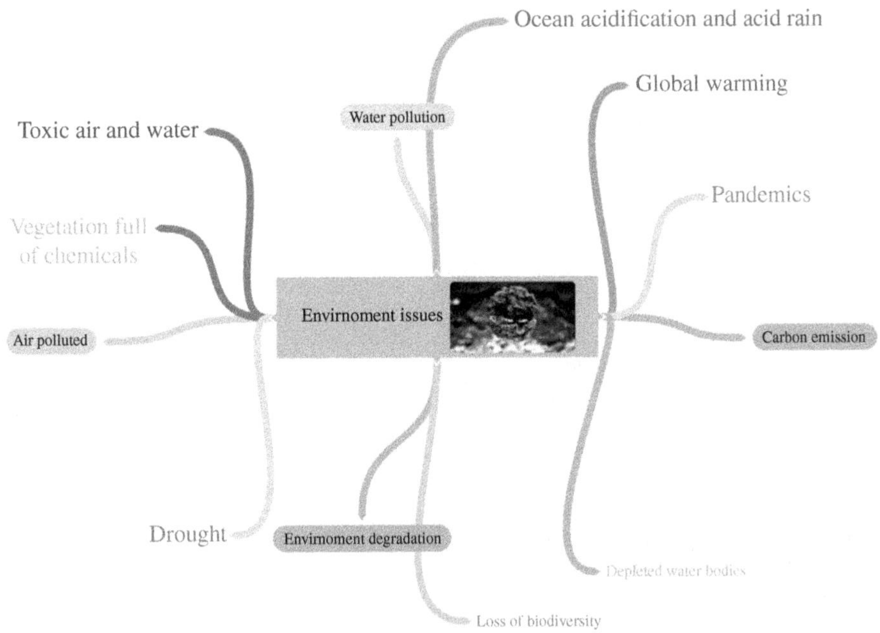

Figure 4.1 Mindmap for various environmental issues. *Source:* Author's creation using Coggle software.

wind energy are expensive in the initial stages but as the country becomes accustomed to its usage, the cost per capita in fact drastically reduces.

A small example as to how the energy sources are impacting world heritage can be seen from the fact how the Taj Mahal – One of the seven wonders of the world has lost its original shine owing to the surrounding industries and harmful fumes. Not only the polluting fumes are damaging the national heritage but also lowering the water table by severely impacting the flow of various rivers and making many of them extinct like river Saraswathi (Misra 2010).

Not only this, the extreme weather condition is proving to be very harsh for various other parts of the world which are facing the environmental rage of heavy rain, increased desertification like the one seen in the Atacama Desert of Chile, increased acidification and a life threat for the beautiful coral reefs and aquatic animals (*Carbon Stock Study* 2018).

The need of the hour is thus to find innovative ways and sources of energy that could balance the issues of economic development and environmental protection.

4.4 Air Quality – Deterioration Leading to Development of another Mars

One of the most disturbing elements of the environmental degradation is the worsening air quality that contributes to millions of deaths worldwide. The situation is worse for the cities that have come up in the ladder of industrialization but have fallen steeply in terms of

Table 4.2 Top most polluted cities of the world (2019).

Rank	City	2019 AVG
1	Ghaziabad, India	110.2
2	Hotan, China	110.1
3	Gujranwala, Pakisthan	105.3
4	Faisalabad, Pakisthan	104.6
5	Delhi, India	98.6
6	Noida, India	97.7
7	Gurugram, India	93.1
8	Raiwind, Pakisthan	92.2
9	Greater Noida, India	91.3
10	Bandhwari, India	90.5

Source: https://www.iqair.com/us/world-most-polluted-cities (Data from World's most polluted cities 2019 (PM2.5), IQAir AirVisual, 2019 World Air Quality Report).

air pollution. The impact is so widely distributed in India that as per the report of WHO, 14 cities of India are among the top 20 most polluted cities of the world (IQAir 2019).

The situation is not only alarming but also devastating to see that in the race of the globalization while polluted cities like Beijing have come down the ladder, the same could not happen with Indian cities where Ghaziabad came at number 1 among the top 20 polluted cities. Table 4.2 shows the data of the most polluted cities of the world as per the IQAir report 2019.

The problem is not only with the toxic air but there is also a huge resource exploitation and environment degradation that has increased the frequency and the chances of the happening of a pandemic. What is required by the leaders of the world is to act cohesively and contribute towards the journey of sustainable development. One way of doing this is to use energy-efficient methods of production.

4.5 Carbon Footprints – Gift of Mankind to Mother Earth

This is an interesting term that is used to measure the carbon dioxide emission in tonnes by an organization or a country. The concept came into frontline because of increased pressure on nature due to increased carbon emission that has increased manifold in the last century. A detail about the countries' contribution to the carbon footprints is given in Figure 4.2 which shows that the most populated and industrialized countries China and Japan are contributing more than 50% of the carbon emission of the whole world and if the United States is added to that then the total contribution becomes three-fourth. India has the fourth position in terms of carbon emission and needs to buck up the efforts to reduce the emission and stop the earth from becoming another Mars.

Increased quantum of carbon emission in the world is in fact above all the emergencies whether it be war, communalism, terrorism or pandemics as the harmful impact of the

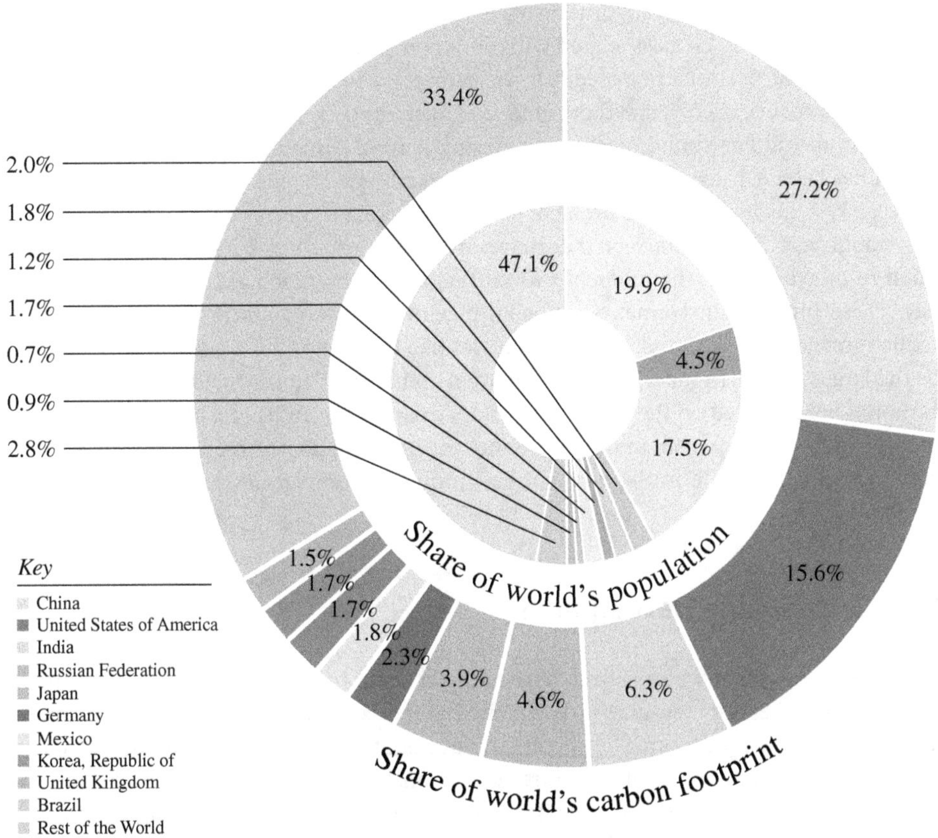

Figure 4.2 Carbon footprints. *Source:* Adopted from Carbon Footprint, 2021 Global Footprint Network Retrived from: https://www.footprintnetwork.org/our-work/climate-change/.

hazardous gas is not only huge but also cannot be contained within the boundaries of any particular country. Moreover, the concentration of the gas in the atmosphere has been rising at an alarming rate. The most dangerous thing about carbon footprint is that every human effort has an associated carbon footprint to it be it travelling, watching shows on television at home, talking to loved ones on phone or even printing a single piece of paper. It is therefore an urgent requirement to bring innovation in the routine activities so that there can be neutrality in carbon emission.

The situation has worsened to such an extent that the global temperature has increased by more than 1° C since pre-industrial era and the impact is visible on the ecological balance, water bodies as well as crops. As per the report of Climate analysis indicator tools (CAIT), the main reason why carbon emission has gone beyond control is the economic development that has been splurging in many countries, ever-growing population and heightened use of energy sources.

Jim Yong Kim, the former World Bank president stated in the year 2017, the only way to reduce the carbon emission and make Paris agreement successful is to make a policy for the trading of carbon credits efficient between the countries so that every country not only tries to reduce the emission but the countries that have gone green are also awarded for the same. The countries which are thus heavy emitters can penalize their businesses who now have the opportunity cost associated with emission which was earlier not paid any attention. The pricing of the carbon credits gives an impetus to the countries to think of innovative ways to reduce carbon emission. One carbon credit permits the holder to emit one ton of carbon dioxide or similar gas by purchasing it from some other company or country which has been carbon negative (producing less carbon). The idea was first thought of in the year 1997 during the Kyoto protocol in Japan wherein it was brought to the notice as to how countries can trade between themselves based on the carbon emissions done by them. Then in the year 2001, the protocol was ratified and carbon credits faced the light of the day. These financial instruments act as an incentive for the countries to devise ways to reduce greenhouse gas emission. Few countries like India and China are in fact expected to be the largest sellers in the Global credit market and domestically also to enhance liquidity carbon is being traded on the Multi commodity exchange of India just like any other commodity. India has also been in the forerunner for imposing tax on the carbon emission by commercial automobile in the name of Environment Compensation charge in the capital Delhi.

4.6 Sustainable Development

The development that we have been harnessing is not going to sustain if the environment is degraded and energy resources are exploited at the same rate. Sustainable development assures that there is economic development but not at the cost of depleting the resources and polluting the environment for future generations. A development which is sustainable can only be considered worth all the efforts. The United Nations Development Programme came up with the Global Goals in the year 2015 to ensure sustainability worldwide. These goals came to be known as the Sustainable development goals and aim to achieve certain targets related to development and sustainability by the year 2030. To attain these objectives, the world leaders adopted a set of goals termed as 'Millennium Development Goals' in the year 2000 with the twin objective of human development and poverty alleviation. However, it was later found out that the richer segment of the society whether it be the developed or developing countries had adopted largely the consumption and production patterns that could not be sustained for long, and hence it is called for emergent and sustainable measures to be adopted at a Global level that could help in achieving the country and individual specific goals apart from being sustainable and environment friendly. Hence were born the 17 Sustainable development Goals in the year 2015 whose foundation was laid down in the year 2012 in Rio de Janerio. The 17 Sustainable Development Goal's (SDG's) had 17 important goals that took care of all the issues of the developing as well as developed countries along with the sustainable factor being kept into consideration. These 17 SDGs are not mutually exclusive as they recognize that the impact of one goal would be there on the others and hence all the goals are integrative. The goals had a target to be achieved by the year 2030 with special emphasis on reducing poverty, empowering women and girls, and addressing the climate emergency United Nations Development Programme (UNDP). The Goals' achievement however is not going to be without any cost as the estimated investment required worldwide is somewhere between $5 and $7 trillion per annum. Another limitation of the goals is that they are not legally binding on any country and hence if countries give importance to any local goal attainment then the SDG's may take a backseat. The 17 SDGs are in fact much broader than the Millennium Goals as it has 169 targets in addition to the 17 broad goals.

4.6.1 Problems Faced by the Country in Implementing Sustainable Development Goals (SDGs)

India has been committed to the achievement of the SDG to make Earth a better planet to live in but few obstacles could delay its achievement if not abandon it. Few such hurdles are listed below:

4.6.1.1 Financial Resources

The 17 SDGs that India adopted after replacing the previously laid down Millennium development goals (MDGs) are not only more ambitious in terms of the objectives to be achieved but also more demanding in terms of the financial requirements. India already being in the grasp of fiscal deficit would have to buck up really strongly to achieve the targets efficiently. As per the estimate of United Nations Conference on Trade and Development (UNCTAD) to achieve the goals at the global level, an approximate amount of 5 to 7 trillion dollars may be required annually, and India already lacking in terms of financial resources is expected to have a deficit of Rs 533 lakh crore for the period 2015–2030 (Bhamra et al. 2015). One solution meeting the financial crunch is to go for green financing in terms of green Foreign Direct Investment (FDI) or green bonds. **Green Foreign Direct Investment (FDI):** with the serious threat being posed to the environment and sustainability, one important issue that is being faced by all the countries is the lack of capital and infrastructure. The advent of technology and modernization has led to development, but it cannot be called Green development till we use clean sources of energy. One such measure that has been garnering the attention of the government worldwide is to attract FDI in the energy-efficient technology. Organisation for Economic Co-operation and Development (OECD) recommended that the countries should focus on inviting those FDI's that use either clean energy or help in reducing the impact of environmental damage. As all countries now are dependent on inward FDI's potential as a source of sustainable development cannot be overlooked. It would in fact help developing countries like India that are capital scarce. **Green Bonds** – At Conference of the Parties (COP15) in Copenhagen in 2009, nations agreed to take measures to reduce the greenhouse gas emissions so as to hold the increase in global average temperature below 2° C (2 °C) above pre-industrial levels and cooperate in achieving the peaking of global and national emissions as soon as possible. For sustainable development, the major requirement is that of finance and one financially engineered product that has come to its rescue is the Green bonds that aim to provide the green finance to the industries to meet their capital requirements. Green bonds are designed to help the environment by directing portions of the capital raised to projects related to clean water, renewable energy, energy efficiency, river and habitat restoration or mitigation of climate change impacts (OECD). The term 'Green Bond' was coined in the year 2008, wherein it differed from the conventional bond market in the sense that the amount that was raised through the Green bonds had to be used in an energy-efficient manner. The first green bond was issued by the World Bank in association with Swedish Bank in the year 2008 and since then the green bond market has crossed $500 billion mark globally but its total share in the bond market is still miniscule (Maltais and Nykvist 2020). What is good for India is that it is among the top few countries where the demand for green bonds is at the peak, as per economic times, in February 2020, India is at number 2 in terms of global demand of Green bonds (*Emerging Market Green Bonds Report 2019* 2020).

4.6.1.2 Social Issues Not Covered

Sustainable development has been accused by many for being lopsided towards the economic and environment issues and not paying heed to the social aspects (Ruth Kattumuri). The goals are considered to be infeasible that may widen the disparity between the developed and the developing countries.

4.6.1.3 Natural Calamities and Pandemics

Natural calamities like earthquakes, floods and pandemics like coronavirus can put a break to the achievement of the SDGs as the resources have to be diverted to the disaster and the impact may be long lasting in developing and other weak economies.

4.6.1.4 Illegal Activities Barring the Achievement of the SDGs

In spite of continuous efforts by the Government of various countries what is lacking is good governance from not only the corporates but also the individuals as there have been instances of illegal mining, cutting of trees that nullify the impact of the positive environment efforts taken by others.

4.6.2 Paris Accord

There are numerous challenges that our Green planet is facing today like extreme poverty, widespread inequality in terms of social and economic infrastructure and degrading environment threatening the very existence of the only life-sustaining planet. Keeping this in mind, there have been various efforts taken at the local level by different countries. However, the scale at which the problem has escalated called for the Global effort taken by the integration of all the countries of the world.

However, because of the difference in the resources available to the countries and the peculiar problems faced by each country, there cannot be the same approach that can be used by all the nations. But the broad guideline that can help in the achievement of the sustainable goals is to bring change in the consumption and production pattern by switching to energy-efficient means of production.

There have been numerous steps taken at the global level to reduce the carbon emission and one such landmark accord has been the Paris agreement where the main objective was to reduce the carbon emission by big industrialist countries to control the increase in the temperature in the century.

The Paris agreement provides a set of comprehensive programmes to monitor the activities related to climate change by the member countries. The agreement is based on the premise that climate change and carbon emission are manmade and natural and are also harmful to living beings. It is not only leading to excessive environmental change but also endangering human and livestock health. A chronology of the various steps undertaken over the last four decades is presented in Table 4.3 below:

In India, one important step that can contribute in the attainment of the SDGs is the mandatory CSR – Corporate Social Responsibility under Section 135 of Companies Act 2013.

Table 4.3 Chronology of the environmental pacts.

Year	Agreement
1979	First ever conference with Climate as Agenda takes place.
1988	IPCC – Intergovernmental panel on climate change is formed.
1990	Second conference on climate change to formulate a global treaty for climate change started.
1992	United Nations Framework Convention on Climate Change (UNFCCC) is formulated.
1997	Kyoto Protocol (Japan) The agreement started in 1997 when it was the first time that developed countries adopted mandatorily the targets for reduction in carbon emission.
2001	Marrakesh Accord to strengthen the adoption of Kyoto Protocol.
2005	First meeting of the parties to Kyoto Protocol took place in Montreal.
2009	IPCC's fourth Assessment report was released.
2009	Copenhagen accord wherein the countries took pledge for non-binding carbon emission projects.
2010	Cancun Agreement wherein the above non-binding pledges were made mandatory and mutually accountable.
2011	Durban agreement wherein countries identified the need to formulate a plan to be implemented after 2020.
2012	Doha agreement that brought various amendments to the initial Kyoto protocol.
2014	Lima, Peru agreement convened.
2015	Paris agreement signed by 195 countries and European Union to reduce the impact of global warming and stop the use of fossil fuels by 2050.
2015	Sustainable Development Goals – Global Goals

Source: A UN75 timeline, 26 JUN 2020, United Nations Sustainable Development, UN environment programme.

India is one of the few countries that has tried to share the responsibility of attainment of social goals by the big corporates. The activities that were earlier undertaken by the corporates voluntarily have now been brought under the ambit of legal regulation. There have been huge funds available under the CSR expenditure that contributes in the attainment of the sustainability goals too.

Following table provides how India has been contributing towards sustainability apart from the goal of development, wherein over the years, crores of rupees have been spent over different sectors to ensure that there is development along with the achievement of sustainability.

To achieve the target of the sustainable goals, countries have come up with innovative methodologies to catalyse the private funds to sustainable development goals thereby coming up with the concept of 'Blended funds', wherein the private funds are gelled with public funds to speed up the sustainability achievement.

As per the report of Economic times, India would need a funding of approximately \$2.64 trillion by 2030 to achieve the target of SDGs out of which atleast 50% can come from the private sector (Report – The Economic Times, 21/01/20). The 17 sustainable development goals that are adopted majorly by all the countries of the world revolve around reducing poverty, inequality, better health and education scenario, economic growth, environmental sustainability among others. India, specifically tried to integrate the national goals with the goals laid down in the sustainable framework to ensure that none of them takes a backseat. Various policies too have been initiated at the national as well as local level to ensure that the country is able to step up the Global sustainable Index. Few such measures that the country has adopted are provided in the next section along with the problems that the country might face in its adoption.

4.6.3 Steps Taken by India to Reduce the Carbon Emission

Sustainable development has been coined by the United Nations which provides that the development should be such that there is no compromise on the ability of the successive generations to make use of the resources. It provides that the utilization of the resources should be such that they are not exploited in the present itself leaving nothing for the future generations.

The three P's – people, planet and prosperity as defined by the United Nations specify that there is a need to integrate the 3 P's if sustainability has to be ensured for future generations.

Energy sources have a paradox, and that is on the one hand, it is considered an essential variable for development, and on the other hand, the more is the utilization of the energy the worse are the effects. The paradox exists irrespective of the type of energy – whether it be coal, oil or natural gas. The effect however is less severe in the case of renewable sources of energy like water and air. India being one of the fastest growing economies has the burden of not only going the green way but also finding a solution to the problem of environmental degradation. Few steps that the country has taken up to step up the nation on the sustainability index are listed below:

4.6.3.1 Sustainability Index

When the world went for a revamp and gained the badge for industrialization, no one thought that it could be disastrous for the environment and would take it beyond the environment boundaries. The carbon emission and toxicity have been at an all-time high and its impact is visible in the increased frequencies of pandemic and climatic catastrophe (Vezzoli et al. 2018). India holds an important place in the achievement of the worldwide Sustainable Development Goals (SDGs) as not only India is one of the most densely populated countries of the world but also because it is one of the fastest growing economies and hence impacting the consumption and production pattern. India is also expected to lead the way for sustainability as it has the advantage of demographic dividend with young and bright minds that can find ways for harnessing clean energy.

Not only this but even the corporates of India have accepted that for long-term existence and profitability they cannot focus on just the demand and supply side but the environmental aspect too has to be kept into consideration. Following this path, the country launched sustainability indices namely Bombay stock exchange (BSE) Carbonex and BSE Greenex. Greenex was launched in the year 2012 and includes 25 stocks of the companies that are using energy-efficient methods of production and thus contributing to the reduction in carbon footprints of the country, while Carbonex launched in the same year has 100 stocks and is a replica of the BSE 100 index and it helps in assessing the efforts taken by the BSE 100 companies in reducing the carbon emission.

4.6.3.2 Mandatory CSR

India is one such country that made it mandatory for the corporates to spend a certain percentage of the profit on some national issues (laid down in Section 135 schedule VII of the Companies Act 2013). The mandatory clause not only helped the corporates in accelerating their philanthropist activities but also helped the government in sharing its burden with the private enterprises. Table 4.4 below provides the quantum of CSR amount diverted

Table 4.4 CSR spending towards different sectors for the year 2014 to 2019.

	Amount spent	Amount spent	Amount spent	Amount spent	Amount spent
	FY 2014–2015	FY 2015–2016	FY 2016–2017	FY 2017–2018	FY 2018–2019
Development sectors	(INR Cr.)	(INR Cr.)	(INR Cr.)	(INR Cr.)	(INR Cr.)
Clean Ganga fund	5.47	32.82	24.37	4.54	5.41
Education, Differently Abled, Livelihood	3188.09	4921.06	5559.13	5822.47	7499.26
Encouraging sports	57.61	140.11	180.32	227.75	295.1
Environment, Animal Welfare, Conservation of Resources	853.99	971.06	1317.70	1362.83	1516.63
Gender Equality, Reducing Inequalities	189.92	342.67	468.74	499.65	517.88
Health, Eradicating Hunger, Poverty and Malnutrition, Safe Drinking Water, Sanitation	2525.92	4633.46	3658.14	3316.25	4958.48
Heritage Art and Culture	117.37	119.16	305.57	284.05	189.89
Other Sectors (Technology Incubator and Benefits to Armed Forces and Admin Overheads)	9.5	37.48	60.95	43.33	119.71
Prime Minister's National Relief Fund	228.18	218.04	158.8	161.84	300.07
Rural Development	1059.34	1376.16	1554.77	1479.80	2308.83
Slum Area Development	101.14	14.09	51.49	35.1	50.23
Swachh Bharat Kosh	113.86	325.52	184.06	213.67	93.8
Any Other Fund	277.09	334.34	419.98	255.62	710.59
NEC/ Not Mentioned	1338.39	1051.15	388.95	1.04	87.54
Grand Total	10 065.93	14 517.21	14 333.04	13 708.02	18 653.48

Source: https://www.csr.gov.in/index19.php (Data from National CSR Data Portal, Ministry of corporate Affairs Government of India.)

towards various economic and environmental issues thereby also supplementing the achievement of the SDGs on a national level. The spending as a whole is not only huge but has also increased tremendously in the FY 2018–2019 as compared to the base year of 2014–2015.

4.6.3.3 Innovative Schooling Ideas

The goal of Sustainable development cannot be achieved without the attainment of other country specific goals. In India, specifically reduction in the inequality among classes of people along with regional imbalance is mandatory for the perusal of SDGs. One efficient way of the attainment of sustainable goals is to make use of renewable sources of energy like solar and wind power which can be easily and successfully tapped in the context of developing countries like India. It will not only lead to sustained development but also helps in achieving other goals of reduction in unemployment as these innovative techniques can generate lots of employment in the country. One great example where people of the country have been taught to make use of solar power, wind energy and other natural sources can be seen in the 'Barefoot college' initiated by Mr Bunker Roy, wherein the students study in the light generated by solar power, cook and heat using solar devices and tap the water and do rain water harvesting. Not only is the teaching restricted to the district of Tilonia village of Rajasthan, but it has spread to many other countries outside India, wherein people have not only become aware about the use of clear sources of energy but also created employment opportunities for themselves and others. Another great example of the same has been the SECMOL Students' Educational and Cultural Movement of Ladakh school in Ladakh established in 1988 by Sonam Wangchuk that uses solar energy and uses 0% of fossil fuel. Another interesting concept was introduced by Bindeshwar Pathak in the year 1970 in the name of Sulabh Shauchalaya, wherein it not only solved the problem of unhygienic sanitation but also introduced the usage of biogas as a clean fuel. India is full of innovative minds which if used properly can help in the attainment of the 17 SDGs within the time frame laid down. Another initiative taken by the government to ensure sustainability is to bring the indices like Carbonex and Greenex to bring sustainability features into the corporate disclosure and investment. The indices include the companies that are using energy-efficient modes of production and hence helping the nation in reducing the carbon footprints. Another important thing that can be done by all the countries of the world is to modify the measure of growth. Currently, most of the countries are using GDP – Gross domestic product or national income per capita as the sole measure of growth. There are few countries however like Bhutan which is the only country to have negative carbon footprints in the whole world. Not only this but Bhutan was the first nation to introduce the concept of Gross national happiness index in the year 2008 instead of the most commonly used measure that is GDP. Various other countries too have followed the footprints of Bhutan like New Zealand. The step has served the dual purpose of not only making people happy but also making the environment go back to its original position.

4.6.3.4 Solar Powered Transportation System

India has been leading the world in terms of using clean energy for various modes of transport and Cochin International airport is in fact the world's first fully solar powered airport. Not only this in fact, the Delhi metro rail corporation (DMRC) also makes use of solar power apart from taking other environmental initiatives like water harvesting. 'DMRC has also helped in the reduction of emission of harmful gases into the city's atmosphere and the United Nations Body administering the clean Development Mechanism (CDM) under the Kyoto Protocol has certified that DMRC has reduced Emissions and thus earned carbon credits worth about Rs.47 crore annually for the next seven years and with an increase in the number of passengers, this figure shall increase (Delhimetrorail n.d.).'

India – an emerging economy too had its baggage of environment degradation wherein quick remedial measures were required at the national as well as the base level. India started with various reforms to promote the agenda of sustainable development. India has been the youngest country to adopt the sustainable development goals propounded by the United Nations summit in 2015 which consist of 17 goals and 169 targets that could be targeted till 2030 to have a better future. India has been at the forefront in launching various environment friendly projects like it was India's Prime Minister who launched the International solar Alliance in 2015 during the Paris summit for UN climate change. The idea is to promote the usage of solar energy by bringing down its price with cooperation among the 121 countries. India started working towards SDG after replacing the previously laid down Millennium development goals (MDG). However, achievement of these goals cannot be done without enough resources, huge commitment and innovative thinking.

India has focussed on achieving the goal of sustainability by focusing on 3R's that are Reduce, Recycle and Reuse. For India, it is even more important as because of the huge population, there is a large use of resources and generation of waste which if not treated properly may lead to various issues for the environment. Proper re-use and disposal of waste ensures that the resources are not only aptly used but also properly disposed off. One such example can be building roads made of concrete and waste plastic as is being done in parts of Indonesia and India too (Jamshedpur for example). Another such project has been 'Satat Bharat – Sanatan Bharat' wherein acting on its nationally determined contributions, India has electrified 100% of its villages, reduced 38 million tonnes of CO_2 emissions annually through energy efficient appliances, provided clean cooking fuel to 80 million poor households and set a target to instal 450 GW of renewable energy and restore 26 million hectares of degraded land by 2030 (United Nations sustainable Development Goals n.d.). Apart from this, the government had also tried to bring the corporate sector in the race of sustainable development by incorporating 'Environment sustainability' as a part of its corporate social responsibility (CSR) clause introduced in the Companies act 2013. Above are the few initiatives taken by the Indian Government, its corporates and its people that if harnessed properly can go a long way in solving the most important crisis facing humankind which is 'Environment Degradation.'

4.7 Coronavirus Pandemic and its Impact on the Carbon Emission

The pandemic that came as a health threat has in fact elated the environmentalists as it has brought a huge decline in energy consumption and carbon emission in the first quarter of the year 2020. As per the report of Carbon Brief, there has been a huge decline in the carbon emission by China and it has reduced by almost 25%. Carbon Brief analysis of the data suggests that the pandemic could cause emissions cuts this year in the range of 2000 m tonnes of CO2 (MtCO2). The pandemic has been the sole reason for the biggest annual decline in carbon emission since the beginning of the industrial revolution. The reduction however is not sufficient to meet the target for a green planet which requires a 7.6% decline in the carbon emission per year, still the pandemic has put a halt on the quantum of carbon footprints globally which if extended permanently through above mentioned steps can have a positive impact on sustainable development.

4.8 Conclusion

The only planet known to mankind that could support life is now at the dearth because of pollution and fast spreading industrialization. The situation has become so bad that even Earth now needs an oxygen mask to survive. Since the beginning of the human race, the inquisitive nature of mankind has led to the development of various machines and technology that has no doubt enhanced the living standard and comfort, but at the opposite end, it has destroyed everything that is required for basic survival. Earth is gasping for fresh air and the pandemics, catastrophe, ice melts, volcanoes have become rampant because of disturbing the ecological balance of the mother earth (Parris and Kates 2003). The carbon emission has increased to such an extent that there has been a need to form a complete forum and devise the protocols to reduce the greenhouse gas emission worldwide. There have been various negotiations and agendas that have been framed and named as 'Global Goals' so as to ensure that the Earth could actually survive and there is sustainable development. The last agreement that took place in Paris in the year 2015 has given way to 17 development goals that need to be attained by all the countries by the year 2030 (United Nations Development Program n.d.). The sustainable development goals are the successor of the Millennium Goals, wherein the target is not only to achieve the goals of economic development but also to sustain the development process by laying emphasis on the energy conservation and environment restoration. Member countries have been given the freedom to set their own targets to achieve the SDG objective by the year 2030, and for this attainment, the countries have set up their separate mechanisms like in India there has been the adoption of a 'Holistic' approach, wherein the Government has tried to combine the goals laid down by SDG with the national goals so as to have a balance between the two and ensuring that no sector is left out. Environmental degradation and reduced level of living forced the leaders of the world to think about whether the development that has been achieved till date can be sustained for the future generation and this idea led to the development of 'Sustainable Development Goals' in the Rio de Janerio meeting of 100 countries in the year 1992. Though the mention of sustainable development was there two decades back too in the UN conference of Stockholm but it was not explicitly mentioned. Even after huge steps taken by the Government, the target is still far behind because of lack of investment required and different natural and environmental catastrophes that the country has to face from time to time that leads to diversion of funds and energy from the SDG to other areas (Lélé 1991; Griggs 2013). What is required to ensure the attainment of the sustainability factor is the effort at the grassroot level which can happen if every individual of the country thinks from the perspective of clean and green energy, wherein maximum emphasis is laid on the 5 R's – Reduce, Reuse, Recycle, Rot and Refuse (Lund 2007). If every individual carries out these five steps then there would not be any need of framing nationwide policies. In a country like India with widespread illiteracy and poverty, this can be attained by providing education about this at the basic elementary level and also incentivizing its adoption.

References

Bhamra, A., Shanker, H., Niazi, Z. et al. (2015). Achieving the Sustainable Development Goals in India A Study of Financial Requirements and Gaps. 4 July 2020, from www.devalt.org.

Carbonbrief.org. (2020). Analysis: India's CO2 emissions fall for first time in four decades amid coronavirus. https://www.carbonbrief.org/analysis-indias-co2-emissions-fall-for-first-time-in-four-decades-amid-coronavirus

Delhimetrorail. (2015). *UN Body Credits Delhi Metro- 6.3 Lakh Carbon Credits for Modal Shift Project*. Retrieved 12 October 2020, from http://www.delhimetrorail.com/press_reldetails.aspx?id=746xECETA6Qlld

European Commission. (2020). Climate change and environmental degradation | Knowledge for policy. https://ec.europa.eu/knowledge4policy/climate-change-environmental-degradation_en

Griggs, D. (2013). Sustainable developmnet goals for people and planet. *Nature*, 495 (7441), 305–307.

India 2020 – *Analysis - IEA*. (2020). https://www.iea.org/reports/india-2020

IQAir (2019). https://www.iqair.com/world-most-polluted-cities.

Lélé, S. M. (1991). Sustainable development: a critical review. *World Development*, 19(6), 607–621. https://doi.org/10.1016/0305-750X(91)90197-P

Lund, H. (2007). Renewable energy strategies for sustainable development. *Energy*, 32(6), 912–919. https://doi.org/10.1016/j.energy.2006.10.017

Macrotrends.net. (2015). India Carbon (CO2) Emissions 1960–2020 | MacroTrends. Retrieved 12 October 2020, from https://www.macrotrends.net/countries/IND/india/carbon-co2-emissions

Maltais, A., & Nykvist, B. (2020). Understanding the role of green bonds in advancing sustainability. *Journal of Sustainable Finance & Investment*, 1–20. doi:10.1080/20430795.2020.1724864

Misra, A. K. (2010). A river about to die: Yamuna. *Journal of water resource and protection*, 2(5), 489.

NASA. (2020). Effects | Facts – Climate Change: Vital Signs of the Planet. https://climate.nasa.gov/effects

Parris, T. M., & Kates, R. W. (2003). Characterizing and measuring sustainable development. *Annual Review of Environment and Resources*, 28, 559–586. doi:https://doi.org/10.1146/annurev.energy.28.050302.105551

Rühl, C., Appleby, P., Fennema, J., Naumov, A., & Schaffer, M. (2015). *Economic development and the demand for energy*

Stern, D.I. (2012). The role of energy in economic growth (27 November 2010). USAEE-IAEE working paper no. 10-055 at SSRN: https://ssrn.com/abstract=1715855 or doi:10.2139/ssrn.1715855

The Guardian. (2019). Human society under urgent threat from loss of Earth's natural life | US news | The Guardian. https://www.theguardian.com/environment/2019/may/06/human-society-under-urgent-threat-loss-earth-natural-life-un-report

United Nations Development Program. (2015). Transforming our world: the 2030 Agenda for Sustainable Development.

United Nations sustainable Development Goals. (2015). *India .:. Sustainable Development Knowledge Platform.* Retrieved 12 October 2020, from https://sustainabledevelopment.un.org/memberstates/india

Vezzoli, C., Ceschin, F., Osanjo, L., M'Rithaa, M. K., Moalosi, R., Nakazibwe, V., & Diehl, J. C. (2018). Energy and sustainable development. In *Designing Sustainable Energy for All* (pp. 3–22). Springer Cham.

Prowessiq. https://prowessiq.cmie.com

CSR. www.csr.gov.in

5

Energy Development as a Driver of Economic Growth

Evidence from Developing Nations

Md Rashid Farooqi[1], *Akhlaqur Rahman*[2], *Md Faiz Ahmad*[3], *and Supriya*[4]

[1] *Department of Commerce and Management, Maulana Azad National Urdu University, Hyderabad, Telangana, India*
[2] *District Institute of Education and Training, Bihar Education Service, Patna, Bihar, India*
[3] *School of Management, Malla Reddy University, Hyderabad, Telangana, India*
[4] *Disaster Management Professional, Sitamarhi, Bihar, India*

5.1 Introduction

The need for renewable energy in developing nations is increasing significantly as population and per capita income increase (Wang et al. 2020a). This chapter attempts to show how renewable energy contributes to economic growth and why energy supplies need to be replenished to fulfil the rising demands. To enhance living standards, progressive nations should attain growth and development. Energy requirement in developed nations is increasing as a result of continuous increase in efficiency and output with which energy can be generated and consumed simultaneously (World Bank 2017). In developed nations, both financial and energy resources are available to meet market requirements. The importance of energy has been increasing every day since past few decades as well as change in climate (Bulkeley 2013; Schuur et al. 2015). In developing nations, the energy sector is poorly developed and energy consumption is relatively very less compared to that of developed nations (Wang et al. 2020b). The average energy consumption in developing nations is very low. Increase in energy consumption is alarming and posing a threat to the ecosystem. Most of the developed nations are establishing frameworks to curb this threat as energy sources are becoming essential to the entire world (Liao and Shi 2018). The basic energy sources that lead to environmental pollution need to be replaced by green energy sources (Dong et al. 2019a); as a consequence, most developed nations are establishing regulations to strengthen the usage of renewable green energy sources, which will act as a safeguard to the environment (Román-Collado et al. 2018). In today's world, protecting the environment will be a major challenge for the forthcoming generation. Almost 82% of the energy sources are derived from fossil fuels, 22% from natural gas, 28% from coal, and 31% from oil (Gulzara et al. 2018). These are the major sources of greenhouse gas emission. Energy is necessary for economic production, and equally important for society and economic growth (Zheng and Lin 2018). Energy consumption is the major cause of environmental pollution in developing nations. In recent years, various policies and

Energy: Crises, Challenges and Solutions, First Edition. Edited by Pardeep Singh, Suruchi Singh, Gaurav Kumar, and Pooja Baweja.

regulations have been implemented to eradicate the environmental issues associated with energy consumption. Efficient use of cost-effective resources, technologies, and upgradation can reduce environmental degradation. Diverse sources of renewable energy such as wind, sun, biomass are available, and they are clean and reliable compared to fossil fuels (Zhou et al. 2018). To handle social and economic problems, energy is not the only resolution for progressive nations such as India, Sri Lanka, and Bangladesh (Pradhan 2010). Modern and sophisticated energy sources are needed for social and economic progress. Lack of clean energy sources makes it difficult for progressive nations to get help from the other countries. In developing nations, expansion and use of upgraded energy is essential to reduce the consumption of traditional fuels, because using fuels such as charcoal causes health issues and also leads to deforestation. Investment in upgraded energy helps to stimulate economic growth (Wang et al. 2018). Dependence on imported energy cannot be lowered by using renewable energy in developing nations; however, it conserves foreign exchange (Fallahi 2011). The fast growth of energy use in developing nations has significant impacts. Energy services are important and essential for economic growth and development and improve standards of living. Economically and environmentally sound approaches for energy development offer greater benefits for developing nations as well as developed nations (Raeeni et al. 2019). Energy largely contributes to economic progress in developing countries, thereby leads to higher standards of living (Joyeux and Ripple 2011), reduces hunger and poverty, and provides a better environment. As mentioned, such approaches are equally beneficial for developed countries. The United States is the leading trade partner for developing nations. Rapid economic growth in these nations could accelerate the US exports, including energy export. Advanced energy technologies can reduce greenhouse gas emission, which is a global benefit. Energy use and development cause regional and local environmental damage (Dong et al. 2019b) in progressive nations (developing countries) and in few urban localities. Many developing nations have enjoyed major economic gains through the increasing developing world energy demand over the last 15 years. Developing nations observed significant growth in economy as well as considerable poverty reduction in the last decade (Neto et al. 2014). Energy is an important factor for economic growth as it is a basic input resource for various consumption and production activities (Alam 2006; Ohler and Fetters 2014). Some experts even argue that increase in energy consumption is directly related to GDP growth. In the development of any country's energy supply infrastructure, energy prices play a crucial role (Aslan et al. 2014).

5.2 Energy and Economic Development

Energy is the oxygen of production and development of economy and lifeblood of growth. From a technical viewpoint, the use of energy brings industrial changes and development and considered the centre of operation of any modern economy (Bhattacharya et al. 2016). It is an attempt to distinguish different relationships between energy consumption and economic growth. A two-way traffic was found between economic development and energy in developing nations (Liu et al. 2017). It is important to know the impact of economic growth in progression of economic development on developing nations. In the progression of economic development, commercial energy moves rapidly compared to economic activities. Modern infrastructures increase transportation of goods, urbanization, and the substitution of commercial for local fuels (Henderson 2003).

Today's energy forms are considered an economic good, which has the potential to improve the living standards of masses. Most people in developing nations are lacking access to energy services and their consumption levels are far lower than that of people in developed nations. This is the pollution arising due to energy production and used termed as economic 'bad' no energy used itself. A reasonable number of technologies are available and capable of resolving the pollution issues at a minimum margin of the total cost of energy supplies (Zheng and Lin 2018). The differences will be identified by policies; the more likely we will be able to fulfil the increasing world energy demands, pollution will be definitely reduced. Energy contributes to economic growth and development (Tang and Tan 2015). First, renewable energy is a sector that is important for the economy that helps to generate jobs, values by transforming, evaluating, and distributing energy goods and services throughout the economy (Vandaele and Porter 2015). Energy is a crucial element for the development of economy and an important factor of production that is required in all economic activities. This basic production input provides a conducive platform for business growth and production; on the other hand, economic growth is very important in developing economies. A poor economic condition leads to poverty and unemployment (EIA 2014a). On the other hand, good economic conditions generate employment and reduce poverty. So, it can be stated that the use of energy services is strongly associated with economic development (Chontanawat et al. 2008; Belke et al. 2011; Esseghir and Khouni 2014; Gulzara et al. 2018). Developing countries have been currently using more than 50% of the world's energy. Figure 5.1 depicts the share of leading three developing nations (i.e. the United States, China, and India) of the entire world primary energy use. The share of these three countries has risen from 39% in 1965

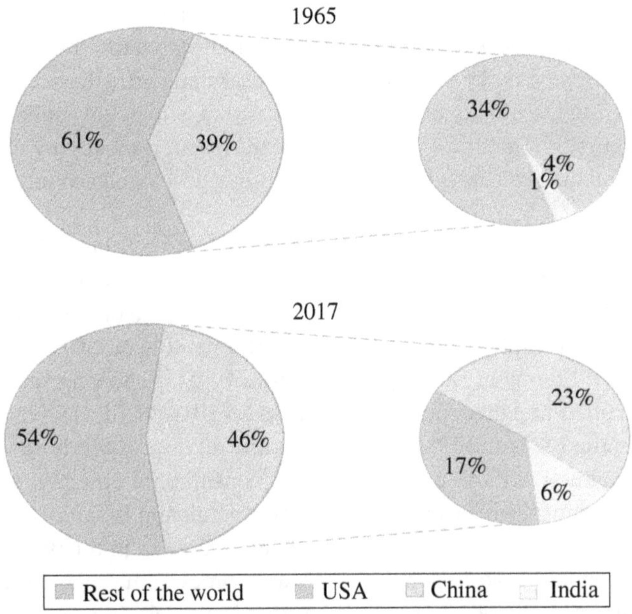

Figure 5.1 Change in the share of primary energy in 1965 and 2017. *Source:* Reference BP. Statistical Review of World Energy, 2017.

to 46% in 2017 (BP 2017); the percentage increase comes primarily from the remarkable surge in China, which was the major user of primary energy in 2017 (BP 2017; World Bank 2017).

5.2.1 The Impact of Economic Development on Energy

The accelerated growth of economy and its emerging activities have a significant impact on the energy sector (Barro and Sala-i-Martin 1995). During 1965–1987, the economy of developing nations grew at an annual average of 5.4% and their commercial energy consumption increased by just 7% (U.S. Congress Report: Energy in Developing Countries 1991). These energy growth rates were higher compared to the industrial nations over the same period. As a result, the developing nations' share of global commercial energy consumption also increased. A high economic growth rate is required on a continuous basis in progressive nations to provide the rapidly growing population with high standards of living (World Energy Conference 1989; Begum et al. 2015).

The literature shows the impact of renewable energy development on the economy (Vandaele and Porter 2015). It maintains the cash flow within the domestic economy and reduces the coal and natural gas imports even if consumers are spending more on electricity. Renewable energy is becoming vital to power generation (Fadaeenejad et al. 2013). Advantages of wind power generation are low water consumption, relatively low greenhouse gas emission, and favourable social impacts. Energy offers reliable power supplies and diversification of fuel, improving energy security and lowering the risk of fuel spills. It is helpful to protect a country's natural resources (Ellabban et al. 2014; Koçak and Sarkgüne¸si 2017).

a) **High Standards of Living:** Population is geometrically increasing in developing nations, so the major challenge with development is to provide high standards of living. Actual standard of living can rise only by means of positive economic growth, which basically comes from increasing the quality of factors of production, which are capital, land, labour, and entrepreneurship. Factors of production are the resources that provide a foundation in creating or manufacturing goods and services in a developed economy (Cansino et al. 2015).

   This task is important due to the decline in per capita income of most of the nations of Latin America and Africa. According to the predicted increase in progressive nations' population, a yearly average of around 1.7% in the next 35 years, an increase in the mean per capita income of 3% per year implies the economic growth rate of about 4.7% per annum (U.S. Congress Report: Energy in Developing Countries 1991). Achieving such growth rates will be difficult. Foreign debts of the progressive world (developing nations) increased greatly in the 1970s and 1980s. Total goods and services export is now twice as that in 1970. The higher a nation's per capita energy consumption, the higher the standard of living and the country will be more prosperous (Apergis and Payne 2012).

b) **Rise in Population:** The past two centuries have witnessed rapid growth in the world population. The major addition to the world population growth comes from progressive nations and it is found from the study of Lou et al. (2016) that in the last 50 years the world's population growth is contributed by progressive nation. The effect of population growth can be positive or negative depending on the situation. High population has the potential to boost economic development. With more people, more output can be achieved; more work will generate more value. Economic growth and development are boon to improve the quality of life rapidly within the middle class than within the lower class. Size of the economy increases, and the fiscal condition is strengthened by rapid growth in GDP. Broadly, shared growth in per capita GDP increases the standard of living (Chen et al. 2018).

The World Bank has projected greatest increases in population to come from two major continents Asia and Africa, which account for around 30 and 58% of the total global increase, respectively (U.S. Congress Report: Energy in Developing Countries 1991). In Africa, the most basic reason is the rapid growth in population, despite of low population increase compared to the entire population from a recent 12 to 19% in 2025 (U.S. Congress Report: Energy in Developing Countries 1991). On the other hand, the growth of population rate is relatively low in Asia. Population growth in China is estimated to be moderate since the existing low rates of growth are sustained. The growth in China's population is projected to be critical since it holds a huge share of the world population.

The global population is likely to increase despite of the constant decline in the fertility rate. The increasing rate of the world population, which was less in the previous days, still shows a significant growth in the human population. According to the World Bank projections, the world population will grow from 5.4 billion in 1990 to 8.5 billion by 2025(U.S. Congress Report: Energy in Developing Countries 1991). All the increments will come from developing nations. Population growth in the progressive world will likely to increase from its current point of 4.1 billion to 7.2 billion in 2025, which will increase the world population from 78 to 88% (Rodolfo et al. 1989–1990; Population and Human Resources Department, World Bank 1988). The rising population uses more energy. If there is 20% growth in the population size, energy consumption will increase by 43.6%. On the other hand, if 20% population declines, it will leads to 44.6% decrease in energy (Huang et al. 2018).

5.2.2 Economic Development and Fluctuations in Energy Consumption

Economic growth and development is critical to gain a higher-quality standard of living for rising population of the progressive world, which will lead to comparable increases in energy consumption (Al-Mulali and Ozturk 2015). The majority of people in progressive nations rely on traditional fuels such as animal dung, crop, and wood for their energy requirements. In fact, traditional fossil fuels are time consuming and difficult to collect. Similarly, most people in progressive nations depend on animal and human muscle power to perform their work in spite of its limited output and low efficiency.

First, as most of developing nations have been building their infrastructures such as transportation, industrial, and commercial sectors, a huge quantity of energy-intensive materials such as cement and steel are needed. Second, urbanization has been observed in developing nations, which has a significant impact on the form of energy used (Shahbaz et al. 2017). Third, today's manufacturing technologies and raw products have significantly reduced the actual cost of several fast-moving products from radios to refrigerators. In developing nations, consumers can purchase multiple products in the development cycle than did consumers in recent developed nations.

Energy intensities are affected by modern technologies used by various progressive nations for their development in the economy. For example, older-generation coal burning technologies still used in developing nations are very inferior to today's technologies. The countries that depend on coal burning are more energy efficient than the countries that rely on gas and oil (Park and Yoo 2014). Factors that influence energy consumption are the country's size and location of natural resources in the context of business, industry, etc.

5.2.3 Energy Consumption in Developing Nations

Rapid rise in population and high economic growth rate (Bastola and Sapkota 2015) provides high standards of living, urbanization, industrialization etc. and they are associated with an accelerated rate of increase in commercial energy consumption (Asafu-Adjaye 2000) in a progressive nation (i.e. Table 5.1 shows the twenty largest commercial

Table 5.1 Largest energy consumers.

Twenty largest commercial energy consumers			
Ranking	**Country**	**Total consumption of commercial energy**	**Per capita consumption of commercial energy**
I	United States of America	68	280
II	Union of Soviet Socialist Republics	55	195
III	China	24	23
IV	Japan	14	111
V	West Germany	10	166
VI	United Kingdom	9	151
VII	Canada	8	292
VIII	India	7	9
IX	France	6	110
X	Italy	6	105
XI	Poland	5	142
XII	Mexico	4	51
XIII	East Germany	4	232
XIV	Australia	3	202
XV	Brazil	3	23
XVI	South Africa	3	84
XVII	Romania	3	137
XVIII	Netherland	3	214
XIX	Czechoslovakia	3	186
XX	Spain	2	148

Table 5.1 (Continued)

Ten biggest progressive Countries' consumption of energy (Consumer)			
Ranking	**Country**	**Total Consumption of commercial energy (%)**	**Per capita consumption of commercial energy (%)**
I	China	24	22
II	India	7	8
III	Mexico	4	50
IV	Brazil	3	22
V	S.A. (South Africa)	3	83
VI	S.K. (South Korea)	2	52
VII	Argentina	2	56
VIII	Venezuela	2	88
IX	Indonesia	1	8
X	Egypt	1	20

Source: U.S. Congress Report: Energy in Developing Countries (1991).

energy consumers and ranking of ten biggest progressive countries consumption of energy). The practices of industrialization and urbanization will lead to high consumption of energy (Anderson 2000).

The World Energy Conference has predicted a low growth rate in commercial energy consumption in progressive nations. In developing nations an annual average growth of three and half percent will lead to be three times more than that in the year 1985 and the use of local (traditional) fuels 26% more. Population growth and rise in standards of living account for about 50 % of the overall growth (U.S. Congress Report: Energy in Developing Countries 1991) (Figure 5.2). Table 5.2 shows the commercial energy consumption and projection of the progressive nations.

5.2.4 The Price of Energy and Management of Demand

Energy price plays an important role in the energy development sector. Energy pricing policies have multiple objectives: appropriate and consistent allocation of resources, ensuring adequate supplies to consumers, promoting regional resources, fair returns to energy producers, and affordable supplies to customers. The weights of these multiple objectives in the formulation of energy policies and strategies vary from nation to nation depending on geographical structure, history, and politics (Al-Mulali and Ozturk 2015). The objectives related to social factors are important in framing energy pricing policies (Wang and Chen 2015). At the primary level of life, energy is a substantial part of whole household expenditures. Governments also play a crucial role in the commercial energy sector. In most of the progressive nations, the electricity sector is governed by the government. In many

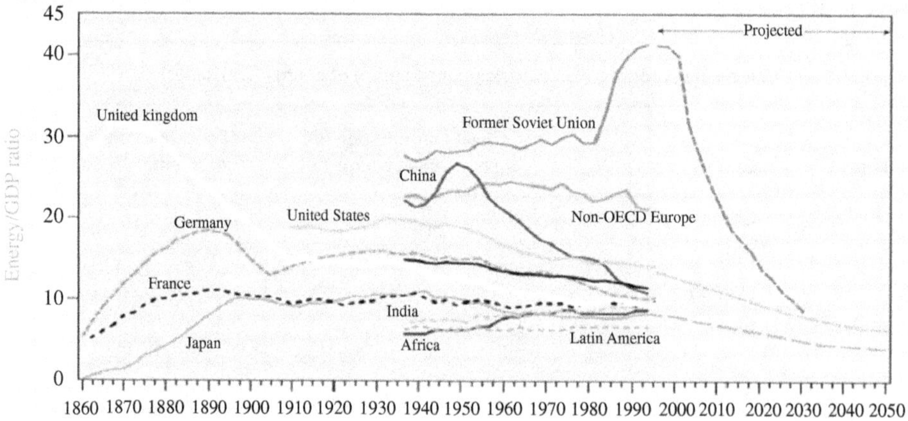

Figure 5.2 Projections and the ratio of energy consumption to GDP for some nations and regions from 1860 to 1996. *Source* CEC (2000), IEA (1997, 1998). Note: Energy consumption is expressed in mega joules and GDP in US dollars in PPP (purchasing power parity). Before 1961, calculations of GDP were based on exchange rates. Data of energy not included from biomass.

Table 5.2 Commercial energy consumption and projection.

	1985	Share of total percent	2010	Share of total percent	Increase in consumption 1985–2010
Progressive Nations	**70**	**23.49**	**176**	**34.51**	**106**
China	22	4.90	60	8.82	38
Organization of the Petroleum Exporting Countries (OPEC)	12	2.70	29	3.70	17
Excluding Organization of the Petroleum Exporting Countries (Non-OPEC)	36	8.10	87	11.11	51
Organization for Economic Co-operation and Development (OECD)	**156**	**52.35**	**215**	**42.16**	**59**
U.S. (United States)	74	16.78	98	12.51	24
Union of Soviet Socialist Republics (U.S.S.R) & East Europe	**72**	**24.16**	**119**	**23.33**	**47**
Total	**298**	**100**	**510**	**100**	**212**

Source: Marine and Schrattenholzer (1989)

other countries, the government manages gas, oil, and coal sectors. Price stability is an important policy objective for the majority of poor people in the population. Economic objectives are also important, such as encouraging the key strategic development sectors, including agriculture and transportation, which come under pricing policies.

A progressive nation to achieve its primary objective of improving its economy must improve the transportation and agricultural sectors with strategic planning and development through policy revisions. Keeping energy prices low will also produce adverse effects. Income from energy sales is not sufficient to cover the cost of energy supply, which is quite a common problem in developing countries. A constant decrease in important financial indicators has been observed as profits from electricity sales lagged the increasing cost. Pricing is one of the tools to control the energy demand. Apart from the various factors such as energy pricing policies, allocation of resources, promoting regional resources, fair returns to suppliers etc., it is important to make aware the consumers how to preserve energy which is cost effective.

5.3 Energy Services in Developing Nations

The objective of this chapter is to understand how energy is consumed in progressive nations. In developing nations, energy consumption is increasing rapidly. In 1960, developing nations used just around 15% of the global commercial fuel (U.S. Congress Report: Energy in Developing Countries 1991). By 1985, this share increased significantly to 24%. Progressive nations' share of global energy consumption was around 29% in 1987 (U.S. Congress Report: Energy in Developing Countries 1991). Progressive nations have been consuming more commercial energy compared to current industrial nations by early in the coming era. The factors that drive this rapid increase in energy use are economic growth, urbanization, and population growth. Overall energy consumption per capita rate in developing nations in 2025 would be just one sixth that of the United States in 1988 (U.S. Congress Report: Energy in Developing Countries 1991).

Various channels are available to provide desired services, each with their own specialities and with varying costs, for example, transportation can be facilitated by means of cycle, rail, bus, car, or airways. Customers will prefer these means based on factors such as convenience, speed, cost, and comfort. In addition to engineering and economics, energy analysis must consider cultural, social, and institutional factors. They are considered in a service framework rather than in a traditional energy supply analysis.

Energy use in rural areas shows a greater different set of parameters than that of the current economy of urban areas. First, traditional energy consumption is an element of a complicated and typical hereditarily system compared to fossil fuels. The biomass, which is one of the important components of this system, provides food for animals and humans, etc.; similarly, the bullock also provides milk, meat, leather, and fertilizer.

The productivity and efficiency of traditional energy technologies in developing nations can be improved significantly (Ladu and Meleddu 2014). For the developing nations, the industrial and commercial sectors contribute the biggest end-use energy market, together accounting for 88% the energy consumed by end users including traditional fuels. Transportation accounts for the remaining 12%. Around 40% of all energy consumed in progressive nations or well over 25% of the total energy is used in progressive nations such as China and India (U.S. Congress Report: Energy in Developing Countries 1991). Cooking is the single largest energy form used in most of developing nations. The transition in cooking fuels is associated with high income, urbanization, and supply availability. For cooking purpose, traditional fuels are used in rural areas and poor households, for

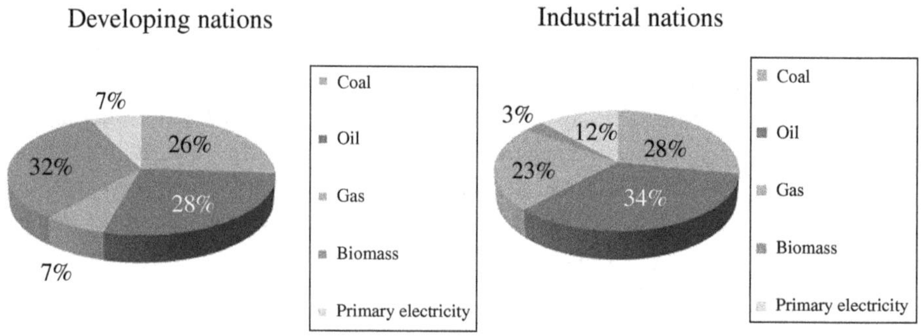

Figure 5.3 Energy consumption, industrial and developing region fuel mix. *Source:* Adapted from U.S. Congress. (1991).

example, dung, crop, and wood, whereas more affluent households prefer LPG (Liquefied Petroleum Gas) and electricity. LPG can use less energy for cooking compared to wood and charcoal. In the progression of cooking fuels, kerosene is the next alternative in the rural and poor households. In most places, kerosene oil prices are subsidized or paid by the government, constituting an efficient cap with respect to the price of charcoal and wood. Customers can switch to these fuels according to availability and price. With increasing household income, electricity is used for lighting. In developing countries, relatively less energy is used for residential space cooling. In commercial buildings and government offices, space cooling has become more significant. However, energy use for space cooling is expected to increase rapidly in the future. Figure 5.3 provides percentage share of traditional energy consumption for developing and industrial nations.

5.4 Energy Supplies in the Developing Nations

The energy supply sector is important in the development of economy (Hidayatullah et al. 2011; Giraldo et al. 2014). The energy supply sector includes a great share of public, private, and foreign investment. The commercial energy supply system such as gas, oil, coal, and electricity needs a good amount of foreign exchange, skilled labour, and efficient management. In the energy sector, operational efficiencies are much lower in most of developing nations than in industrial nations. Progressive nations generate much energy than their customers, and a significant amount of gas and oil are exported from less developed nations. So, considerable differences can be seen – only a few developing nations export energies.

The utilization of energy which will produce more light by progressive nation can be observed space due to growth of economy which leads to increase in energy and motor consumer fossils fuels. On one hand, prosperity in developed nations leads to more global energy consumption; on the other hand, traditional biomass (wood, agriculture waste, dung) is the main source of energy in less developed nations. Besides economic development and environmental benefits, renewable energy has the potential to provide increased security and financial stability (Islam et al. 2011). The use of renewable energy sources will reduce the dependency on expensive fossil fuel imports and help many countries to increase their balance of payments.

a) **Coal:** The coal production and consumption in developing nations are limited in few countries such as China, South Korea, and India, and these countries are accountable for 96% of the total coal production in progressive nations (U.S. Congress Report: Energy in Developing Countries 1991). The coal production rate is growing rapidly. Coal quality in developing nations varies from nation to nation. In China and India, the coal has a high ash content but relatively low sulphur. The coal industry significantly contributes to the overall commercial energy investment requirement in lower-income progressive nations. The Asian countries such as China and India are the biggest coal users. In developing countries, the government is solely accountable for distribution, transport, and coal mining (Wang and Song 2019).
b) **Oil:** Oil plays an important role in the energy sector of developing nations (Behmiri and Manso 2014). Oil consumption is expected to continue rising every year. Almost the entire developing world uses around 25% less oil compared to the United States (U.S. Congress Report: Energy in Developing Countries 1991). Most progressive nations depend heavily on oil imports. Oil and gas, which are considered natural resources, are treated as state property.
c) **Natural Gas:** It is a non-renewable hydrocarbon used as a source of energy for cooking and producing electricity and as a fuel for motor vehicles. Natural gas offers splendid opportunities for industrial development in many developing nations, which can launch their economies on a reliable path of rising income and poverty elevation, so natural gas is always an engine of growth and development (Bildirici and Bakirtas 2014; Chen et al. 2018). In the energy supply system, natural gas plays a relatively minor role in developing nations (EIA, 2014b). It provides merely 7% of the aggregate usage of energy in progressive nations. However, natural gas is a key energy source for few developing nations (Liu et al. 2017). Five countries Argentina, Venezuela, Mexico, China, and Algeria account for 58% of the progressive world oil and gas consumption. Natural gas is used for electricity generation (U.S. Congress Report: Energy in Developing Countries 1991).
d) **Biomass Fuels:** Biomass fuels like fossil fuels are mostly occurring and used in rural areas of developing nations where a majority of the world population lives. In progressive nations, it is considered a critical energy source and supplies three fourths of the total energy consumed in developing nations (Koçak and Sarkgüne¸si 2017). Biomass fuel consumption is difficult to measure. Biomass fuels are widely used in rural areas and in poor urban areas also. Private sectors are the sole suppliers of biomass fuels.
e) **Electricity**: In developing nations, electricity is produced from coal. A few countries consider oil for electricity generation. Electricity plays an important role in social and economic development of any country (Apergis and Payne 2012). In the present scenario, social, environmental, and economic resources are largely consumed by the electric sector. Electricity production in the developing nations is increasing at rapid pace. For electricity generation, different kinds of fuels are used in the United States. The functional capacities of electricity-generating plants are smaller in progressive nations than in the United States (Al-mulali et al. 2014). A few developing nations were generating electricity from nuclear power plants. A comparison of the United States and other developing world can be seen in Figure 5.4, in terms of coal, oil, natural gas, biomass, and others. The Table 5.3 shows the world population consumption of fuels.

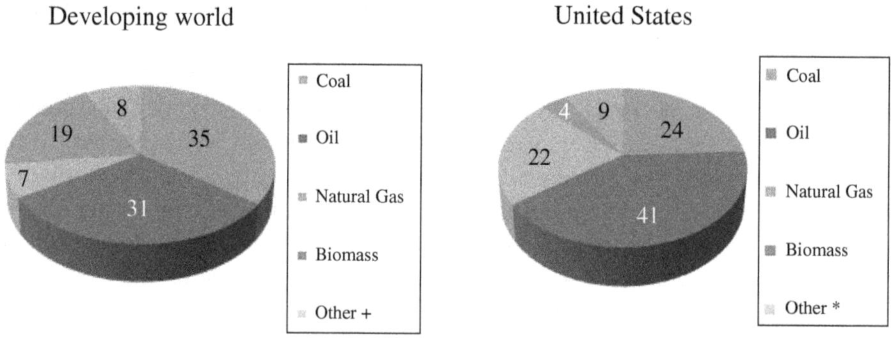

Figure 5.4 Comparison of supply mix in developing countries and the United States. *Source:* IEA, Paris OECD (1989), IEA Paris OECD (1987) and IEA Paris OECD (1984).

Table 5.3 World population consumption – cooking fuels.

	Percentage of people who use fuels		
Area	**Fossil energy**	**Fuel wood**	**Dung and crop waste**
Africa South Sahara	10	63	27
India	10	47	43
Remaining South Asia	12	46	42
E. Asia (developing Pacific)	36	41	23
Asia (centrally planned economies)	22	51	27
M. East and N. Africa	53	17	30
L. America and Caribbean	71	26	3
N. America, OECD Pacific	100	0	0
West Europe	100	0	0
Europe (centrally planned economies)	100	0	0
Total	47	33	20

Source: Hughart (1979)

5.5 Energy and the Environment in Developing Nations

At present, the major challenge for the future generation is how to protect the environment. More energy consumption poses a threat to the global ecosystem. With increasing energy consumption, carbon dioxide in the environment also increases (Shao et al. 2016). So, energy consumption is one of the reasons for environmental pollution in developing nations such as India, Sri Lanka, and Bangladesh.

Many developing nations are experiencing serious environmental degradation. Rural (i.e. countryside) areas, increase in population density, and low productivities of agricultural produce are the root causes that force the people to use ecologically fragile lands (Bilgili 2015). There is rapid growth in population in most of the urban areas in progressive nations. Also, poor infrastructure has led to a dramatic increase in pollution, levels of sulphur dioxide, ozone, and nitrogen oxide in technologically advanced nations.

Energy production and use cause environmental degradation in progressive nations. In addition to them, other major factors are industrialization, population growth, and government policies (Bilgili et al. 2016). To get the role of energy in whole environmental degradation, it is, therefore, important to consider all these factors. Energy use provides many environmental benefits in progressive nations. Energy consumption is highest in developing nations and countries with growing populations due to the focus on industries and high usage of household appliances.

5.6 Conclusion

The study results have highlighted some important facts regarding the economic growth and energy consumption in developing nations. The chapter presented several outcomes such as that increase in economic growth leads to increase in energy consumption. Renewable energy is criterian for economic, social and technological environment in progress where in manufacturing, labour and capital can be used. However, the linkage between these two can be reduced by many factors including shifting towards higher-quality fuels and technologies that lead to increase in economic productivity. The study reveals that in order to do business and gain competitive advantage, energy is the most basic requirement. Governments should adopt prudent economic energy policies that improve the efficient use of energy, which helps economy to grow compared to energy consumption. The study has shown that energy consumption and GDP are positively correlated. Policy makers should take multiple actions to improve energy efficiency. For controlling overuse of energy, government regulators must identify the areas that have minimum energy requirements. The sectors that consume more energy are recommended to modernize their technologies to be energy efficient. Failing to follow the recommendations may adversely affect the growth of GDP. The findings also suggest taking necessary steps so that it improves trade with foreign nations, which will introduce efficient resources in host countries. By improving trade with foreign nations, innovation in technologies will grow and aid in modernizing the industries.

One may reserve renewal energy from various sources such as wind, solar, and biomass. These sources are reliable and clean. In developing nations such as India, Africa, and Bangladesh, development in today's technology has become significant to minimize the usage of conventional fuels, because using fuels such as charcoal causes health issues and also leads to deforestation. Urbanization also has a positive correlation with economic growth. Two major reasons highlight the significance of the energy supply sector: first, the economic growth relies on the presence of reliable energy sources and second it requires a large investment. The pivotal role of energy in the progress of economic growth cannot be overlooked. Most of developing nations have abundant renewable energy resources, which include solar energy, biomass, etc., and also the potential to produce labour-intensive systems. Using such energy sources, developing nations will be able to ease their dependency on oil and natural gas. Therefore, it can be concluded that use of renewable energy sources is helpful in sustaining incessant development of economy; hence, the more the use of renewable energy sources, the more sustained economic development can be achieved by the nations.

References

Alam, M.S. (2006). Economic growth with energy, MPRA Paper, No. 1260.

Al-Mulali, U. and I. Ozturk (2015). "The effect of energy consumption, urbanization, trade openness, industrial output, and the political stability on the environmental degradation in the MENA (Middle East and north African) region." *Energy* 84: 382–389.

Al-mulali, U., Fereidouni, H.G., Lee, J.Y. (2014). Electricity consumption from renewable and non-renewable sources and economic growth: evidence from Latin American countries. *Renewable and Sustainable Energy Reviews* 30, 290–298.

Anderson, D. (2000). Energy and economic prosperity. "World Energy Assessment: Energy and the Challenge of Sustainability", Chapter 11: 393–413.

Apergis, N. and Payne, J.E. (2012). Renewable and non-renewable energy consumption growth nexus: evidence from a panel error correction model. *Energy Economics* 34 (3),733–738.

Asafu-Adjaye, J. (2000). "The relationship between energy consumption, energy prices and economic growth: time series evidence from Asian developing countries." *Energy Economics* 22(6): 615–625.

Aslan, A., Apergis, N., Yildirim, S. (2014). Causality between energy consumption and GDP in the US: evidence from wavelet analysis. *Front. Energy* 8 (1), 1–8.

Barro, R.J., and X. Sala-i-Martin. 1995. Economic Growth. Cambridge, MA: MIT Press.

Bastola, U. and P. Sapkota (2015). "Relationships among energy consumption, pollution emission, and economic growth in Nepal." *Energy* 80: 254–262.

Begum, R. A., K. Sohag, S. M. S. Abdullah and M. Jaafar (2015). "CO2 emissions, energy consumption, economic and population growth in Malaysia." *Renewable and Sustainable Energy Reviews* 41: 594–601.

Behmiri, N. B. and J. R. P. Manso (2014). "The linkage between crude oil consumption and economic growth in Latin America: the panel framework investigations for multiple regions." *Energy* 72: 233–241.

Belke, A., F. Dobnik and C. Dreger (2011). "Energy consumption and economic growth: new insights into the cointegration relationship." *Energy Economics* 33(5): 782–789.

Bhattacharya, M., Paramati, S.R., Ozturk, I., Bhattacharya, S. (2016). The effect of renewable energy consumption on economic growth: evidence from top 38 countries. *Applied Energy* 162, 733–741.

Bildirici, M. E. and T. Bakirtas (2014). "The relationship among oil, natural gas and coal consumption and economic growth in BRICTS (Brazil, Russian, India, China, Turkey and South Africa) countries". *Energy* 65: 134–144.

Bilgili, F. (2015). Business cycle co-movements between renewables consumption andindustrial production: a continuous wavelet coherence approach. *Renewable and Sustainable Energy Reviews* 52, 325–332.

Bilgili, F., Koçak, E., Bulut, Ü., 2016. The dynamic impact of renewable energy consumption on CO2 emissions: a revisited environmental Kuznets curve approach. *Renewable and Sustainable Energy Reviews* 54, 838–845.

BP. (2017). Statistical review of world energy 2017. https://www.connaissancedesenergies.org/sites/default/files/pdf-actualites/bp-statistical-review-of-world-energy-2017-full-report.pdf

Bulkeley, H., 2013. Cities and Climate Change. Routledge.

Cansino, J.M., Sánchez-Braza, A., Rodríguez-Arévalo, M.L., 2015. Driving forces of Spain's CO2 emissions: a LMDI decomposition approach. *Renewable and Sustainable Energy Reviews* 48, 749–759.

CEC (Commission of European Community). (2000). 1999 Annual Energy Review. Energy in Europa. Special Issue. Brussels.

Chen, J., Zhou, C., Wang, S., Li, S. (2018). Impacts of energy consumption structure, energy intensity, economic growth, urbanization on PM2.5 concentrations in countries globally. *Applied Energy* 230, 94–105.

Chontanawat, J., L. C. Hunt and R. Pierse (2008). "Does energy consumption cause economic growth: evidence from a systematic study of over 100 countries." *Journal of Policy Modeling* 30(2): 209–220.

Dong, F., Li, J., Wang, Y., Zhang, X., Zhang, S., Zhang, S. (2019a). Drivers of the decoupling indicator between the economic growth and energy-related CO2 in China: a revisit from the perspectives of decomposition and spatiotemporal heterogeneity. *The Science of the Total Environment* 685, e631–e658.

Dong, F., Yu, B., Pan, Y. (2019b). Examining the synergistic effect of CO2 emissions onPM2. 5 emissions reduction: evidence from China. *Journal of Cleaner Production* 223, 759e771.

EIA. (2014a). Kenya: Country Analysis Note. Washington DC: U.S. Department of Energy.

EIA. (2014b). Natural gas: sources and uses. In Electricity Explained. Washington, DC: Department of Energy.

Ellabban, O., Abu-Rub, H., Blaabjerg, F. (2014). Renewable energy resources: current status, future prospects and their enabling technology. *Renewable and Sustainable Energy Reviews*39, 748–764.

Esseghir, A. and L. H. Khouni (2014). "Economic growth, energy consumption and sustainable development: the case of the Union for the Mediterranean Countries." *Energy* 71: 218–225.

Fadaeenejad, M., Saberian, A. M., Fadaee, M., Radzi, M. A., Hizam, H., AbKadir, M. Z. 2013. The present and future of smart power grid in developing countries. *Renewable and Sustainable Energy Reviews*, 29, 828–834.

Fallahi F (2011). Causal relationship between energy consumption (EC) and GDP: a Markov-switching (MS) causality. *Energy*; 36:4165–70.

Giraldo, J., Mojica-Nava, D., Quijano, N. 2014. Synchronization of isolated microgrids with a communication infrastructure using energy storage systems. *Electrical Power and Energy Systems*, 63, 71–82.

Gulzara T, Huaping S, Muhammad H, Hafiz M J, and Yushen K (2018). "Energy consumption and economic growth: evidence from four developing countries". *American Journal of Multidisciplinary Research*, 7(1): 100–107.

Henderson, V. (2003). "The urbanization process and economic growth: the so-what question." *Journal of Economic Growth* 8(1): 47–71.

Hidayatullah, N. A., Blagojce, S., and Kalam, A. 2011. Analysis of distributed generation systems, smart grid technologies and future motivators influencing change in the electricity sector. *Scientific Research*, 2, 216–229.

Huang H, Du Z, He Y. (2018). The effect of population expansion on energy consumption in Canton of China: a simulation from computable general equilibrium approach. *International Journal of Engineering Sciences & Management Research*, 5(1): 19–26.

Hughart, D. (1979). Prospects for traditional and non-conventional energy sources in developing countries, World Bank staff working paper No. 346,132 pp.

IEA (International Energy Association). 1997. Energy Policies of IEA Countries. Paris: OCED.

IEA (International Energy Association). 1998. Balances of OECD Countries 1995–1996. Paris: OECD.

International Energy Agency (IEA). Energy Balances of Developing Counties 1971–1982 (Paris: OECD, 1984).

International Energy Agency (IEA). Energy Balances of OECD Countries 1970–1985 (Paris: OECD, 1987)

International Energy Agency (IEA). World Energy Statistics and Balances 1971–1987 (Paris: OECD, 1989)

Islam, F., Shahbaz, M., and Alam, M. (2011). Financial development and energy consumption Nexus In Malaysia: a multivariate time series analysis. MPRA Paper 28403. @ http://mpra.ub.uni-muenchen.de/28403

Joyeux, R. and Ripple, R. (2011) "Energy consumption and real income: a panel cointegration multi-country study." *The Energy Journal*, Vol. 32, pp. 107–142.

Koçak, E. and Sarkgüne¸si, A. (2017). The renewable energy and economic growth nexus in black sea and Balkan countries. *Energy Policy*, 100, 51–57

Ladu, M. G. and M. Meleddu (2014). "Is there any relationship between energy and TFP (total factor productivity)? A panel cointegration approach for Italian regions." *Energy* 75: 560–567.

Liao, X. and Shi, X. (2018). Public appeal, environmental regulation and green investment: evidence from China. *Energ Policy* 119, 554–562.

Liu, J., Cheng, Z.H., Zhang, H.M. (2017). Does industrial agglomeration promote the increase of energy efficiency in China? *Journal of Cleaner Production* 164, 30e37.

Lou CR, Liu HY, Li YF, Li YL (2016). Socioeconomic drivers of PM2.5 in the accumulation phase of air pollution episodes in the Yangtze River Delta of China. *International Journal of Environmental Research and Public Health*; 13:928.

Marine, A.S. and Schrattenholzer, L. (1989). International energy workshop: overview of poll responses. Stanford University International Energy Project.

Neto, A. F., Perobelli, F. S., Bastos, S. Q. (2014). Comparing energy use structures: an input–output decomposition analysis of large economies. *Energy Economics*, 43, 102–113.

Ohler, A., Fetters, I. (2014). The causal relationship between renewable electricity generation and GDP growth: a study of energy sources. *Energy Economics*, 43, 125–139

Park, S.-Y. and S.-H. Yoo (2014). "The dynamics of oil consumption and economic growth in Malaysia." *Energy Policy* 66: 218–223.

Pradhan, R. P., (2010). Energy consumption-growth nexus in SAARC countries: using cointegration and error correction model||, *Modern Applied Science*, Vol.4, No.4, pp.74–90.

Raeeni, A.A.G., Hosseini, S., Moghaddasi, R. (2019). How energy consumption is related to agricultural growth and export: an econometric analysis on Iranian data. *Energy Reports* 5, 50–53.

Rodolfo A. Bulatao, My T. Vu, and Eduard Bos Europe (1989). Middle East, and Africa (EMN) region population projections, Policy, planning, and research working papers; WPS 328.

Román-Collado, R., Ordoñez, M., Mundaca, L., (2018). Has electricity turned green or black in Chile? A structural decomposition analysis of energy consumption. *Energy* 162, 282–298.

Schuur, E.A., McGuire, A.D., Sch€adel, C., Grosse, G., Harden, J., Hayes, D.J., Hugelius, G., Koven, C.D., Kuhry, P., Lawrence, D.M., (2015). Climate change and the permafrost carbon feedback. *Nature* 520 (7546), 171e179.

Shahbaz, M., A. Chaudhary and I. Ozturk (2017). "Does urbanization cause increasing energy demand in Pakistan? Empirical evidence from STIRPAT model." *Energy* 122: 83–93.

Shao S, Li X, Cao JH, Yang LL. (2016). China's economic policy choices for governing smog pollution based on spatial spillover effects. *Economic Research Journal*; 9:73–88.

Tang, C. F. and B. W. Tan (2015). "The impact of energy consumption, income and foreign direct investment on carbon dioxide emissions in Vietnam." *Energy* 79: 447–454.

U.S. Congress. Office of Technology Assessment, Energy in Developing Countries, OTA-E-486 (Washington, DC: U.S. Government Printing Office, 1991).

Vandaele, N and Porter, W (2015). Renewable energy in developing and developed nations: outlooks to 2040. *Journal of Undergraduate Research*15(3) 1–7.

Wang, Q., Chen, X. (2015). Energy policies for managing China's carbon emission. *Renewable and Sustainable Energy Reviews* 50, 470–479.

Wang, Q., Song, X. (2019). India's coal footprint in the globalized world: evolution and drivers. *Journal of Cleaner Production* 230, 286–301.

Wang SJ, Li GD, Fang CL. (2018). Urbanization, economic growth, energy consumption, and CO2 emissions: empirical evidence from countries with different income levels. *Renewable and Sustainable Energy Reviews* 2018; 81:2144–59.

Wang, Q., Jiang, X.-t., Yang, X., Ge, S., (2020a). Comparative analysis of drivers of energy consumption in China, the USA and India e a perspective from stratified heterogeneity. *Science of the Total Environment* 698, 134117.

Wang, Q., Song, X., Liu, Y., (2020b). China's coal consumption in a globalizing world: insights from multi-regional input-output and structural decomposition analysis. *Science of the Total Environment* 711, 134790.

World Bank. (1988). Review of world bank lending for electric power. Industry and Energy Department Working Paper, Energy Series paper No. 2 March 1988.

World Bank. (2017). The World Bank Data, GDP Growth (Annual %).

World Energy Conference. (1989). Global energy perspective 2000–2020. *14th Congress, Montreal 1989*, Paris

Zheng, Q.Y. and Lin, B.Q. (2018). Impact of industrial agglomeration on energy efficiency in China's paper industry. *Journal of Cleaner Production* 184, e1072–e1080.

Zhou CS, Chen J, Wang SJ. (2018). Examining the effects of socioeconomic development on fine particulate matter (PM2.5) in China's cities using spatial regression and the geographical detector technique. *Science of the Total Environment* 2018;619–620:436–45)

6

Pathways of Energy Transition and Its Impact on Economic Growth

A Case Study of Brazil

Pooja Sharma

Daulat Ram College, University of Delhi, New Delhi, India

6.1 Introduction

Climate Change has taken centre stage among all global challenges which both developed and developing nations are facing today. Each developed and developing economies is determined to adopt newer, cleaner and efficient ways of producing energy to mitigate the harmful and disastrous consequences (Paris Agreement, COP 21). It is for the first time in 20 years that COP21 will aim to achieve a legally binding along with international climate agreements to keep the average temperature below 2 °C (UNFCCC, COP 21). India is going to play a crucial role in meeting the targets. 'India aims to lower its emission intensity by 33% to 35% by 2030 below 2005 levels', (Times of India 2 October 2016). In this context, most of the advanced countries have established a legal framework with the view to encourage the shift from fossil-based energy to a non-conventional clean energy system. This attempt requires a well-planned conscious effort to accomplish global objectives of energy efficiency and transition towards clean energy alternatives.

India is also committed to expand the contribution of non-fossil-based capacity of electricity generation to 40% of installed electric power capacity by 2030 (The Economic Times, 2 October 2018). The efforts are substantially towards decarbonization which refers to the efforts towards the direction of reducing carbon-emitting conventional fuels and substituting them by renewable and zero-carbon sources of energy. India is struggling with its twofold objectives of energy equity and sustainable energy growth. Under such circumstances when the country is struck by multitude problems of unemployment, poverty, inequality, lack of enough infrastructure which directly highlights an acute need for cheap sources of energy, it would be a real challenge for the nation to undergo energy transition in favour of alternate sources of energy. It is highly challenging for a developing country and developing nations to adhere to the targets as the conventional sources of energy are very cheap, and the alternative sources are expensive as they involve technological innovations (Ahuja and Tatsutani 2009).

Energy: Crises, Challenges and Solutions, First Edition. Edited by Pardeep Singh, Suruchi Singh, Gaurav Kumar, and Pooja Baweja.

With emerging concerns of Climate Change, the Climate regime has formulated a very comprehensive or holistic energy regime. All the developed and developing countries are determined to undergo a transition in their energy system to attain climate-resilient growth so that they are successfully able to trade-off between emission and economic growth. In this context, Brazil constitutes a remarkable example in achieving carbon-free growth laying down benchmark in energy innovation and transition pathway.

'Brazil, the largest country in the region of South America and fifth-largest in the world', (John and Monique 2011). It is the most prominent country in terms of international trade in South America. Brazil's institutional regime, policy framework and choices has made it possible to achieve successful carbon-intensive growth. "About 45% of primary energy demand is met by renewable energy" (IEA 2012). "The Brazilian ambitious targets were bio-power 19.3 GW, small-scale hydropower '7.8 GW' and wind power '15.6 GW by 2021' (Renewable Energy Policy Brief BRAZIL, June 2015)".

This phenomenal success has been an outcome of a considerable amount of investment in Research and Development and innovation in renewable technology in Brazil. The country has provided a well-coordinated environment of institutional structure, financing system and political structure to foster the innovation and make it commercially viable. In the process of understanding the energy pathways of Brazil, one needs to explore the three-fold objectives, first how the energy mix of Brazil should coexist to meet the energy requirement of poor who are deprived of primary energy access. This answers the question for whose benefit the policies cater to their needs in the context of investment in renewable sources of energy. The second most crucial issue needs to be addressed in the policy framework of decarbonized, climate-resilient growth is the institutional aspect dealing with the institution of structural regime existing for funding the Research and Development and innovation. Finally, to comprehend the financial system that backs up technology and production of other alternative clean energy sources.

Finance or source of funds constitutes a fundamental factor in determining the success of any innovation. Schumpeter (1982) emphasized the significance of the banking system in the process of economic development. As a result, examining the existing system of sources of finance in Brazil holds a crucial significance while studying the carbon-free pathways chosen by the political system, ultimately promoting the energy transition and energy diversification. The study would overview the Regime system in Brazil, which provides systematic framework financing innovations and Research in the energy transition, mainly renewable energy. An attempt is made to understand the regime of institutions which subsequently reduce the uncertainties associated with any research or innovation ready to break through the interface between the institution, power and end-user or consumers. The chapter emphasizes the role of policies in the decisions and incentives provided to the firms to invest in a particular technology (de Martino Jannuzzi and Melo 2013).

Concerning the significance of the policy framework pointed above, there exists a need to analyse the political economy in Brazil, determining the role of renewable energy in achieving a low carbon pathway. This analysis would revolve around three broad discussions, first in context of the regime system and vested interests that shape the energy policy in Brazil, second who decides the clean energy targets and who benefits from the new Policy, concerning the issues related to accessibility, affordability and acceptability. Finally, to analyse the system of power that exists and governs finance and technology either government, non-state actors or donors. By analysing these issues, one can understand the factors that shape the energy structure in Brazil (de Martino Jannuzzi and Melo 2013).

Newell et al. (2014) analyse the case of Kenya and conclude that both Government and donors together participate in financing the renewable energy in moving on the path of climate compatible development both supporting geothermal energy sources. Hall and Lerner (2010) pointed out that new energy policies and initiatives have facilitated Brazil in achieving an economically, socially sustainable development, reducing energy dependency on other countries.

In Brazil, renewable energy has one of the most significant proportions of total renewable energy generated in the world. The Brazilian Government has played a leading role in promoting biofuels, it began substituting oil for ethanol in the 1970s. Between 1960 and 1990, federal Government worked in the arena of national and regional interest while states limited their scope to generation, transmission and distribution of electricity within their territories (Roehrkasten 2015).

This phenomenal success has been an outcome of a considerable amount of investment in Research and Development and innovation in renewable technology in Brazil. The country has provided a well-coordinated environment of institutional structure, financing system and political structure to foster the innovation and make it commercially viable. Maji (2015) investigated effect of clean energy deployment on economic growth and deduced a mixed relationship in Nigeria, however, not significant.

Apergis and Danuletiu (2014) examined the relationship between deployment of renewable energy and economic growth, suggesting that use of renewable energy sources causally affect the real GDP. This empirical study confirms the interdependence of the consumption of renewable energy and economic growth. Lutz et al. (2012) used economy – energy-environment model PANTA RHEI to conclude that an increase in investment expands demand in the short – run and further minimizes expenditure on energy in the long – run. But in the regional level, renewable energy along with efficiency creates an additional value-added and employment in the economy. The observed transition from conventional energy to energy-saving techniques and renewable energy consumption supports the German economy, opening export opportunities for German industries. Blazejczak et al. (2014) analyse the economic impact of renewable energy deployment in Germany and concluded that renewable energy increases economic growth.

Apergis et al. (2010) 'investigated the inter-relationship between consumption of renewable energy, CO_2 emissions, consumption of nuclear energy and economic growth for selected groups of countries both developed and developing countries, between 1984 and 2007'. A statistically significant negative relationship has been observed between carbon emissions and deployment of nuclear energy and a significant positive correlation between carbon emissions and consumption of renewable energy. This work suggests that nuclear energy plays a critical role in reducing CO2 emissions in short-run, while the consumption of renewable energy does not lead to emissions. The consumption of nuclear energy reduces carbon emissions, suggesting the critical role played by nuclear energy in mitigating carbon emissions (Menyah and Wolde-Rufael 2010).

Nela V. and Saša Ž. (2010) concluded that 'the real GDP growth has causal impact on oil consumption, primary energy production and net energy imports in a bivariate model'.

Apergis et al. (2010) reveals 'a long-term relationship between real GDP, renewable energy consumption, real gross fixed capital formation and labour force'. Ramprasad Sengupta (2007) analyse the instrumental role played by energy efficiency, energy mix and technology in identifying the extent of inter-relationship between consumption of energy and intensity of emissions.

Simsek and Simsek (2013) contended that the significance of clean and renewable energy sources has increased on the account of severe threat not only to ecosystems but also world economies. Non-conventional energy sources are required more and more as a solution to the climate change challenge and exponentially rising energy demand. Consequently, a transition from primary energy sources deteriorating environment to clean energy sources is a central concern of literature related to energy led growth. A legal framework has been established by most of the advanced countries to encourage and deploy renewable energy sources advocating institutions such as the 'International Energy Agency' (IEA).

In this context, the chapter considers Brazil as a benchmark in the renewable energy breakthrough. Considering Brazil as a successful model, it is imperative to overview the policy regime and financial institutional system in Brazil and then try to analyse whether the Brazilian energy reform can deliver a climate-resilient growth analysing the social consequences on energy accessibility and affordability. A detailed account of energy policies, various state and non-state actors followed by a variety of programmes and laws initiated for the energy transition. The sources of finance for electricity, heating and transport, describing the agencies and causes that contribute to Research and Development have been discussed. Presenting the critical findings of the symbiosis of policy regime and financial structure that favoured a particular energy mix, affecting climate change and socio-economic consequences.

Considering energy as a critical component of economic growth, the critical role of renewable energy in economic growth has been examined. Since BRICS includes all the fastest growing economies of the world, these nations are prone to emit more and more emissions. It is crucial to explore the role of renewable energy sources in these nations; therefore, the chapter investigates the contribution and impact of renewable energy technologies in economic growth and development.

The chapter is organized into various sections. Section 6.2 presents the rationale for public investment in Research and Development in the Energy sector. Overviewing the electricity sector in Brazil with a detailed account of energy policy in Section 6.3, the market structure is described in Section 6.4. Brazilian Government has formulated various programmes and laws for fostering energy transition, presented in Section 6.5. The sources of finance, which is a crucial factor for the deployment of the energy transition, has been discussed in Section 6.6. The energy transition must achieve the sole objective of environmental sustainability as described in Section 6.7, as well as energy security in the context of availability, affordability and accessibility, discussed in Section 6.8. The process of energy transition incorporates an integral component of political economy as described in Section 6.9, followed by Section 6.10 that identifies the inter-relationship between consumption of renewable energy and growth of economy. Section 6.11 presents the key finding of the chapter concluding the critical appraisal on the energy transition pathways adopted by Brazil.

Energy policy or regime plays an instrumental role in the status of the energy sector and in defining pathways for the energy transition. Energy is a public good and therefore entails public provisioning. Energy sector comprises of power utility companies and the industries involved in exploration and development of oil, natural gas, refinery, etc. This sector caters to public services and represents a case of public provisioning and government investment. The following section discusses the rationale for public investment in the energy sector.

6.2 The Rationale for Public Investment in Research and Development in Energy Sector

Energy sector predominantly entails public service in the form of electricity generation, transmission, distribution and final end-use. Public services are largely characterized by market failure in the form of externalities, case of monopoly, information asymmetry and the notion of merit good. Negative externalities associated with electricity or energy systems are in terms of pollutants creating social and health damage or consequences. While positive externalities are in the form of provision of electricity in remote rural areas, opening up opportunities for productive growth.

Further, the information asymmetries among the suppliers and providers, this difference results in the unequal access of electricity or energy and therefore, the market fails. The energy sector is characterized as a case of natural monopoly involving high infrastructure cost and economies of scale, thus constituting a strong case for state provisioning of energy utility by state further facilitating innovation in technology through funding Research and Development.

Research & Development in the energy sector is an evolving process enabling the basic Research by which end-use energy, energy supply and eventually carbon management technologies undergo a transition (Dooley 2000, PNNL-13158).

Given the above reasons for market failure, the state must invest in this sector and also achieve the current decarbonized development by initiating and promoting appropriate technological innovations to attain climate-compatible energy developments. In the process of understanding the energy pathways of Brazil, one needs to explore the threefold objectives, first how the energy mix of Brazil should coexist to meet the energy requirement of poor who are deprived of primary energy access. This answers the question for whose benefit the policies cater to their needs in the context of investment in renewable sources of energy. The second most crucial issue needs to be addressed in the policy framework of decarbonized, climate-resilient growth is the institutional aspect dealing with the institution of structural regime existing for funding the Research and Development and innovation. Finally, to comprehend financial structure that constitutes the foundation for technology and varied production methods for clean energy sources.

6.3 Overview of the Electricity Sector in Brazil

'Brazil became part of BRICS and also one among the fastest-growing economies in the world'. Brazil is subsequently recognized globally in terms of its economic reforms. It is the tenth-biggest energy consumer in the world and the second largest producer of ethanol. The prominent players in the energy sector in Brazil are the state-owned companies, Petrobras and Eletrobras.

Being highly dependent on hydroelectricity, Brazil possesses the most abundant water storage system in the world (OECD, Project Closing Report 2005). Brazil has 97% access to electricity, but since it is dependent majorly on hydroelectricity, the country becomes vulnerable to power supply shortage at the time of droughts.

6.3.1 Energy Policies in Brazil

This section overviews the energy policies in Brazil concerning all energy sources and sectors. This section would analyse the policies framed under the existing Energy regime in Brazil (Meisen and Hubert 2010). The reform process of the energy sector, specifically the electricity sector, can be categorized into different phases.

6.3.1.1 Energy Sources and Associated Policies

Brazilian energy policies favouring hydropower, biofuel, energy efficiency and electricity have to be examined in terms of their acceptability by society or community at large.

- Biodiesel: Brazilian Government uniquely favours biodiesel production with 20% biodiesel content by 2020. Large-scale production of biodiesel will contribute to economic growth and also job creation. Since the technology has been available in this energy source for a long time; therefore, economies of scale would further bring their costs down.
- Ethanol fuel using sugarcane: Sugarcane is the most planted crop across Brazil, possessing the world's second largest producer of ethanol fuel (Renewable Fuel Association 2018). Over the years, the production costs have gone down tremendously. Large-scale production would further create employment for people. Brazilian Car manufacturing companies highly promoted the ethanol fuel by making vehicles which could run on any combination of ethanol fuel and the Government also made it mandatory the blending of ethanol fuel with gasoline, and now there is no vehicle which runs in Brazil on pure gasoline.
- There was evidence of ethanol crises in the past when sugarcane production went down; then this made sugar prices more attractive than produce ethanol fuel; this was coupled with the economic crises of 2008. The prices of ethanol fuel became unattractive, and ethanol fuel production went down.
- Large hydropower plants are in the form of large dams displacing large population facing severe consequences of resettlement. Such an energy source has significant environmental, ecological, social consequences, so Brazilian Policy favoured small hydro with Brazil having considerable potential between 7 and 14 GW. This small investment would attract new players, increase in investment would lower costs of regional supply reducing transmission and distribution costs, also lowering environmental and social negative impact.

- LPG gas: This fuel is mostly used for cooking in the residential sector. This type of fuel has an environmental advantage over firewood. The prices in this fuel have almost doubled since 1992. It is widely accepted.
- Electricity rates have been rising consistently since the 1970s. The target of universal access has been prevalent since 2008.

6.3.1.2 The First Phase of Reforms in the Electricity Sector: 1990s

Opening of economy through Liberalization of 1990s provided a conducive environment for investment in Brazil (Baumann 2002). As a result of which public enterprises started to undergo privatization. Some of the companies became global, enhancing their investment abroad, for example, Petrobras worth approximately $90 billion in 2008. The electricity market underwent a drastic transformation by becoming free and open. Therefore, the electricity market became more attractive to private companies. Electrobras is currently the leading power company in Brazil, initially a national public company and later privatized with the Government holding 52% company stock.

6.3.1.3 Second Reform of the Electricity Market: 2004

The emerging electricity crises led to the regulation of the electricity market. Consequently, Energy Research Company was formulated to assist the Ministry of Energy. From 2004 onwards, energy trade to be carried out in terms of long-term contracts. Two trade contracts were designed in the form of

- Regulated Environment (ACR): Energy is bought in public auctions by the distribution companies.
- Contracting Environment (ACL): A contract that allows Distribution companies to buy power in public sales.

6.3.2 Climate Change: National Policy 2009

Brazilian Government implemented 'National Policy on Climate Change' which advocated targets of national emission reduction from 36.1 to 38.9% by 2020 (World Resource Institute). These targets are to be realized by using energy efficiency policies in construction, farming and industry.

6.3.3 Prioritization of Policies in Choice of Energy Mix (International Atomic Energy Agency, 2006)

- In case of **natural gas**, high priority was given to the expansion of natural gas use in buses and internalized sulphur emissions costs associated with fuel and less weightage delivered to adopting minimum efficiency standards in thermal power plants.
- **Renewable energy sources** used for electricity gave more priority to adopt renewable energy portfolio standards, and less preference was given to stimulate use of PV solar in remote and off-grid areas, off-grid areas.
- **Ethanol, other bioenergy sources** 'A very high priority has been allotted in policy structure to increase output and consume ethanol, promoting efficient CHP systems etc.'

Further develop and promote new bioenergy sources. While a very low weightage given to R&D capacity building for hydrogen and fuel cells
- **Electricity Conservation:** A considerable attention was given to 'establish a new national energy efficiency agency, adopt demand-side bidding', implementing standard laws for appliance efficiency, while less priority paid towards adopting energy codes for new commercial buildings.
- **Energy conservation and energy transition:** Priority was awarded for adopting CO_2 emission standards for new passenger vehicles. Less priority is given to improving the efficiency of freight transport, passenger transport and further to shift to less energy-efficient modes.
- Energy transition pathways of energy efficiency and adopting renewable energy together provide financing mechanism for the companies.

This is very clear from above that it is entirely in the hands of policymakers and in this case Brazilian Government to promote the energy source of their choice and further choose the pathway to achieve a decarbonized growth.

6.4 Market Structure

With the reforms in the electricity sector, satisfying demand and then engage in external trade is the present policy. The current market structure is that most of the conventional and renewable energy companies are now privatized. Brazil currently has a two-faced energy market

- Petrobas: the primary objective is to engage and control the drilling of oil and gas.
- Electrobras: owns the majority of hydropower, nuclear and wind power companies.

The above companies hold monopoly of Brazil energy involved in the energy sector transition to sustainable energy alternatives. Public companies mostly hold electricity generation while distribution is effectively privatized.

Brazil National Bank for Development (BNDES) is responsible for privatizing and regulating electricity distribution. Fixed costs, electricity prices and taxes are fixed as rates, while variable costs are indexed as inflation. This endeavour will benefit the hydropower sector, by enabling them to offer lower prices for electricity as the sector faces immense competition from natural gas.

6.4.1 Government Players

The stakeholders governed by the state or the Government in Brazil are listed below.

- 'Ministry of Mines and Energy': Implementation of renewable and conventional energy and laws for attaining government targets of decarbonization (Global Energy Network Institute).
- 'Brazilian Institute of Environment and Renewable Natural Resources' (IBAMA): The aim is to preserve natural resources in Brazil.

- 'The Brazilian Electricity Regulatory Agency' (ANEEL), in Portuguese, is an autonomous body of the Government of Brazil linked to the Ministry of Mines and Energy. It aims to ensure favourable conditions to facilitate a congenial environment among the agents.
- The Energy Research Company (EPE): established in 2004, assists the 'Minister of Mines and Energy' in various activities such as sector planning and expansion auctions (Global Energy Network Institute)
- Operador Nacional do Sistema Elétrico (ONS): Managing and controlling the market is the main aim of this operating system company,
- The National Nuclear Energy Commission (CNEN) established in 1956, under the 'Ministry of Science and Technology'. It is responsible for the nuclear regulation and implementing laws related to protection from radiation.

6.4.2 Private and Public Players

There are certain private and public players in the market, as discussed below.

- Electrobras: Around 60% of the total power in Brazil is provided by this company. It regulates generation, transmission and distribution and works with several specialized subsidiaries.
- Tractebel Energia: It is a private company with 'an installed capacity of 11 GW generated' from hydropower plants.
- Petrobras: A significant oil producer and a major distributor of oil products is the second largest public listed company in the world.

6.5 Programmes and Laws Under the Government of Brazil

The programmes and laws are described in Table 6.1, with the goals and initiatives.

- PROCEL: A national electricity conservation programme was established by the Government of Brazil at the end of 1985. The main aim of this initiative is to finance energy efficiency projects undertaken by both state and local utilities, universities, research institutes and private companies. For various initiatives such as energy efficiency low interest is offered by state and local utilities
- PROINFA: It is 'Incentives Program for Alternative Sources of Electric Energy' ANEEL. The objective is to promote investments in research and development and to determine laws and resolutions (Prado 2008).

6.6 An Overview of the Sources of Finance in the Energy Sector: Brazil

'In 2010, the second phase of Growth Acceleration Programme (PAC-II) consisting of new investment plans for the periods 2011–2014 and post – 2014' (Energy Research Knowledge Centre). 'The first phase (2007–2010) called for an investment of €269 billion, out of which

Table 6.1 Energy Programmes and initiatives undertaken by the energy sector.

Programme	Year	Goals
1. Alternative Energy Source Incentive Programme (PROINFA)	2002–2010	It is created by the Ministry of Mines. The primary purpose is to develop clean and alternative sources of electricity. The electricity was generated from wind power, biomass and small hydroelectric plants.
2. National Energy Conservation Programme (PROCEL)	1985	This programme aimed to reduce wastage of electricity. Energy efficiency, promoting technical development, energy security, economic efficiency and reduced environmental impact are some of the objectives of the programme.
3. 'National Programme for the Rational Use of Oil and Natural Gas' (CONCEPT)	1991	The purpose is to encourage energy efficiency in all the sectors of the economy.
4. 'National Biodiesel Production and Usage Programme' (PNPB)	2004	The aim is to regulate Biodiesel produced from oilseeds.
5 "Light for All "(Luz para Todos)	November 2003 extended to period 2011–2014	This initiative was undertaken to initiate electricity inclusion in Brazil.
6. Hydropower Programme	Ten-year energy plan	This programme aims at '71 new hydro plants by 2017 with a potential output of 29 GW'.
7. Wind Energy Programme		It initiates most significant 'wind energy capacity in Latin America and Caribbean'.

63.3% took place'. The phase II of Growth Acceleration Programme (PAC-II) was launched with following initiatives.

The above allocation reveals that post-2014, a considerable amount is allocated for the energy sector, as observed in Table 6.2. The details of sources of funding and its nature along with the profile for investment have been compiled in Table 6.3.

Table 6.2 'PAC – II Investments'(€ billion).

PAC – II Initiatives	2011–2014	Post - 2014
Urban Infrastructure	24.1	—
Citizens + Community	9.7	—
Housing	117.4	—
Light for All	12.8	—
Transportation	44.1	1.9
Energy	196.6	264.8
Total	404.7	266.7

Table 6.3 Funding organizations at the national level.

Funding organization	Nature or Type	Profile
1. 'Studies and Projects Financing Agency' (FINEP)	It is a Public company managed by the 'Ministry of Science Technology and Innovation'	This agency promotes development of socio-economic parameters.
2. 'National Council for Scientific and Technological Development' (CNPq)	It is a Federal government organization under the 'Ministry of Science Technology and Innovation'	To provide aid and training of Brazilian researchers.
3. 'Petroleo Brasileiro SA' (Petrobras)	The largest company, State-controlled oil and gas producer,	'It submits the annual budget for approval by Ministry of Planning, Budget and Management, the Ministry of Mines and Energy and the Brazilian Congress'.
4. 'ElectrObras'	Privatized in the 1990s, supports Federal Government,	To design, build and operate power stations are the key objectives of this company. After privatization, it is also an energy distributor.

Source: Modified from Robson Braga de Andrade, FINANCING INNOVATION IN BRAZIL, World Intellectual Property Organization (WIPO).

6.6.1 The Regime for Funding Agency (World Energy Outlook 2013)

A well-developed Financial structure backed up by Institutional Regime is the foundation for combating the risks and uncertainties associated with technological innovation (OECD, World Energy Outlook 2011). A multilevel initiative is required that plays a catalytic role in transforming the energy sector. Moreover, energy is a sector vulnerable to market failure due to reasons like natural monopoly, information asymmetry, externalities, and the Government plays a decisive role in incentivizing this sector and making it more competitive keeping the social issues in consideration. From the 1990s onwards, Brazil instruments of public R&D financing for the private sector was better placed, resulting in expansion and diversification of funding in order to ensure a better resource allocation. The funding Regime of Brazil can be schematically expressed as visualized in Chart 6.1.

1. Federal Funding Agency

- "Science and technology agencies from the Ministry of science, Technology and Innovation (MCTT) such as the National Innovation Agency (FINEP) National Council for Scientific and Technological Development (CNPq)"
- 'Regional Development Agencies such as Amazon Development Agency (Sudam), The Superintendence for the Development of the North East(Sudene)'

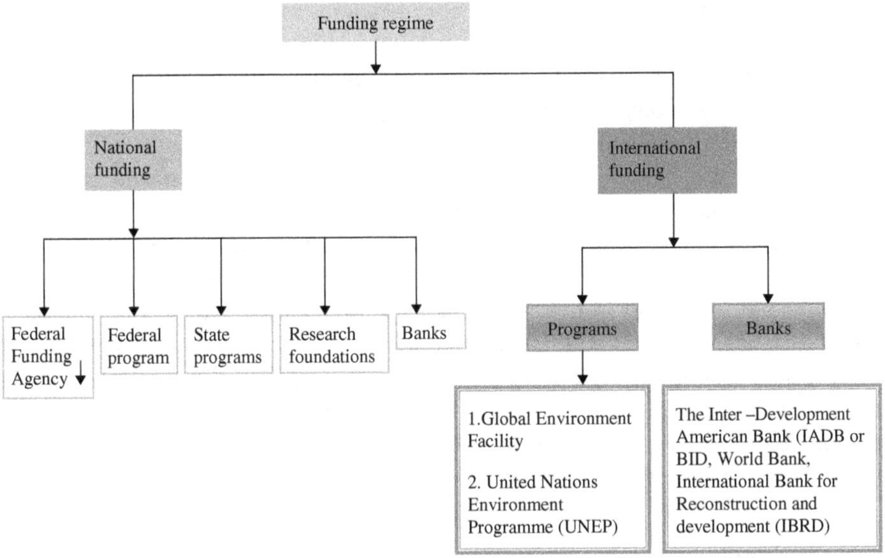

Chart 6.1 Regime for funding agency

2. **Federal Programmes**
   Regulatory Agencies like:
   - "The Brazilian Electricity Regulatory Agency (Aneel)
   - National Agency of Petroleum
   - Natural Gas and Biofuels (ANP), etc.
   - Technological Export Programme for exports (Progex)
   - Constitutional Fund for financing the Centre West Region etc."
3. **State Programmes**
   'State funding to support Research such as Fundes (RJ), Fomento (GO) and other research funds'.
4. **Research Foundations**
   - "Banco do Brasil Foundation (FBB)
   - States Foundations such as Sao Paulo Research Foundation (Fapesp)
   - Research and Development Foundation (FADESP)
   - Foundation to Support Research and Extensions (Fapex), etc."
5. **Banks**
   - "Banco do Brasil
   - Brazilian Social and Economic Development Bank (BNDES) etc."

6.6.1.1 Regime Structure and Legal Regulatory: Key Takeaways

Brazilian institutional regime has been open enough to accommodate private industry. The policy regime has been flexible enough to provide a market-driven approach, specifically in the electricity sector. A well-defined legal framework backs up the policies implemented.

Brazil has invested majorly in energy efficiency and energy sources like wind, biofuel and solar in terms of designing policies, providing incentives and channelizing funds through the funding regime. Brazilian Government takes appropriate initiatives and programmes like 'PROINFA and Technological Development Program for Biodiesel'.

Public finance initiatives in Brazil have been the backbone in the energy sector reforms. Government has taken essential steps to improve further the funding opportunities for projects, Research and Development and innovation. In this process, the main challenge lies in coordinating and integrating both private and public knowledge-generating agencies and further ensuring a constant flow of funds through financial institutions. There has been stability and consistency in the flow of funds in Brazil as the substantial funding is linked to the National Budget. Still, it has been observed less incentive and weightage is given to the solar energy as a clean energy source. This phenomenal breakthrough in renewable sources of energy has to be analysed in terms of energy security and accessibility in remote areas. Brazil has proved to be the most extensive clean grid in the world, but the country still has to invest in making it more sustainable.

6.6.2 Source of Funding and Trends in Research and Development

'The Ministry of Science and Technology' (MCT) is responsible for Brazil's innovation policy. It has identified several strategic areas for energy R&D like funding for biofuels, transmission, distribution, hydrogen and renewable energies, oil, gas, coal and nuclear (de Azevedo Rezende 2019). Brazil has also set up the Interministerial Commission on Climate Change (ICCC 2008) to increase energy efficiency and decrease electricity consumption by 10%. Maintaining high levels of renewable electricity supply and increase the use of biofuels in the transport sector are other objectives.

The electricity production from various sources in Brazil pertains to 2% from coal, 3% from gas, 3% from oil, 3% from nuclear and 84% from hydro in 2009 according to the World Bank data. The sources of R&D funding are significant as far as the other sources are considered, other than the government. Besides, the spending on R&D is mainly on Fossil fuel technology as compared to other technologies. It is also evident that government spending on R&D funding has significantly reduced over the period.

6.6.2.1 Finance and Innovation in Renewable Energy: Key Takeaways

'The proportion of R&D expenditure in GDP has been increasing since 2000, more prominently after 2006' (World Bank data). The Brazil policy framework has provided a conducive environment for all state, non-state actors, private companies and NGOs to invest in Research and Development about various energy sources.

By observing the pattern of R&D expenditure by the Government between 2000 and 2008, it is evident that the Policy has been favouring renewable energy in terms of expenditure on R&D over the fossil fuels (Belfer Centre). Government is purposefully using the institutional regime and funding agencies to invest and fund Research and Development in energy-efficient technologies and innovation in transmission, distribution and storage as this is the biggest challenge of sustaining the renewable energy sources.

But coincidently the expenditure on R&D by other sources is more in fossil energy as compared to energy technologies and negligible for other sources of energy. It is mainly because it involves a lot of risk and uncertainty to invest in new technology and investment in R&D in fossil fuels would make the fuel even more competitive and carbon-free thus accruing the investors more profits.

The data for patents also reflects the fact that several patents in renewable energy sources have been rising to show a positive outcome of growing R&D expenditure in renewable energy in terms of biofuel, wind energy, hydro – conventional and energy storage. The policies promoting various types of funds have been concentrated towards energy efficiency, biomass and hydropower through multiple programmes like INNOVA ENERGIA, PROEOLICA 2001 programme, PROINFA, etc. As R&D expenditure to improve energy efficiency, wind energy and biofuel increases, the number of patients increase in wind energy, biofuel and energy storage devices.

A smaller number of patents being filed in solar energy and tidal sources. The data further shows an evident rise in patents in 2011 owing to second phase reforms in Heating industry through a network of policy framework involving institutional and legal regime, ensuring a stable funding structure. The number of patents initially rise steeply in Brazil and the decline sharply immediately after 2013.

The Electricity sector is a significant part of the infrastructure development of a country and a successful electricity production meeting the demands of all sectors of the economy is a predominant requirement of a well-placed energy system. Some of the challenges faced by this sector are the issue of efficient and less carbon-intensive production using renewable, losses involved in transmission and distribution and finally accessibility and affordability in remote areas. The issue of accessibility and affordability is also a social and political issue which will be tackled towards the end.

6.7 Climate-Resilient Growth: Environmental Consequences

Brazilian Policy and legal regime structure is designed to deliver a climate-resilient growth, reducing the carbon emissions most cost-effectively. This section would highlight the status of carbon emissions and its relation to the use of renewable energy as promoted by the policy regime in Brazil. Appendix Figure 6.A.1 indicates that though the carbon dioxide emissions have been rising in Brazil, there is some interval of years when they have stabilized showing the control over emissions using clean energy technologies.

A negative relationship between carbon dioxide emissions and renewable energy consumptions, reflecting that with the use of more and more renewable energy, carbon dioxide emissions will reduce (World Bank data 1990–2012).

6.7.1 Environmental Consequences: Key Takeaways

Carbon dioxide emissions have been consistently rising with some intervals. 'There exists a trade-off between carbon emissions and Renewable energy consumption' (World bank data). Carbon emissions through combustion are also rising consistently with a slight decline during the period 2008–2010, suggesting a positive impact of increased use of renewable sources of energy which accounts mainly due to well-designed Policy and stable and reliable sources of finance. PM2.5 is an indicator of air quality status which has been rising tremendously since 1990. Still, the rate of increase has been curtailed 2010 onwards establishing the success of efforts made on behalf of a policy framework to deliver a resilient carbon growth, according to World Bank data.

6.8 Social Consequences: Availability, Affordability and Accessibility

The energy system has been transformed in Brazil from the middle of twentieth century and beginning of twenty-first century. 'The rural areas which used to rely majorly on fuelwood for energy use have effectively transformed themselves to adapt to the urban, modern commercial form of energy utilities like electricity, fossil fuel and modern biomass', (International Atomic Energy Agency, Brazil, 2006).

Despite the remarkable transition experienced in energy utility across the country, it is observed that still, it is not economical to expand the power grid in most of the remote and isolated areas in Brazil.

6.8.1 Social Consequences: Key Takeaways

In terms of access to electricity, there was a substantial gap between the rural and urban population till 2000, from 2000 onwards the accessibility of electricity started improving and catching up with the urban population. And further increased in 2012 with only some marginal difference with urban areas. In case of access to non-solid fuels like petrol, diesel, kerosene, there is a substantial gap between rural and urban areas, and the gap remained the same in 2010 and 2012 (World Bank database). Consequently, the Gross national income in Brazil has been rising consistently. The Gini Index reflects the degree to which the distribution and consumption of individuals deviates from equal distribution. There is observed a rise in Gini index suggesting disparities across regions in terms of various household amenities including the availability of water internet connections, infrastructure, availability of electricity and fuel, etc. with the appropriate reforms carried out under policy framework along with the energy system regime the Gini Index came down in 2011 (World Bank data).

The main challenge for the policymakers is to find the balance to cover investment costs involved in creating infrastructure across remote areas and at the same time deliver energy utility services at affordable prices. This is achieved by subsidizing the benefits for low-income groups. If the prices are set such that they are not affordable by low income, then the repercussion would be that the poor people would continue using conventional sources with inefficient energy technologies.

Gasoline prices have averaged 1.13US dollar /litre from 1992 until 2016 reaching highest in 2010 and lowest in 1992. The gasoline prices started rising from 2008 all-time peaked prices in 2010 and then started declining, reaching very low in 2016 and further started rising (Numbeo, Gas Prices, Brazil). Brazilian Policy has used direct transfer effectively ever since general subsidies were abolished during the oil market liberalization. 'The step was taken to offset the negative economic impacts of liberalization' (International Atomic Energy Agency, Brazil, 2006).

6.9 The Political Economy of Energy Transition: A Brazilian Experience

Brazil is ranked one in the production of renewable energy with only 24% share of fossil fuel in electricity production along with carbon dioxide emission in the combustion of only 0.2% (Energy Atlas, Brazil). From the paper, it is evident that the Brazilian policy regime has successfully acted on the well-laid foundation of the legal framework promoting many non-state actors. The policies are well inclined towards achieving climate-resilient growth targets. Brazilian Government controls and governs the flow of funds and ensures consistent and reliable finance for funding all significant initiatives taken by the policy regime to achieve environmentally sustainable growth. It is also observed that the Government has taken appropriate steps in opening up and providing the market – based approach to Research and Development in the innovations required in renewable energy sources. This has positively worked in favour of successfully carrying out R&D with a consistent flow of funds. The number of patents achieved by Brazilian residents has phenomenally increased in those energy sources where both state and non-state actors have invested.

Further examining the climate resilience of the policy regime and their outcomes, Brazil has been able to stabilize its CO_2 emissions and emissions due to combustion, there is also a reduction in indicator of air quality suggesting a clean growth. Further, Brazilian energy transition effort made through successful R&D and innovations has resulted in reducing the disparity between the rural and urban areas related to accessibility and also reduced the income inequality through the remarkable efforts made by the policy regimes.

The issue of affordability and accessibility of commercial energy is also coupled with a dilemma of providing efficient energy and affordability since generally efficient energy sources are available at higher prices. To resolve such a dilemma, the Brazilian Government had successfully made use of subsidy and direct transfers to lower-income groups, especially at the time when the gasoline was privatized.

The Brazilian policies have favoured some of the energy sources which have a positive impact on socially desirable and acceptable goals, generating employment opportunities at the same time delivering an environmentally sustainable energy source. Last but not the least despite the huge amount being spent by the Brazilian Government on R&D and innovation the investment in grid connectivity has not come down making it still expensive to reach out to rural and remote areas leaving many people without the access of electricity.

6.10 Interlinking Economic Growth and Energy Use: A Theoretical Construct

Energy plays an indirect role in the production and economic growth. The oil crises of 1973–1974 ensured that the process of production is entirely energy-dependent. In economics, the neoclassical production function is expressed as a function of labour, capital and technology as inputs determining the output.

'In some of the recent growth models, human capital is a crucial variable which determines growth, mainly in case of endogenous growth models'. Human capital is defined as the population and labour force. Romer (1990), Aghion and Howitt (1996) and Grossman and Helpman (1990, 1991) emphasized the impact of Research and Development on economic growth.

In most of the empirical studies, human capital is quantifiable through proxies like R&D expenditure, health expenditure, etc. Some of the studies have evaluated the relationship between economic growth and energy. 'However, these studies differ in their results drastically despite witnessing a strong interdependence between economic growth and energy consumption'. Energy is a crucial input along with labour and capital as reflected by some of the studies, while others have concluded a neutral impact on growth.

Nela Vlahinić-Dizdarević2, Saša Žiković3 (2010), (Nicholas and Payne 2010) suggested a considerable causality between consumption of renewable energy and economic growth. 'Payne et al. (2010) assert an inverse relationship between consumption of nuclear energy, emissions, while a significant positive relationship between renewable energy use and emissions'. 'It was observed that Nuclear energy could help to mitigate CO_2 emissions; however, renewable energy could not make any significant contribution as emphasized by' (Wolde-Rufael and Menyah 2010). With the onset of climate change as a global challenge, the role of clean and renewable energy sources has increased significantly (Simsek and Simsek 2013).

6.10.1 Renewable Energy Consumption, per Capita GDP Growth, CO_2 Emissions, Research and Development Expenditure: A Comparison of BRICS

Brazil, Russia, India, China and South Africa, constituting a group of countries known as BRICS, are among the fastest developing countries in the world. The growth is eventually accompanied by the cost of emissions that a country emits while growing. Therefore, by observing the data related to their growth, CO_2 emissions, use of renewable energy and expenditure on Research and Development would deliver a clear understanding of each country's performance when it comes to growth and emissions. The data for concerned variables are obtained for all BRICS countries from 1990 to 2013 in the Appendix Figure 6.A.2.

The above graphs indicate that CO_2 emissions are highest in the case of Canada and lowest in the case of India in Appendix Figure 6.A.1. The carbon emissions in Brazil are also very low but higher than in India. In contrast, the proportion of renewable energy in total energy consumption is highest for Brazil and India, which perhaps is the reason that these countries are emitting lower CO_2. The data on Research and Development indicate that Canada spends the maximum amount on Research and Development. At the same time, India and South Africa have a lesser proportion of their GDP being spent on Research and Development. This points out that the research and development expenditure may not be directed towards the innovation of renewable energy technologies Figure 6.A.3.

Comparing the data from 1960 to 2016, it is observed that GDP per capita fluctuates substantially in both the countries and becomes negative at most of the times in both nations. The data of Brazil indicate a steady movement in renewable energy consumption and GDP per capita growth rates. CO_2 emissions per capita have been consistently rising in Appendix Figure 6.A.4. Similarly, the data between 1990 and 2012 reflect a steady decline in proportion of renewable energy consumption, while the GDP growth rates fluctuate. The trend in CO_2 emissions in Brazil and India depicts that in both the countries, the emissions have been rising consistently since 1961, shown in Appendix

Figure 6.A.4. Brazil CO_2 emissions are higher than the emissions in India. The data indicate that Brazil spends a considerably larger proportion of their GDP on Research and Development as in Appendix Figure 6.A.5. The renewable energy consumption as a percentage of total energy consumption in India has been much higher than Brazil till 2006, as shown in Appendix Figure 6.A.6. Still, the data indicate that 2006 onward renewable energy consumption in Brazil started rising.

6.11 Conclusion

Energy plays one of the predominant roles in the process of development of all developing countries. All the growing countries require energy to meet the demand for their growth. In the light of the climate change which is now a global issue, all the countries need to curtail their emissions and therefore move towards energy transition from conventional energy to more clean and renewable sources of energy.

Brazil's institutional regime, policy framework and choices have made it possible to achieve successful carbon-intensive growth. The phenomenal success has been an outcome of a considerable amount of investment in Research and Development and innovation in renewable technology in Brazil. The country has provided a well-coordinated environment of institutional structure, financing system and political structure to foster the innovation and make it commercially viable. In the process of understanding the energy pathways of Brazil, the chapter explored the twofold objectives. First, the policies were critically examined and analysed to examine whether the financial regime, along with the political system, can deliver a climate-resilient, efficient and equitable energy system. Second, the chapter analyses the relationship between renewable energy and economic growth for the BRICS countries in general and Brazil in particular.

It is realized that the institutional aspect dealing with the institution of structural regime existing for funding the Research and Development and innovation is one of the most critical issue in the discourse of energy transition. Finally, to understand the financial regime that constitutes the foundation of technology for clean and alternative energy sources.

Despite strong interdependence between economic growth and energy consumption, various recent studies have explored the relationship between energy and economic growth. However, there is no uniformity across the findings regarding their conclusion. A robust significant negative correlation between use of renewable energy and carbon emissions highlighting the role of renewable energy investment on the improvement of environmental quality. Further, renewable energy would be able to contribute successfully to economic growth. The proportion of GDP spent on Research and Development may not be on renewable energy technologies. Renewable energy consumption has been persistently decreasing, while Brazil energy consumption is rising consistently. Brazil spends a substantial amount of its income on Research and Development as compared to India. Brazil is a growing country emitting more CO_2 emissions as compared to India. Consequently, the rationale and genesis of investment on R&D, renewable energy technologies, as well as the role of the state in developing countries have been established substantially.

References

Aghion, P., & Howitt, P. (1990). A model of growth through creative destruction (No. w3223). National Bureau of Economic Research.

Aghion, P., & Howitt, P. (1996). Research and development in the growth process. *Journal of Economic Growth*, 1(1), 49–73.

Aghion, P., Howitt, P., Howitt, P. W., Brant-Collett, M., & García-Peñalosa, C. (1998). *Endogenous growth theory*. MIT press.

Ahuja, D., & Tatsutani, M. (2009). Sustainable energy for developing countries. *SAPI EN. S. Surverys and Perspectives Integrating Environment and Society*, 2 (1).

Apergis, N., & Danuletiu, D. C. (2014). Renewable energy and economic growth: evidence from the sign of panel long-run causality. *International Journal of Energy Economics and Policy*, 4(4), 578.

Apergis, Nicholas, James E. Payne, Kojo Menyah, and Yemane Wolde-Rufael. "On the causal dynamics between emissions, nuclear energy, renewable energy, and economic growth." *Ecological Economics* 69, no. 11 (2010): 2255–2260.

D. de Avila Vio; "AnOverview of PROINFA - A Brazilian Incentive Programme for Alternative Electricity Sources". OGEL 1 (2011), www.ogel.org, URL: www.ogel.org/article.asp?key=3079

de Avila Vio, D. (2011). An overview of PROINFA-A Brazilian incentive programme for alternative electricity sources. *Oil, Gas & Energy Law Journal (OGEL)*, 9(1). www.ogel.org/article.asp?key=3079

de Azevedo Rezende, L., Bansi, A. C., Alves, M. F. R., & Galina, S. V. R. (2019). Take your time: examining when green innovation affects financial performance in multinationals. *Journal of Cleaner Production*, 233, 993–1003.

Baumann, R. (2002). Brazil in the 1990s: an economy in transition. *En: Brazil in the 1990s: an economy in transition-London: Palgrave/St. Antony's College, 2002-p.* 1–38.

Blazejczak, J., Braun, F. G., Edler, D., & Schill, W. P. (2014). Economic effects of renewable energy expansion: a model-based analysis for Germany. *Renewable and Sustainable Energy Reviews*, 40, 1070–1080.

Dooley, J. J. (2000). *A Short Primer on Collecting and Analyzing Energy R&D Statistics* (No. PNNL-13158). Pacific Northwest National Lab.(PNNL), Richland, WA (United States).

Grossman, G. M., & Helpman, E. (1990). Trade, innovation, and growth. *The American Economic Review*, 80(2), 86–91.

Grossman, G. M., & Helpman, E. (1991). Trade, knowledge spillovers, and growth. *European economic review*, 35(2–3), 517–526.

Grossman, G. M., & Helpman, E. (1996). Rent dissipation,free riding, and trade policy. *European Economic Review*, 40(3–5),795–803.

Hall, B. H., and Lerner, J. (2010). The financing of R&Dand innovation. *In Handbook of the Economics of Innovation* (Vol.1, pp. 609–639). North-Holland.

Howitt, P., & Aghion, P. (1998). Capital accumulation and innovation as complementary factors in long-run growth. *Journal of Economic Growth*, 3(2), 111–130.

John J.C. and Monique V. (2011). Brazil Today: An Encyclopaedia of Life in the Republic (pp.23). South Dakota State University.

John J.C. and Monique V. (2011). Brazil Today : An Encyclopaedia of Life in the Republic, South Dakota State University. P.23. ISBN 978-0-313-34673-6

Lutz, C., Lehr, U., and Pehnt, M. (2012, June). Economic impacts of energys efficiency and renewable energy in Germany. In *Abstract for the 20th International Input--Output Conference.*

Maji, I. K. (2015). Does clean energy contribute to economic growth? Evidence from Nigeria. *Energy Reports*, 1, 145–150.

de Martino Jannuzzi, G., & de Melo, C. A. (2013). Grid-connected photovoltaic in Brazil: policies and potential impacts for 2030. *Energy for Sustainable Development*, 17(1), 40–46.

Meisen, P., & Hubert, J. (2010). Renewable energy potential of Brazil. *Global energy network institute (GENI).*

Menyah, K., & Wolde-Rufael, Y. (2010). Energy consumption, pollutant emissions and economic growth in South Africa. *Energy Economics*, 32(6), 1374–1382.

Newell, P., Phillips, J., Pueyo, A., Kirumba, E., Ozor, N., & Urama, K. (2014). The political economy of low carbon energy in Kenya. *IDS Working Papers*, 2014(445), 1–38.

OECD (2005). https://www.oecd.org/about/34711139.pdf.

Prado, M. M. (2008). The challenges and risks of creating independent regulatory agencies: a cautionary tale from Brazil. *Vanderbilt Journal of Transnational Law*, 41, 435.

Renewable Fuel Association (2018). https://ethanolrfa.org/.

Roehrkasten, Sybille (2015):"Global Governance on Renewable Energy." *Global Governance on Renewable Energy*. Springer Fachmedien Wiesbaden, 2015. 73–116

Romer, P. M. (1990). Endogenous technological change. *Journal of Political Economy*, 98(5, Part 2), S71–S102.

Schumpeter, J. A. (1982). The “crisis” in economics-fifty years ago. *Journal of Economic Literature*, 20(3), 1049–1059.

Sengupta, R. (2007). *High economic growth, equity and sustainabAle energy development of India* (No. 213).

Simsek, H. A., & Simsek, N. (2013). Recent incentives for renewable energy in Turkey. *Energy Policy*, 63, 521–530.

Vlahinić-Dizdarević, N., & Žiković, S. (2010). The role of energy in economic growth: the case of Croatia. *Zbornik radova Ekonomskog fakulteta u Rijeci: časopis za ekonomsku teoriju i praksu*, 28(1), 35–60.

Wolde-Rufael, Y., & Menyah, K. (2010). Nuclear energy consumption and economic growth in nine developed countries. *Energy Economics*, 32(3), 550–556.

Websites

Energy Research Knowledge Centre, ERKC. European Union. https://setis.ec.europa.eu/energy-research. https://ec.europa.eu/jrc/sites/jrcsh/files/fta2014-posters-tools-energy-research-knowledge-centre.pdf.

Global Energy Network Institute http://www.geni.org

ICCC. (2008.)https://www.nccs.gov.sg/

IEA, OECD. (2012).World energy outlook 2012. *Vienna: IEA* (2012). Renewable Energy Policy Brief BRAZIL, IRENA, JUNE 2015

International Atomic Energy Agency. (2006). Brazil: A Country Profile for sustainable Energy Development, Brazil 2006. http://www-ub.iaea.org/MTCD/publications/PDF/Pub1247_web.pdf

Meisen, P. and Jordi, H. (2010). Renewable Energy Potential of Brazil Or how to foster a transition to sustainable, environmentally friendly ways of producing energy in Brazil. Global Energy Network Institute (GENI) http://www.geni.org/globalenergy/. . ./renewable-energy-potential-of-brazil/re-pot-of-brazil.

Observatory of Renewable Energy in Latin America and The Caribbean, available at http://www.renenergyobservatory.org
OECD. (2011). https://www.oecd-ilibrary.org/energy/world-energy-outlook-2011_weo-2011-en
Organisation for Economic Co-operation and Development – OECD, World Energy Outlook. (2011). Fact Sheet: global energy trends. http://www.worldenergyoutlook.org
Renewable energy policy Brief, BRAZIL, June 2015. http://www.irena.org/. . ./IRENA_RE_Latin_America_Policies_2015_Country_Brazil.pdf
Renewable Fuel Association (2018). https://ethanolrfa.org/
World Energy Outlook. (2013). https://www.iea.org/reports/world-energy-outlook-2013
www.ogel.org, URL: http://www.ogel.org/article.asp?key=3079
https://afdc.energy.gov/data/10331
https://www.wri.org/blog/2010/03/brazils-global-warming-agenda
https://www.wipo.int/edocs/pubdocs/en/wipo_pub_gii_2020-chapter12.pdf e Andrade, R. B. FINANCING INNOVATION IN BRAZIL.
https://www.belfercenter.org/program/science-technology-and-public-policy.
https://www.iaea.org/publications/reports/annual-report-2006

Appendix A

Author's compilation

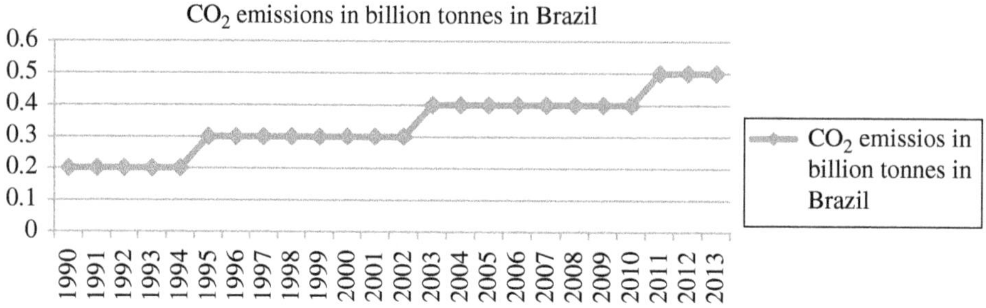

Figure 6.A.1 CO2 Emissions in billion tonnes in Brazil, World Bank.

Author's Calculations, World Bank Data

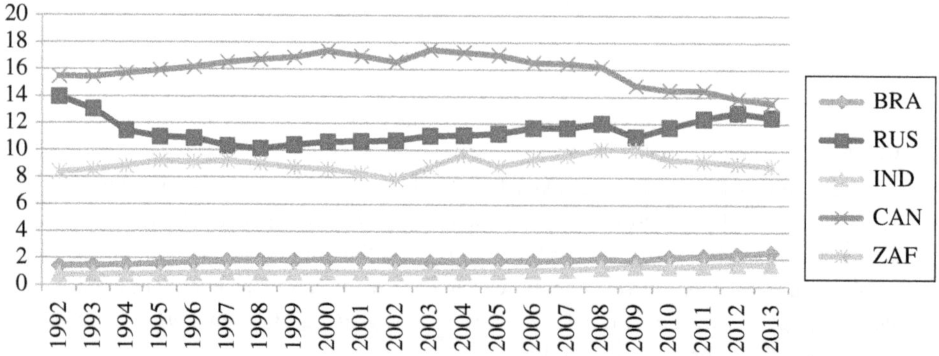

Figure 6.A.2 CO2 Emissions (metric tons per capita): A comparison of BRICS.

Author's Calculations. World Bank Data

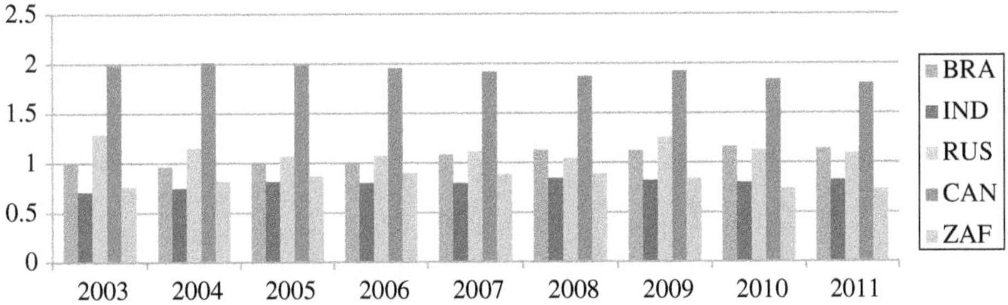

Figure 6.A.3 Research and development expenditure (% of GDP): A comparison of BRICS.

Author's calculation, World Bank database

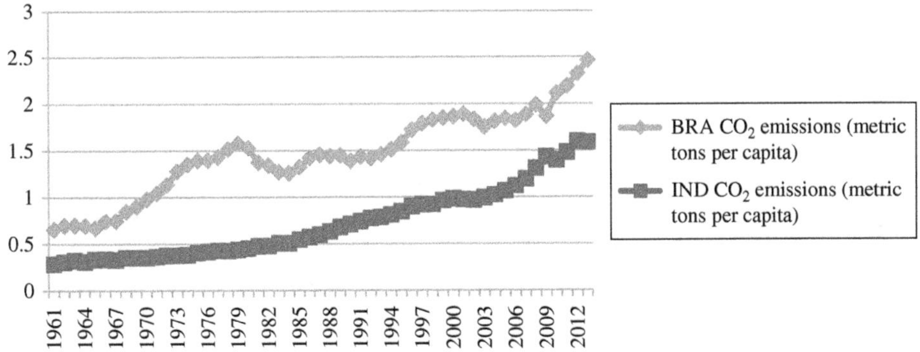

Figure 6.A.4 Trends in CO2 emissions in Brazil and India.

Author's calculation, World Bank database

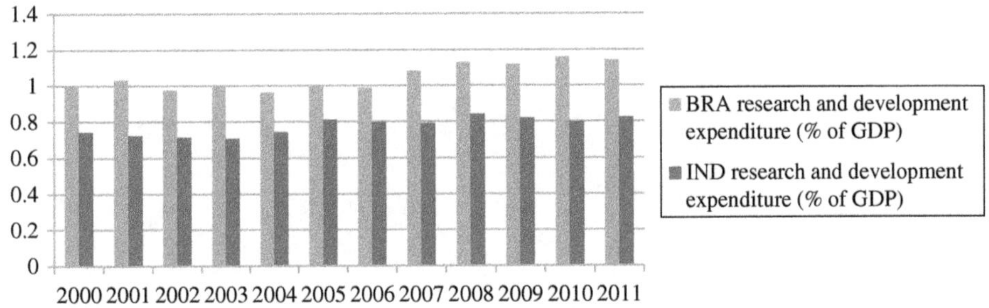

Figure 6.A.5 Expenditure on Research and Development as a percentage of GDP in Brazil and India.

Author's calculation, World Bank database

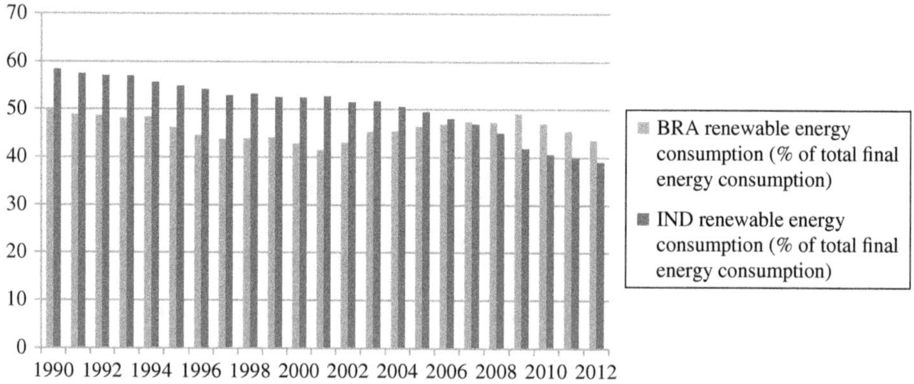

Figure 6.A.6 Renewable Energy Consumption as a percentage of total final energy consumption in Brazil and India.

7

Renewable Energy

Sources, Importance and Prospects for Sustainable Future

Shachi Agrawal and Renu Soni

Department of Botany, Gargi College, University of Delhi, New Delhi, India

7.1 Introduction

Past few decades have experienced exponential increase in use of energy all over the world from 8588.9 million tonnes (in 1995) to 13 147.3 million tonnes (in 2015) (Dong et al. 2020). The energy demand in India is also increasing very drastically, and it is predicted that by 2030 the whole energy demand of India may be more than twofold and electricity demand will be almost threefold than that of existing value (IRENA 2017). This global energy demand is to be fulfilled from various sources which can be mostly divided into two classes – the traditional non-renewable and modern renewable sources. As the name suggests, non-renewable sources cannot be replenished and get exhausted, whereas renewable resources are regenerative in some form or the other and do not deplete over time. Non-renewable resources, majorly fossil fuels, take millions of years to form. As per the World Energy Forum, at the current consumption level across the world, coal, fossil-based oil and gas reserves can last another century as they account for almost 80% of the primary energy consumed at the current global level (Zerta et al. 2008; Zhang et al. 2017; Jamshidi and Askarzadeh 2019; Ahmad and Zhang 2020). In addition, usage of non-renewable resources negatively affects the carbon balance of the earth's ecosystem and is majorly responsible for the climate worry worldwide.

Renewable energy (RE), on the contrary, leverages energy sources constantly refilled by nature itself e.g. Sun, water, wind, plants and Earth's heat (Ellabban et al. 2014). The sources of RE are inexhaustible, and their availability is not limited to small pockets but across the world in various proportions. For many upcoming centuries, the Sun's existence is not questionable, the wind will not stop blowing or water will not exhaust. By using various technologies, most of the RE forms may be transformed into other energy forms e.g. electricity, heat, chemicals and mechanical power (Tromly 2001; Infield and Freris 2020). Heat is an essential need of mankind. Solar, geothermal and biomass energies can be used to fulfil this demand. Renewable energy can also be converted into transport fuels albeit with some difficulty and cost (García-Olivares et al. 2018). As this resource is naturally distributed in

Energy: Crises, Challenges and Solutions, First Edition. Edited by Pardeep Singh, Suruchi Singh, Gaurav Kumar, and Pooja Baweja.

remote areas too, so can provide energy in such areas without using any extensive energy transport system and thus provide additional advantage. Sun-based water heating and wind-controlled water siphoning are good instances of frameworks which do not need power at all for their appropriate working. However, RE can increase its major contribution by supplying people's needs in electrical form only. Solar energy, one of the largest natural RE sources, alone carries the potential to fulfil all the energy demands of the entire world using the modern technologies (Kabir et al. 2018). Currently mankind is able to harvest only a miniscule 1% of the total solar energy that reaches earth annually. The technologies available today and those to be developed in future can leverage these unending natural sources of energy to satisfy the ever-increasing demand of progressive mankind and an effectively sustainable solution to long-standing energy deficits to various economies across the world.

As a result of anthropocentric developmental activities, nations around the world are struggling to cope with ever-increasing worry of global warming which makes the significance of RE in the energy-cum-electricity mix paramount. Share of renewables is taking prominence in the overall energy portfolio for all geographies across the globe, irrelevant of whether they are developing or developed. It has become the only ray of hope to sustain a pristine environment for future generations to flourish as it improves environmental sustainability (Alola et al. 2019). And India has been showing the path to its contemporaries for socio-economic growth without affecting the environment adversely. Whether it is augmentation of grid power or access to energy or avoiding the over-dependence on availability of fossil fuels, India is leading humanity to pursue a low-carbon development path (MNRE Annual Report 2019-20).

This chapter focuses on various sources of RE, their importance, role in sustainable development. Its limitations and future prospects are also discussed in brief.

7.2 Sources of Renewable Energy

RE sources are natural resources; therefore are inexhaustible, and energy released from them continuously flows in our environment. They represent a keystone to control the direction of the energy system towards sustainability. They include: solar energy, wind energy, geothermal energy, hydropower, bioenergy and tidal energy (ocean) (Figure 7.1).

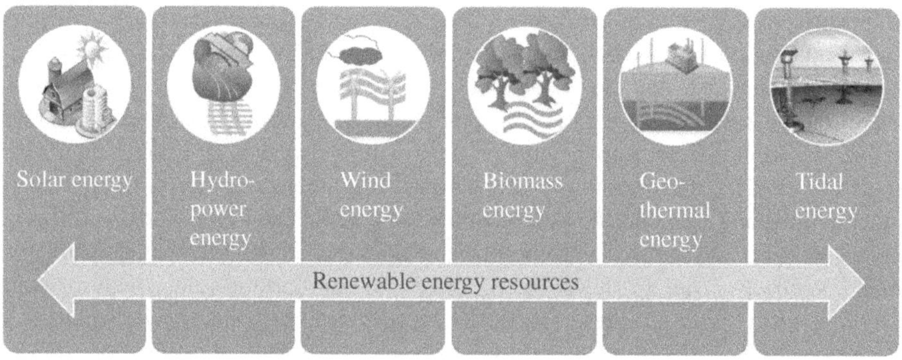

Figure 7.1 Overview of renewable energy resources.

7.2.1 Solar Energy

The Sun is the main source of unlimited free solar energy (kabir et al. 2018). Solar energy has the highest possibility for producing safe, unpolluted and reliable energy from the Sun. It is used for light and also to produce heat and electricity. Earth continent receives solar energy more than 200 times the whole marketable energy consumption yearly by individuals (Ashokverghese 1998). On Earth, Sun's energy is the greatest source of everlasting energy and can be used directly (solar radiation) and in indirect (wind, hydro, biomass, ocean etc.) forms (Kumar et al. 2010). The concept of whole solar energy is related to harvesting and consumption of heat and/or light energy produced by the Sun and technologies (active and passive) involved in attaining goals. (Besarati et al. 2013). Solar energy technologies are classified into two types:

7.2.1.1 Active Solar Energy Technology

The solar radiations are electromagnetic in nature and are used in producing electrical energy. These systems collect solar energy and convert it into heat and electrical energy by using electrical and mechanical equipment (fan, pump, solar water heater, etc.). These technologies are arranged into two classes: (i) photovoltaic technology and (ii) solar thermal technology (Herrando and Markides 2016).

Photovoltaic technology comprises of wafer-based cells (conventional crystalline arsenide gallium or silicon), business meagre thin-film cells (amorphous silicon, cadmium telluride) and technologies (perovskites, quantum dots, organic materials) (Alharbi and Kais 2015). This technology utilizes sun-based cells which convert the daylight energy into power which can be utilized for different uses, such as communications, pumping, lighting and energy supply (Kumar et al. 2010).

In solar thermal technology, solar energy is converted into heat energy for commercial and/or domestic applications such as heating, drying, cooking, cooling, etc. (Seddegh et al. 2015). Use of solar heat energy is an advantage towards carbon emissions and sustainability. In the industrial sectors, use of solar heat collectors is perceived as promising possibility for an alternative energy resource due to power security, financial achievability and natural advantages over fossil fuels. Industrial use of heat processes is classified on the basis of solar accumulator technologies and heat requirement. These technologies have many industrial applications and are used for sustainable development (Kumar et al. 2019).

7.2.1.2 Passive Solar Energy Technology

It makes both direct and indirect use of heat energy obtained from the Sun. In this technology, accumulation of solar energy is involved and used without transforming into any other form. Here solar energy is gathered, stored and conveyed as warmth for additional utilization (Sun and Wang 2016). Passive utilization of energy is conceivable just in building (or) structures. In latent frameworks, structures are worked with windows and warmth fascinating surfaces set up to enhance sun-fuelled warming in winter (Alrikabi 2014).

New innovations are being locked in, to deliver power assembled from the solar energy. These are broadly drilled everywhere in the world as sustainable alternatives. In this way, the acknowledgement of the Sun-based innovations would significantly reduce and moderate issues related with energy security, joblessness, environmental change and so forth (Kabir et al. 2018).

7.2.2 Wind Energy

Wind energy is driven by atmospheric air and is broadly distributed, commercially available energy resource. It is produced by complex mechanisms such as rotation of the Earth, Sun's heat energy, temperature difference between sea and land, the cooling effects of the polar ice caps and oceans and the physical effects of mountains and other obstacles (Kumar et al. 2010). It does not cause water or air pollution; therefore is environmentally safe and cost-effective (Nix 1995).

According to power density of wind, which is directly related to the cube of wind velocity, seven classes are categorized: beginning from class 1 (the most reduced) to class 7 (the best). For effective applications, class 4 (average density ranges from 320 to 400 W/m_2) or higher classes are generally required (Nix 1995).

Wind energy is based on a simple technology converting moving (kinetic) energy into mechanical energy which is directly used (e.g. water pumping or milling) or transmitted to an electrical generator. It consists of a set of turbine blades which are moved by the wind and turn a mechanical shaft coupling to an energy generator which produces electric power. In different regions, the availability of wind varies. Based on the axis of blade rotation, wind turbines are either vertical-axis or horizontal-axis machines (mostly used). Turbines are unable to work when wind speed is lower than 13 km per hour. Location of turbines is also an important factor for proper functioning of machines. These wind turbines need to be installed on top of buildings or high towers (Alrikabi 2014).

Wind velocity is determined by several factors such as the time of day, weather, geographical locations, season, landforms and height above the earth's surface. By knowing wind characteristics, we can optimize the design of wind turbines, develop techniques used for measuring wind and also help in selection of wind farm sites (Tong 2010).

Significant achievements with movement of wind turbine rotor size and their appraised energy yield since year 2000 with developing future patterns are shown in Figure 7.2. Hence, the establishment of sustainable power source advancements is of incomparable interest. Presently enhanced movement of wind energy prediction and its utilization around the world have been watched (Aubrun et al. 2013).

The most genuine negative impacts of wind-turbine-based innovation are habitat loss for fauna and flora, deforestation, soil erosion, noise pollution and shadow effects (Figure 7.3). To summarize, it is basic to comprehend the environmental-related impacts of wind turbines. So, it can be concluded that if power is to be produced, it is important to construct power in a manner that has minimal effect on the environment. Considering the exploration needs, leaders, designers and different partners should work to near normalize the arrangements to limit the negative ecological effect of this natural energy source (Nazir et al. 2020).

The overall objectives of wind energy research are to decrease the cost of the instruments, reduce maintenance, improve energy harnessed, increase reliability and enhance

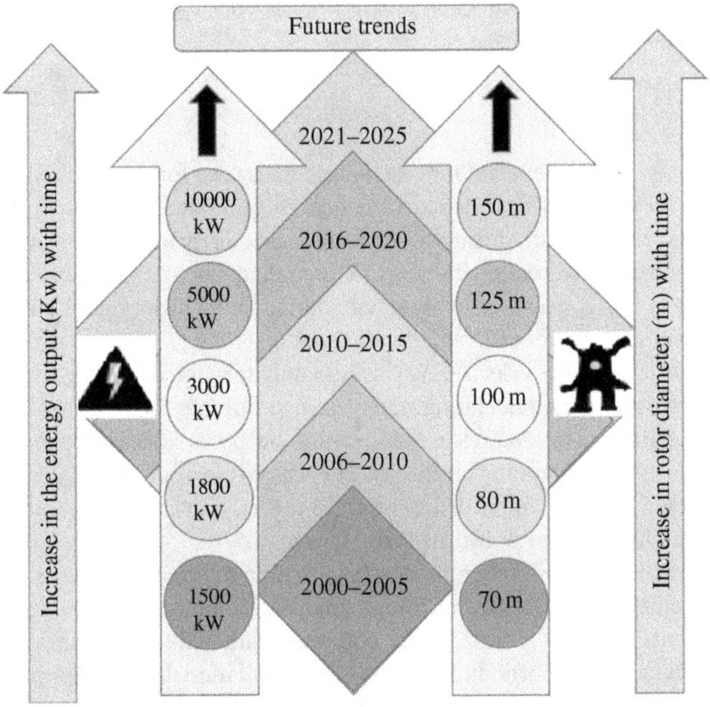

Figure 7.2 Major revolution in the wind energy output in relation to turbine rotor size. *Source:* Modified after Nazir et al. (2020).

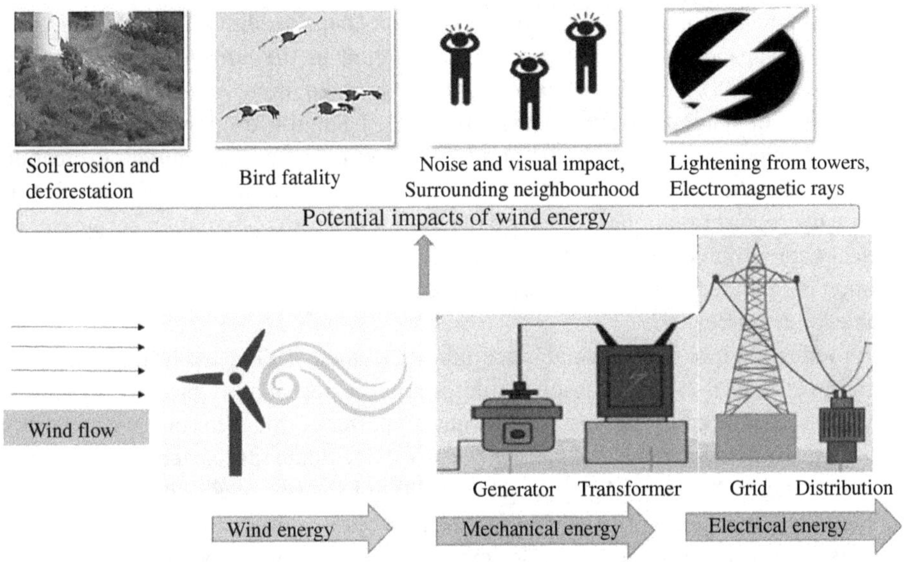

Figure 7.3 Development of wind energy and its potential impacts. *Source:* Modified after Nazir et al. (2020).

the system, also focussing on the environmental and aesthetics effects at the same time. (Asumadu-Sarkodie and Owusu 2016).

7.2.3 Hydropower

Hydropower is one of the largest and ancient renewable resources used for the production of electricity (Abolhosseini et al. 2014). It denotes the energy generated from water, by converting energy (kinetic or potential) of water into mechanical energy as used in watermills, textile machines, etc. or as hydroelectric energy (Kumar et al. 2010). Hydropower is a sustainable type of energy that is delivered by water circulating in the hydrological cycle, which is eventually determined by the sun's radiation (Kumar et al. 2011).

The kinetic energy of water produces electricity. This is mainly through damming rivers to form large water reservoirs and then liberating water through turbines to generate electricity. Stored water from dams can be later released for various purposes such as irrigation, industrial and domestic use and generating electricity. The water reservoir behaves like a battery, by releasing storied water when needed to produce power. As water is the initial source, therefore is termed hydropower or hydroelectric power.

Hydropower has become a key possibility for extensive usage in different nations as an alternative energy source because of its various social, financial and natural advantages. The primary advancements utilized in hydropower comprise dammed supply, running waterway, storage capacity, stream innovation and innovation gravitational vortex. Because natural availability, financial and ecological advantages of hydroelectric energy which make it a significant supplier of future energy, especially in the non-industrial nations. Hydropower plant gets well known in certain countries because of bumpy territories and being secured with various waterways (Chala et al. 2019),

The development of enormous hydropower stations requires the huge capital expense and causes ecological issues making it less fascination in the current world economy (Kabo-bah et al. 2018). This circumstance has prompted the improvement of the tremendous capability of small-scale hydropower plant (SHP). Small hydropower delivers under 2 MW power; micro-hydro is for creation under 500 kW; and pico-hydro alludes to creation under 10 kW (Balkhair and Rahman 2017). Accordingly, energy from water in scaled down/miniature hydropower plant has picked up the most noteworthy fascination because of its ecological approachable action and is suitable for rural regions' electricity creation. The innovation is extended to be future reasonable as it creates spotless, modest and natural agreeable force sources.

In non-industrial nations, sustainable sources add to about 47GW, and little hydropower plant shares around 1–2% of the absolute limit. Nations using supportable energy featured that hydroelectric force has become a significant supporter of things to come world energy blend, especially in the agricultural nations because of its innate specialized, monetary and natural advantages (Manzano-Agugliaro et al. 2017). It likewise got consideration because of its high thickness, unsurprising and ceaselessly accessible force. In addition, there are enormous wellsprings of hydropower, for example, wave energy, flowing energy and sea energy (Behrouzi et al. 2016).

Moreover, hydropower energy can be decentralized as it can legitimately be utilized for grain processing, battery charging and so on in little businesses, giving bounty advantages to the nation (Haidar et al. 2012). Despite its high development costs, the hydroelectric force plant has a low breakeven point because of its low working and upkeep costs with the ideal normal yearly precipitation, making it savvier when contrasted with other green force sources. Hydropower can likewise help to control flood, can be utilized as modern and water system and can be formed into recreational travel industry (Bildirici and Gökmenoğlu 2017). Miniature hydro is one of the most ecologically agreeable energy innovations accessible these days. In addition, little hydropower plant turns into an ideal energy arrangement in numerous nations worldwide because of its less expensive cost, and it is customizable after the topographical, monetary and social conditions (Yaakob et al. 2014; Yah et al. 2017).

Hydropower does not produce any emissions into the atmosphere; however, the development of dams on a river can produce important environmental problems such as water quality, prominent impact on population displacement, fish migration, loss of biodiversity and wildlife habitat. (Edenhofer et al. 2011; Kumar et al. 2011). The principle difficulty for the reception of this innovation is enormous hydropower plant requests for immense deforestation. This would desperately bring about the destruction of basic biological systems, movement of the enormous populace downstream and other disastrous results, for example, changing stream morphology and directions, streak floods and avalanche etc. (Yah et al. 2017).

7.2.4 Geothermal Energy

Geothermal energy is the main environmentally friendly power source delivered normally from earth itself from the warm energy created in the centre, mantle and crust of the earth (Kumar et al. 2010). Geothermal energy is economical and clean. Geothermal energy assets reach to boiling water from the shallow ground to hot stone present a couple of miles underneath the earth's surface and magma (molten rock) found in more profound layers with the immensely high temperatures. The highest 6 miles of the earth's crust layer are a possible source of the geothermal energy (Alrikabi 2014). As of now geothermal energy is utilized in various areas around the globe to create power at a cost-effective and give energy legitimately to food, space warming, refrigeration, industrial use and aquaculture.

In geothermal power plants, high temperatures present deep inside the earth used to produce hot water or steam. These pools can be used for a variety of uses, such as heating buildings or to produce electricity. Superficial earth's stable temperature can be used for cooling and heating buildings by using geothermal heat pumps (GHPs). GHPs are more effective than energy-efficient heating systems (furnace). Geothermal power plants use high temperatures from the underground hot water and stream from drilled wells, which drive steam turbines to generate electricity. The RE cycle is completed when used water is reverted back to ground to recharge the pools (Alrikabi 2014).

There are three types of geothermal power plants: flash steam, binary cycle and dry steam. In both flash steam and binary cycle plants, energy is drawn from pools of hot water while in dry steam plants, they draw energy from reservoirs of steam. In case of flash steam plants, temperatures of water are greater than 360 °F while in case of binary cycle, water temperature is lower around 225°–360 °F.

Geothermal energy can be used directly and indirectly. In direct use, applications require lower geothermal temperatures between 70°and 302 °F for power creation. Whereas in a direct-use system, a well is drilled into a geothermal pool, which makes available a sturdy hot water stream (Tromly 2001).The use of geothermal energy for power generation has been expanding consistently, and this fast advancement has been joined by its negative effects on the neighbourhood climate. These potential natural effects may have broad consequences for the local people's well-being, on environment ecology and economy, which may influence investments and social tolerability (Meller et al. 2018).

The quantitative evaluation of ecological dangers brought about by geothermal energy advancement can be performed through various models based on intensity and assessment data. It is encouraged to set up a geothermal information sharing stage, and multidisciplinary demonstration should be engaged later on (Chen et al. 2020).

7.2.5 Biomass

Biomass is a one of the most encouraging sources of natural energy as it is obtained from various sources such as agricultural waste, industrial waste, bark and wood (Lim et al. 2012). It consists of materials in solid or liquid form obtained from plants or animal waste, products and leftovers of forest and agricultural production. After partial processing of these materials, biomass can be in liquid state (bioethanol and biodiesel), solid state (pellets, briquettes) or gas (hydrogen, biogas and syngas) (Dodić et al. 2012). High energy efficiency can be obtained through liquid fuels processed through biomass (Suntana et al. 2009).

Biomass is naturally occurring carbon resource which is used for energy generation in burning process; therefore is used in large quantity as an alternative to fossil fuel (Kelly-Yong et al. 2007). Wide obtainability of biomass makes it a growing source of RE with high potential growth. In waste conversion plants, direct burning of biomass produces electricity (Li et al. 2017), whereas in boilers it generates heat at residential and industrial level (Perea-Moreno et al. 2018). It can significantly improve the environment and provide energetic and economical security. For energy requirements, solid biomass has important properties such as renewability, versatility, carbon neutrality, reactivity, inexpensive, availability, low ash and sulfur contents.

Various methods converting biomass into bioenergy are biological, thermochemical and chemical methods. Mainly five thermochemical ways are available for transformation of biomass into heat and combined heat power (CHP): combustion, gasification, pyrolysis, hydrothermal processing and hydrolysis. These technologies are primary transformation technologies, which transform solid biomass into fuels and heat. Heat produced during combustion is either used directly or transforms into electricity. For production of electricity, secondary transformation technologies are involved such as steam engines, stirling engines, turbines (steam, micro, gas) etc. (Malico et al. 2019).

7.2.6 Tidal Energy

Tides are long-period waves that arise through sea or oceans in response to inter-attractive gravitational attraction by the Sun and Moon. Tidal energy is produced by the relative movement of the celestial bodies – Earth, Moon and Sun, which interact through gravitational forces. The magnitude depends on changing positions of the Sun and Moon in relation to the Earth, Earth's rotation effects and the shape of the coastlines and sea floor. Potentiality of tidal energy production directly depends on the velocities of tidal current or the height of water level (Rashid et al. 2012; Vikas et al. 2016).

On the basis of occurrence of tides, there are three types of tides: diurnal (once per day), semidiurnal (twice daily) and mixed (Hagerman et al. 2006). Tidal energy is one of the forms of hydro energy which is artificially generated from tidal motions by converting energy produced from motion into electricity or other forms.

Tidal power is the vast pool of RE sources as our earth's surface is two-thirds covered by the ocean. In the coastal front, kinetic energy of water movement is changed into electrical power. It is a renewable, inexhaustible, energy source and a unique developing technology for research and development (Tousif and Taslim 2011). As the tidal currents or tides are both reliable and predictable, tidal power has an advantage over both solar and wind power systems. Tidal power generation can be precisely calculated in advance, letting for easy incorporation with prevailing electricity grids (Vikas et al. 2016).

7.3 Advantages and Disadvantages of Various Renewable Energy Resources

There are various advantages as well as disadvantages of above described RE resources which are summarized in Table 7.1.

Table 7.1 Advantages and disadvantages of various renewable energy resources.

Energy resource	Advantages	Disadvantages
Solar energy	Has the potential to supply energy infinitely. Causes neither air pollution nor water pollution.	Reliable only when sunlight is available. Sometimes not economical. Needs systems for storage as well as for backup.
Wind energy	Setup of wind farms is relatively economical. Causes no air or water pollution. Land around the wind farms may be used for other purposes.	Constant and significant amount of wind is must. Significant amount of land is required. Sometimes significant visual impact on landscapes.
Hydropower energy	Plentiful, clean and safe source of energy. Can be stored easily in reservoirs. Relatively inexpensive method of electricity production. Recreational benefits e.g. boating and fishing are also there.	Hydropower energy plants can be setup only near large water bodies. Sometime there are chances of flood in surrounding landscapes and communities. Major ecological impacts on local hydrology due to dams. May have a negative impact on the environment too. Suitable sites for dams are very limited.

(Continued)

Table 7.1 (*Continued*)

Energy resource	Advantages	Disadvantages
Geothermal energy	Unlimited energy supply. Causes neither air pollution nor water pollution	High start-up/development costs. Due to corrosion, maintenance costs are also there.
Biomass energy	Abundant source of energy. Waste products may be used to burn to get energy.	Causes air pollution. Not cost-effective.
Tidal/Marine Energy	Very good for an island country. Trap energy that would otherwise not be collected	Construction can be costly Have a negative impact on wildlife and the environment. Shipping infrastructure needs loads of open area to operate

7.4 Importance of Renewable Energy

Increased use of RE resources promises multiple advantages, ranging from economic growth and job creation to tackling the problem of climate change and ultimately in the sustenance of environment quality. RE production is essential to handle the problem of climate change and to reverse the biodiversity loss associated with it (Sonter et al. 2020; Spillias et al. 2020). Because of the exponentially increasing acknowledgement and concern for global warming as a looming threat to climate-dependent mankind, many geographies across the world have espoused RE goals to achieve the threshold percentages of total energy and thereby focus on climate change in a timely fashion (Spillias et al. 2020).

RE resources do not produce any pollutants and therefore do not deteriorate the quality of environment. As they are used as substitute of conventional fuels obtained from fossils and thus reduce the emissions of harmful gases coming out of fossil fuels (Bilgili et al. 2016; Vo and Le 2019). RE resources are never going to deplete unlike conventional fossil fuels (Tsoutsos et al. 2005; Akella et al. 2009) and, therefore, never become burden over the environment by increasing the harmful-free elements coming out from extraction and mining activities. Additionally energy production by RE sources does not cause thermal pollution, whereas it is always there in case of energy production through conventional energy sources (Akella et al. 2009).

Thus, RE sources are widely accepted as pristine sources of energy, and efficient utilization of these energy resources will result in negligible environmental impacts, produce near-zero secondary wastes and might be considered long-term sustainable on the basis of present and future economic as well as social requirements of the society. Renewable-energy-based technologies offer excellent window to mitigate the effects of greenhouse gas emissions and to lessen the global warming. These environment-friendly technologies need aggressive investments to ultimately achieve environmental sustainability (Nathaniel and Khan 2020). The infinite potential in renewable options will provide sustainable long-term solutions to persistent energy issues concerning the developing nations across the world. With limited harm to the environment than the conventional fossils along with abundant supply, the RE sources are hygienic in nature meeting the domestic energy demands with the potential to provide services to energy sector with near-zero emissions of harmful particles such as air pollutants and greenhouse gases, thanks to the zero-carbon footprint (García-Olivares et al. 2018).

A system developed around RE will make most crucial tasks possible to resolve e.g. (i) improvement in the reliability of energy supply; (ii) more efficient organic fuel economy; (iii) solving issues related to energy and water supply at local level; (iv) increasing opportunities for employment of the local area population and as a result, the standard of living; (v) assuring sustainable development of the remote areas with extreme climatic conditions such as desert and mountain; (vi) execution of the liabilities of the countries regarding fulfilment of the environmental protection-related international agreements (Zakhidov 2008). This will have a positive effect in reducing migration of people from rural areas to urban areas (Bergmann et al. 2008). As RE can be harvested in a decentralized manner, it gives excellent opportunity to fulfil the energy needs at rural and small scale that would be more reliable, cost-effective and sustainable for the environment (Reddy and Subramanian 1980; Ravindranath and Hall 1995). In the field of RE, majority of the investments are made on materials and human resources to construct and run the energy production units, whereas cost associated with energy import is very less.

7.5 Benefits of Renewable Energy Production to the Society

RE provides a range of benefits to society. These benefits may be environmental, economic, technological, political or social which are summarized in Figure 7.4. Decline in carbon dioxide (CO_2) emissions is, of course, one of the major benefits of RE. Additionally, various policies related to RE contribute towards achieving goals which include creation and development of local environmental and health infrastructure; facilitation of the access of energy,

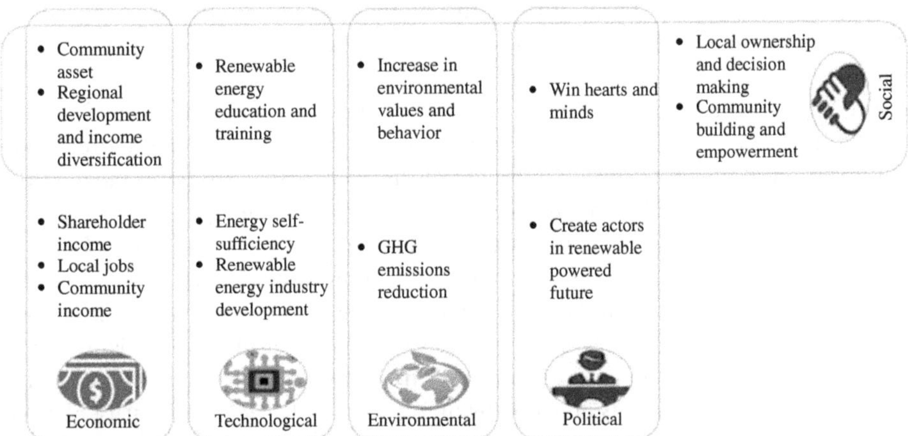

Figure 7.4 Benefits of renewable energy production at global level (Abbreviation GHG = Green House Gases).

especially for rural areas; achieving energy security targets by expanding the portfolio of energy resources and related technologies; and improving socio-economic standards by providing good opportunities for jobs (Angelou et al. 2013).

7.6 Renewable Energy and Sustainable Development Goals

RE plays a very significant role in sustainable development of both developed and developing countries (Güney 2019). Moreover, compared with non-renewable energy, renewable energy affects the sustainable development very substantially. Thus, RE must be included in policy and in order to reach the energy sustainability goals on a global scale, specifically in all emerging countries (Armin Razmjoo et al. 2020). Increased production of RE can contribute to achieve many Sustainable Development Goals (SDGs) adopted by UN General Assembly in September 2015. To achieve the SDGs, serious effort should be made to generate a high demand for RE for various reasons. First, and most evident of them all, is the availability of energy, which provides foundation for achieving a number of objectives of any nation, developing or developed. Energy is a must for financial growth and thus to eradicate the issue of poverty (Eggoh et al. 2011; Barnes et al. 2014). It has the capacity to make agro produce more productive, hence contributing to the goals of food security (Mushtaq et al. 2009). It reduces the dependency in fuel imports, hence, contributing towards nations becoming self-sustainable in terms of energy sources.

Second, currently widespread alternatives, such as traditional biomass and coal, cause side effects of local damage which is absent in the case of RE. This contributes towards ensuring healthy standards of human lives by removing the source of indoor air pollution (Torres-Duque et al. 2008), hence improving local air quality index from co-emitted air pollutants (West et al. 2013). This promotes the long-term sustainable and efficient use of surrounding and available ecosystems, in particular natural reserves such as forests, by replacing the need of firewood (Odihi 2003). Due to the fact that RE can be deployed in a decentralized way in rural and remote regions, making it economically competitive, it is more apt to replace the sources of indoor air pollution (Deichmann et al. 2011).

Third, and most imperative, reason is combating climate change which itself is one of the defined goals of SDGs. The RE generates hardly any CO_2. This has a direct positive impact on incrementally achieving long-term prospect of attaining most of the remaining SDGs. As per IPCC report, the damages of climate change are so varied and detrimental that any progress made unsustainably on the SDGs could be ephemeral (Barros and Field 2014). Damage of any scale to agriculture activities will be a direct threat, not just food security, but also a major negative impact to economic growth on a society that most often relies on agriculture produce. In addition, it is a proven hypothesis that any surge in food prices will lead to more poverty (Ivanic and Martin 2008). This counteracts effort on bringing down poverty and disparity. Various other impacts of climate change have been demonstrated to influence human health very adversely (Haines et al. 2009).

7.7 Limitations in Renewable Energy

In the past decade, the use of RE has increased considerably. Still there is a long way to go to achieve its full potential. In any country, one of the main reasons behind its slow development is the lack of public awareness. The other limitations or constraints in the field of RE are policies of national and international interest, availability of real estate for RE plant setups, financial and technical issues and political and social awareness. In comparison to non-renewable plants, the capital expenditure required to instal RE-related plants is very high (Sharma et al. 2012). The commercial bank is the only option for the financing of renewable energy projects (Rathore et al. 2019). In every RE resource, availability of land and its clearance are other big problems e.g. the wind plants are usually set up on farmland, the good habitat of birds and thus leads to their habitat destruction. Their setup on these sites leads to deforestation too (Leung and Yang 2012). In hydropower generation, areas of the enormous potential of water need to be evacuated for the construction of dams. This leads to the destruction of animal habitats of that area. Lack of technical knowledge and awareness among consumers is one of the major technical barriers. From the customer's perspective, for example, solar technologies are not trustworthy as they are available in plentiful amount only during sunny days. There is lack of awareness in the society of the fact that solar technology may be highly reliable when used in combination with hybrid photovoltaic systems. Research and development initiatives require high level of economic and logistical support from the government which is presently unavailable (Halkos and Gkampoura 2020). Insufficient number of training institutes, insufficient guidance and technical content for the engineers are some of other barriers in research and development of RE resources (Beck and Martinot 2004). The government, in addition to private energy agencies, must drive towards public awareness as well as regulatory policies related to it and should focus on the economic consumption of RE (Elavarasan et al. 2020).

7.8 Current Status and Future Perspectives

The transition from conventional sources to RE portfolio is slowly gaining pace worldwide, including India. The advanced technologies and forthcoming new innovations impart substantially in enhancing the efficacy of RE and its execution everywhere. RE sources can produce more than 3000 times the present energy demand at global level (Ellabban et al. 2014). In this world of power generation, the contribution of RE sources, including hydropower sources, has been increasing expeditiously since the last two decades. Energy production by solar and wind resources has gained impetus, favoured by forceful environmental strategies implemented by European Union, United States of America, Asia including India, China, Australia and Japan. In the recent past, wind and solar development costs have been dropped many folds, which now enables the developing countries to enhance their aptitude in RE sector (Ahmad and Zhang 2020). Currently one-third of the total energy mix is being produced by renewables in the European Union, one-sixth in the United States of America, India, Japan and one-fourth in China. The proportion of renewable resources in the global electricity supply has been increased to 27% in 2019, which is the maximum value recorded till now (IEA 2020). Hydropower showed 60% of the RE generation at global level in 2019. In 2019, renewable electricity generation increased by 6%, and 64% of this increase was contributed by wind and solar PV technologies. Hydropower, solar PV and wind power altogether contributed for more than 85% of renewables growth, accompanied by bioenergy (IEA 2020).

With only 7% increase annually over 2019–2030, the share of renewables must be expanded significantly to fulfil the Sustainable Development Scenario (SDS) share of approximately 50% of generation by year 2030. It is expected that the share of RE in total primary energy supply would increase from 14% in 2015 to 63% in 2050 (Gielen et al. 2019). The RE share in power sector would also rise from 25% in 2015 to 85% in 2050 (Gielen et al. 2019). Global renewable electricity generation is expected to increase 2.7 folds between 2010 and 2035 (Ellabban et al. 2014). Consumption of biofuels is expected to increase more than three times over the same time period (2010–2035). Almost all of the biofuels are being used in road transport; however, the usage of biofuels in aviation is expected to make an advancement towards 2035. The use of modern renewable resources in heat production is expected to be almost double over this time period. (Ellabban et al. 2014).

In India too, the infrastructure and inclination towards renewable energy are increasing day by day and encroaching rapidly into the energy mix. The target of India is to achieve 175 GW of total energy from RE sources by 2022, to fulfil the energy demand of 1.36 billion people (Elavarasan et al. 2020). India has set a new goal to produce 40% of its total installed electricity capacity using non-fossil fuels by 2030. Each state formulates policies and frameworks specific to their needs and targets. Karnataka, Gujarat and Tamil Nadu are the front runners in the race to adopt RE generation in the country. As of May 2019, 44% of total installed capacity of RE production of the nation gets fulfilled by these three states alone. Energy policies for each state are framed on the basis of its physical state and geological location for the development of suitable RE. These developments will lead India towards a sustainable green future.

All over the world, the COVID-19 outbreak has triggered many problematic issues in the RE sector too e.g. delayed supply chain, many challenges in tax stock markets and risk of not getting any advantage from government incentives as earlier expected in the year end (Eroğlu 2020). Therefore, countries need to reveal very significant incentives in the field of clean RE.

Under specific circumstances, it is likely that scenarios which forecast a significant rise in the volume of RE will materialize. However, this is possible only when there is extensive involvement of various government bodies in the promotion of RE technologies (Karatayev et al. 2019).

7.9 Conclusion

Due to the exhaustible resources diminishing fast, and different ecological issues brought about by the greenhouse gas emissions, traditional fossil fuel–based non-renewable resources are usually taken into account as unsustainable from long-term perspective. Energy resources such as solar, hydropower, wind etc. are creative and rich alternatives for power generation and their immense capacity for power generation in principle can fulfil the world's energy demand very easily. These alternative clean resources can impart positive impact considerably in the country's total energy mix bringing down overall environmental, economical and societal costs. Thus, nations must switch to the utilization of renewable energy sources as far as they are able to do. The demand, production and consumption of energy generated from pristine renewable resources should be one of the principal objectives to meet energy demand in a secure, economic and sustainable-in-long-term manner and achieve stable financial growth, in addition to maintaining the environment degradation at a very low level, ultimately leading towards a sustainable future.

References

Abolhosseini, S., Heshmati, A. and Altmann, J., 2014. A Review of Renewable Energy Supply and Energy Efficiency Technologies. IZA-Institute of Labor Economics, Germany

Ahmad, T. and Zhang, D., 2020. A critical review of comparative global historical energy consumption and future demand: the story told so far. *Energy Reports*, 6, pp.1973–1991.

Akella, A.K., Saini, R.P. and Sharma, M.P., 2009. Social, economical and environmental impacts of renewable energy systems. *Renewable Energy*, 34(2), pp.390–396.

Alharbi, F.H. and Kais, S., 2015. Theoretical limits of photovoltaics efficiency and possible improvements by intuitive approaches learned from photosynthesis and quantum coherence. *Renewable and Sustainable Energy Reviews*, 43, pp.1073–1089.

Alola, A.A., Bekun, F.V. and Sarkodie, S.A., 2019. Dynamic impact of trade policy, economic growth, fertility rate, renewable and non-renewable energy consumption on ecological footprint in Europe. *Science of the Total Environment*, 685, pp.702–709.

Alrikabi, N.K.M.A., 2014. Renewable energy types. *Journal of Clean Energy Technologies*, 2(1), pp.61–64.

Angelou, N., Elizondo Azuela, G., Banerjee, S.G., et al. (2013). *Global tracking framework* (No. 77889, pp. 1–289). The World Bank.

Armin Razmjoo, A., Sumper, A. and Davarpanah, A., 2020. Energy sustainability analysis based on SDGs for developing countries. *Energy Sources, Part A: Recovery, Utilization, and Environmental Effects*, 42(9), pp.1041–1056.

Ashokverghese, N. (1998). Renewable energy resources. *Proceedings of International conference on Alternate Energy resources, Asian Institute of Technology, Thailand* (pp. 403–410).

Asumadu-Sarkodie, S. and Owusu, P.A., 2016. The potential and economic viability of wind farms in Ghana. *Energy Sources, Part A: Recovery, Utilization, and Environmental Effects*, 38(5), pp.695–701.

Aubrun, S., Loyer, S., Hancock, P.E. and Hayden, P., 2013. Wind turbine wake properties: comparison between a non-rotating simplified wind turbine model and a rotating model. *Journal of Wind Engineering and Industrial Aerodynamics*, 120, pp.1–8.

Balkhair, K.S. and Rahman, K.U., 2017. Sustainable and economical small-scale and low-head hydropower generation: a promising alternative potential solution for energy generation at local and regional scale. *Applied Energy*, 188, pp.378–391.

Barnes, D.F., Samad, H. and Banerjee, S.G., 2014. The development impact of energy access. *Energy poverty: Global challenges and local solutions*, 54.

Barros, V.R. and Field, C.B. (2014). Climate change 2014: impacts, adaptation, and vulnerability. Part B: regional aspects.

Beck, F. and Martinot, E.E. (2004). Renewable energy policies and barriers, forthcoming in Encyclopedia of Energy, Cutler J. Cleveland Elsevier Inc, pp 365–338

Behrouzi, F., Nakisa, M., Maimun, A. and Ahmed, Y.M., 2016. Global renewable energy and its potential in Malaysia: a review of hydrokinetic turbine technology. *Renewable and Sustainable Energy Reviews*, 62, pp.1270–1281.

Bergmann, A., Colombo, S. and Hanley, N., 2008. Rural versus urban preferences for renewable energy developments. *Ecological Economics*, 65(3), pp.616–625.

Besarati, S.M., Padilla, R.V., Goswami, D.Y. and Stefanakos, E., 2013. The potential of harnessing solar radiation in Iran: generating solar maps and viability study of PV power plants. *Renewable Energy*, 53, pp.193–199.

Bildirici, M.E. and Gökmenoğlu, S.M., 2017. Environmental pollution, hydropower energy consumption and economic growth: evidence from G7 countries. *Renewable and Sustainable Energy Reviews*, 75, pp.68–85.

Bilgili, F., Koçak, E. and Bulut, Ü., 2016. The dynamic impact of renewable energy consumption on CO2 emissions: a revisited environmental Kuznets curve approach. *Renewable and Sustainable Energy Reviews*, 54, pp.838–845.

Chala, G.T., Ma'Arof, M.I.N. and Sharma, R., 2019. Trends in an increased dependence towards hydropower energy utilization – a short review. *Cogent Engineering*, 6(1), p.1631541.

Chen, S., Zhang, Q., Andrews-Speed, P. and Mclellan, B., 2020. Quantitative assessment of the environmental risks of geothermal energy: a review. *Journal of Environmental Management*, 276, p.111287.

Deichmann, U., Meisner, C., Murray, S. and Wheeler, D., 2011. The economics of renewable energy expansion in rural sub-Saharan Africa. *Energy Policy*, 39(1), pp.215–227.

Dodić, S.N., Vasiljević, T.Z., Marić, R.M., Kosanović, A.J.R., Dodić, J.M. and Popov, S.D., 2012. Possibilities of application of waste wood biomass as an energy source in Vojvodina. *Renewable and Sustainable Energy Reviews*, 16(5), pp.2355–2360.

Dong, K., Dong, X. and Jiang, Q., 2020. How renewable energy consumption lower global CO2 emissions? Evidence from countries with different income levels. *The World Economy*, 43(6), pp.1665–1698.

Edenhofer, O., Pichs-Madruga, R., Sokona, Y., Seyboth, K., Kadner, S., Zwickel, T., Eickemeier, P., Hansen, G., Schlömer, S., von Stechow, C. and Matschoss, P. eds., 2011. Renewable Energy Sources and Climate Change Mitigation: Special Report of the Intergovernmental Panel on Climate Change. Cambridge University Press.

Eggoh, J.C., Bangaké, C. and Rault, C., 2011. Energy consumption and economic growth revisited in African countries. *Energy Policy*, 39(11), pp.7408–7421.

Elavarasan, R.M., Shafiullah, G., Padmanaban, S., Kumar, N.M., Annam, A., Vetrichelvan, A.M., Mihet-Popa, L. and Holm-Nielsen, J.B., 2020. A comprehensive review on renewable energy development, challenges, and policies of leading Indian states with an international perspective. *IEEE Access*, 8, pp.74432–74457.

Ellabban, O., Abu-Rub, H. and Blaabjerg, F., 2014. Renewable energy resources: current status, future prospects and their enabling technology. *Renewable and Sustainable Energy Reviews*, 39, pp.748–764.

Eroğlu, H., 2020. Effects of Covid-19 outbreak on environment and renewable energy sector. *Environment, Development and Sustainability*, pp.1–9.

García-Olivares, A., Solé, J. and Osychenko, O., 2018. Transportation in a 100% renewable energy system. *Energy Conversion and Management*, 158, pp.266–285.

Gielen, D., Boshell, F., Saygin, D., Bazilian, M.D., Wagner, N. and Gorini, R., 2019. The role of renewable energy in the global energy transformation. *Energy Strategy Reviews*, 24, pp.38–50

Güney, T., 2019. Renewable energy, non-renewable energy and sustainable development. *International Journal of Sustainable Development and World Ecology*, 26(5), pp.389–397.

Hagerman, G., Polagye, B., Bedard, R. et al. (2006). Methodology for estimating tidal current energy resources and power production by tidal in-stream energy conversion (TISEC) devices. EPRI North American tidal in stream power feasibility demonstration project, 1.

Haidar, A.M., Senan, M.F., Noman, A. and Radman, T., 2012. Utilization of pico hydro generation in domestic and commercial loads. *Renewable and Sustainable Energy Reviews*, 16(1), pp.518–524.

Haines, A., McMichael, A.J., Smith, K.R., Roberts, I., Woodcock, J., Markandya, A., Armstrong, B.G., Campbell-Lendrum, D., Dangour, A.D., Davies, M. and Bruce, N., 2009. Public health benefits of strategies to reduce greenhouse-gas emissions: overview and implications for policy makers. *The Lancet*, 374(9707), pp.2104–2114.

Halkos, G.E. and Gkampoura, E.C., 2020. Reviewing usage, potentials, and limitations of renewable energy sources. *Energies*, 13(11), p.2906.

Herrando, M. and Markides, C.N., 2016. Hybrid PV and solar-thermal systems for domestic heat and power provision in the UK: techno-economic considerations. *Applied Energy*, 161, pp.512–532.

IEA. (2020), Renewable Power, IEA, Paris https://www.iea.org/reports/renewable-power

Infield, D. and Freris, L., 2020. *Renewable Energy in Power Systems*. John Wiley & Sons.

IRENA. (2017), Renewable Energy Prospects for India, a Working Paper Based on REmap. The International Renewable Energy Agency (IRENA), Abu Dhabi. http://www.irena.org/remap

Ivanic, M. and Martin, W., 2008. Implications of Higher Global Food Prices for Poverty in Low-Income Countries. The World Bank.

Jamshidi, M. and Askarzadeh, A., 2019. Techno-economic analysis and size optimization of an off-grid hybrid photovoltaic, fuel cell and diesel generator system. *Sustainable Cities and Society*, 44, pp.310–320.

Kabir, E., Kumar, P., Kumar, S., Adelodun, A.A. and Kim, K.H., 2018. Solar energy: potential and future prospects. *Renewable and Sustainable Energy Reviews*, 82, pp.894–900.

Kabo-bah, A.T., Diji, C.J. and Yeboah, K.A., 2018. Hydropower generation in West Africa—the working solution manual. In Sustainable Hydropower in West Africa (pp. 197–207). Academic Press.

Karatayev, M., Movkebayeva, G. and Bimagambetova, Z., 2019. Increasing utilisation of renewable energy sources: comparative analysis of scenarios until 2050. In Energy Security (pp. 37–68). Palgrave Macmillan, Cham.

Kelly-Yong, T.L., Lee, K.T., Mohamed, A.R. and Bhatia, S., 2007. Potential of hydrogen from oil palm biomass as a source of renewable energy worldwide. *Energy Policy*, 35(11), pp.5692–5701.

Kumar, A., Kumar, K., Kaushik, N., Sharma, S. and Mishra, S., 2010. Renewable energy in India: current status and future potentials. *Renewable and Sustainable Energy Reviews*, 14(8), pp.2434–2442.

Kumar, A., Schei, T., Ahenkorah, A. et al. (2011). Hydropower. *IPCC special report on renewable energy sources and climate change mitigation.*

Kumar, L., Hasanuzzaman, M. and Rahim, N.A., 2019. Global advancement of solar thermal energy technologies for industrial process heat and its future prospects: a review. *Energy Conversion and Management*, 195, pp.885–908.

Leung, D.Y. and Yang, Y., 2012. Wind energy development and its environmental impact: a review. *Renewable and Sustainable Energy Reviews*, 16(1), pp.1031–1039.

Li, Y., Rezgui, Y. and Zhu, H., 2017. District heating and cooling optimization and enhancement–towards integration of renewables, storage and smart grid. *Renewable and Sustainable Energy Reviews*, 72, pp.281–294.

Lim, J.S., Manan, Z.A., Alwi, S.R.W. and Hashim, H., 2012. A review on utilisation of biomass from rice industry as a source of renewable energy. *Renewable and Sustainable Energy Reviews*, 16(5), pp.3084–3094.

Malico, I., Pereira, R.N., Gonçalves, A.C. and Sousa, A.M., 2019. Current status and future perspectives for energy production from solid biomass in the European industry. *Renewable and Sustainable Energy Reviews*, 112, pp.960–977.

Manzano-Agugliaro, F., Taher, M., Zapata-Sierra, A., Juaidi, A. and Montoya, F.G., 2017. An overview of research and energy evolution for small hydropower in Europe. *Renewable and Sustainable Energy Reviews*, 75, pp.476–489.

Meller, C., Schill, E., Bremer, J., Kolditz, O., Bleicher, A., Benighaus, C., Chavot, P., Gross, M., Pellizzone, A., Renn, O. and Schilling, F., 2018. Acceptability of geothermal installations: a geoethical concept for GeoLaB. *Geothermics*, 73, pp.133–145.

MNRE Annual Report. (2019–2020). Ministry of New and Renewable Energy, Govt. Of India.

Mushtaq, S., Maraseni, T.N., Maroulis, J. and Hafeez, M., 2009. Energy and water tradeoffs in enhancing food security: a selective international assessment. *Energy Policy*, 37(9), pp.3635–3644.

Nathaniel, S. and Khan, S.A.R., 2020. The nexus between urbanization, renewable energy, trade, and ecological footprint in ASEAN countries. *Journal of Cleaner Production*, 272, p.122709.

Nazir, M.S., Ali, N., Bilal, M. and Iqbal, H.M., 2020. Potential environmental impacts of wind energy development: a global perspective. *Current Opinion in Environmental Science & Health*, 13, pp.85–90.

Nix, R.G. (1995). *Wind energy as a significant source of electricity* (No. NREL/TP-441-8162; CONF-951173-1). National Renewable Energy Lab., Golden, CO (United States).

Odihi, J., 2003. Deforestation in afforestation priority zone in Sudano-Sahelian Nigeria. *Applied Geography*, 23(4), pp.227–259.

Perea-Moreno, M.A., Manzano-Agugliaro, F. and Perea-Moreno, A.J., 2018. Sustainable energy based on sunflower seed husk boiler for residential buildings. *Sustainability*, 10(10), p.3407.

Rashid, F.L., Mohamad, A.H. and Hashim, A., 2012. Power production using tidal energy. *British Journal of Science*, 3(2).

Rathore, P.K.S., Chauhan, D.S. and Singh, R.P., 2019. Decentralized solar rooftop photovoltaic in India: on the path of sustainable energy security. *Renewable Energy*, 131, pp.297–307.

Ravindranath, N.H. and Hall, D.O., 1995. Biomass, Energy and Environment: A Developing Country Perspective from India. Oxford University Press.

Reddy AKN, Subramanian DK. The design of rural energy centers. *Indian Academy of Science, Bangalore* 1980:109–30.

Seddegh, S., Wang, X., Henderson, A.D. and Xing, Z., 2015. Solar domestic hot water systems using latent heat energy storage medium: a review. *Renewable and Sustainable Energy Reviews*, 49, pp.517–533.

Sharma, A., Srivastava, J., Kar, S.K. and Kumar, A., 2012. Wind energy status in India: a short review. *Renewable and Sustainable Energy Reviews*, 16(2), pp.1157–1164.

Sonter, L.J., Dade, M.C., Watson, J.E. and Valenta, R.K., 2020. Renewable energy production will exacerbate mining threats to biodiversity. *Nature Communications*, 11(1), pp.1–6.

Spillias, S., Kareiva, P., Ruckelshaus, M. and McDonald-Madden, E., 2020. Renewable energy targets may undermine their sustainability. *Nature Climate Change*, pp.1–3.

Sun, D. and Wang, L., 2016. Research on heat transfer performance of passive solar collector-storage wall system with phase change materials. *Energy and Buildings*, 119, pp.183–188.

Suntana, A.S., Vogt, K.A., Turnblom, E.C. and Upadhye, R., 2009. Bio-methanol potential in Indonesia: forest biomass as a source of bio-energy that reduces carbon emissions. *Applied Energy*, 86, pp.S215–S221.

Tong, W., 2010. Fundamentals of wind energy. Wind Power Generation and Wind Turbine Design, Wei Tong, pp.1–44. WIT Press, Southampton, Boston

Torres-Duque, C., Maldonado, D., Pérez-Padilla, R., Ezzati, M. and Viegi, G., 2008. Biomass fuels and respiratory diseases: a review of the evidence. *Proceedings of the American Thoracic Society*, 5(5), pp.577–590.

Tousif, S.M.R. and Taslim, S.M.B., 2011. Tidal power: an effective method of generating power. *International Journal of Scientific and Engineering Research*, 2(5), pp.1–5.

Tromly, K. (2001). *Renewable energy: An overview* (No. DOE/GO-102001-1102; NREL/BR-810-27955). National Renewable Energy Lab., Golden, CO (US).

Tsoutsos, T., Frantzeskaki, N. and Gekas, V., 2005. Environmental impacts from the solar energy technologies. *Energy Policy*, 33(3), pp.289–296.

Vikas, M., Rao, S., and Seelam, J.K. (2016). Tidal energy: a review. *Proceedings of International Conference on Hydraulics, Water Resources and Coastal Engineering (Hydro2016).*

Vo, D.H. and Le, Q.T.T., 2019. CO2 emissions, energy consumption, and economic growth: new evidence in the ASEAN countries. *Journal of Risk and Financial Management*, 12(3), p.145.

West, J.J., Smith, S.J., Silva, R.A., Naik, V., Zhang, Y., Adelman, Z., Fry, M.M., Anenberg, S., Horowitz, L.W. and Lamarque, J.F., 2013. Co-benefits of mitigating global greenhouse gas emissions for future air quality and human health. *Nature Climate Change*, 3(10), pp.885–889.

Yaakob, O.B., Ahmed, Y.M., Elbatran, A.H. and Shabara, H.M., 2014. A review on micro hydro gravitational vortex power and turbine systems. *Jurnal Teknologi*, 69(7).

Yah, N.F., Oumer, A.N. and Idris, M.S., 2017. Small scale hydro-power as a source of renewable energy in Malaysia: a review. *Renewable and Sustainable Energy Reviews*, 72, pp.228–239.

Zakhidov, R.A., 2008. Central Asian countries energy system and role of renewable energy sources. *Applied Solar Energy*, 44(3), pp.218–223.

Zerta, M., Schmidt, P.R., Stiller, C. and Landinger, H., 2008. Alternative world energy outlook (AWEO) and the role of hydrogen in a changing energy landscape. *International Journal of Hydrogen Energy*, 33(12), pp.3021–3025.

Zhang, X., Estoque, R.C. and Murayama, Y., 2017. An urban heat island study in Nanchang City, China based on land surface temperature and social-ecological variables. *Sustainable Cities and Society*, 32, pp.557–568.

8

Clean Energy Sources for a Better and Sustainable Environment of Future Generations

Aparna Nautiyal[1,2] and Ayyagari Ramlal[3]

[1] *Department of Botany, Deshbandhu College, University of Delhi, New Delhi, India*
[2] *i-4 Centre, Deshbandhu College, University of Delhi, New Delhi, India*
[3] *Division of Genetics, ICAR – Indian Agricultural Research Institute (IARI), Pusa, New Delhi, India*

8.1 Introduction

Energy as one of the primary needs for the sustenance of life is a prerequisite which drives all other functioning and processes within the biosphere (Østergaard et al. 2020). Energy is an essential and vital component to meet the basic requirements of humans in the fields such as agriculture, healthcare, medicine and technology development etc. (Halkos and Gkampoura 2020). The energy demand is increasing day by day at an alarming rate due to the increase in the population and growing demand in various sectors. As a major impact, the degradation of the environment is very much evident nowadays. Micro and macro hazardous air pollutants have become recently harmful and toxic to the environment. The rate of tremendous increase in the pollutants such as NO_x, SO_2, CO and particulates has caused much trouble and raised the utmost concern for conservation of the environment. The various kinds of environmental pollution include water pollution, maritime pollution, land use and its impact, radiation and radioactivity and its disposal, solid waste disposal, pollutants and others (Dincer 2000). Such problems of environment can easily be minimized with the usage and application of renewable energy resources which do not cause air pollution or greenhouse gas emissions (Demirbas 2005). The daily basic needs of human beings such as health, cooking, the comfort of space, lighting, communication and mobility are some of the areas in which most of the energy is consumed (Owusu and Asumadu-Sarkodie 2016).

According to one of the reports of the US Energy Information Administration (EIA), the energy demand will increase by 50% by 2050 (Halkos and Gkampoura 2020). Energy sources such as geothermal, solar, wind, ocean, hydropower and biomass are the primary and important sources of renewable origin (Owusu and Asumadu-Sarkodie 2016). Around 1750, the first commercial coal mining that started in Richmond, Virginia, made coal as the most popular and preferred fueling agent for steam engines owing to its more energy carrying capacity than the biomass-based fuels (Owusu and Asumadu-Sarkodie 2016). During

Energy: Crises, Challenges and Solutions, First Edition. Edited by Pardeep Singh, Suruchi Singh, Gaurav Kumar, and Pooja Baweja.

that time, coal was much economical and cleaner fuel but soon with the increased energy demand and overgrowing population, the fossil-fuel-based production of energy ruled and resulted in an increase in various environmental-related issues by generating huge amounts of carbon dioxide (CO_2). Renewable resources are those resources that are non-exhaustible and replenish of their own (naturally) without being exhausted from the earth (Owusu and Asumadu-Sarkodie 2016). On the contrary, non-renewable resources are those which are exhaustible and do not replenish easily like coal and oil (Bharucha 2005). There is a serious need for the significant development which promises the use of energy resources in an efficient way to meet human demands while preserving the environment for the present as well as for the future generations on a sustainable basis. According to Tester (2005), sustainable development means 'a dynamic harmony between the equitable availability of energy-intensive goods and services to all people and preservative of the earth for future generations' (Tester et al. 2012; Owusu and Asumadu-Sarkodie 2016). An index called as the 'Human Development Index' is used as a tool to ensure the aim of sustainable development and health status of an ecosystem that includes increased longevity, income sources and knowledge (Bharucha 2005). Renewable resources being the inexhaustible, non-polluting, domestic in origin are also referred to as alternative energy resources (Panwar et al. 2011). There are two kinds of clean technologies, one mainly the usage of alternative energy sources such as hydro, solar etc. and the other one is how to extract and increase the efficiency of the energy derived using virtual power plants and smart meters etc. (Abolhosseini et al. 2014). There is an unequal distribution of natural resources among the developed and developing countries as the consumption of resources by the developed countries is 50 times more than the usage by others (per capita). Apart from this, the developed countries produce 75% more greenhouse gases (GHGs) and industrial wastes than others (Bharucha 2005).

This chapter gives a brief outline of some of the important and conventional available renewable resources followed by the modern available resources along with their advantages and disadvantages. It also focuses on the prospects of a few alternative sources of energy being used as a substitute for the already available resources.

8.2 Conventional Sources of Energy

The energy resources have been generally classified into three categories i.e. renewable resources, fossil fuels and nuclear resources. Renewable resources such as wind, solar, geothermal are used to meet the needs of the growing human population (Panwar et al. 2011). As per the data studied so far, fossil fuel consumption by the developed countries was found much more than the developing countries. For instance, amongst all countries, the United States consumes most of the energy produced globally. Some resources such as water and biomass (living organisms) are considered as renewable resources; however, they are available for a certain time period and need to be energized (Bharucha 2005).

These resources that are generally used for domestic purposes cause very less environmental pollution and do not emit GHGs (Panwar et al. 2011). Such resources can regenerate, unlike other non-renewable resources which do not have the capacity to multiplicate or regeneration (Bhoi et al. 2020). According to the Paris Climate Conference (COP21) held in December 2015, the earth's temperature has increased after the nineteenth century by about 0.85 °C (Khan 2018). Natural gas, underground gas reserves, lakes and run-off water hydropower plants are all included in the category of non-renewable resources, as once they are used up, cannot be reused again. To overcome this, newer ways of producing energy should be explored which at the same time can resolve the prevailing problem of depleting conventional sources such as coal, natural gas and oil (Bhoi et al. 2020). Emitted carbon from these sources reacts with the oxygen forming carbon monoxide and carbon dioxide, thereby causing both environmental and health problems (Bhoi et al. 2020). Some of the most important sources from which renewable energy is produced include solar, biomass, wind, geothermal and water (Panwar et al. 2011). The various types of renewable energy resources are shown above (Figure 8.1).

8.2.1 Hydro Energy

Amongst all the renewable sources of energy, the energy produced by water is considered to be best as most (~90%) is converted into electricity (from water to wire) with 97% efficiency electricity (Bharucha 2005). About 16.3% of electricity is generated throughout the world using hydropower (Xu et al. 2015). Many countries such as Canada, China, Brazil and the United States have utilized hydropower generation widely owing to their largest hydropower production capacity. China is the richest producer of electricity using water

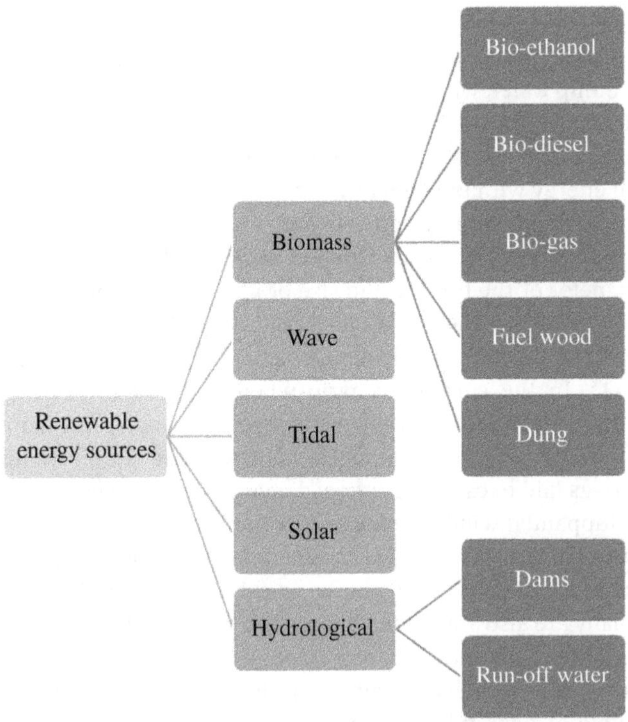

Figure 8.1 Diagrammatic representation showing different types of renewable energy sources.

and can potentially generate 694 GW (Giga Watt) (Mohtasham 2015). The largest hydropower plant in the world is Itaipu Dam. It was constructed on the river Parana and was developed by Paraguay and Brazil jointly, whereas the largest hydropower plant in India is Koyna hydroelectric plant on the river Koyna, Maharashtra, India. Itaipu produced 340 TWh (Tera Watt hour), 2994 of electricity in 1950 and 2005 respectively (Abolhosseini et al. 2014). As of 2009, the dam produces 700 MW (Mega Watt) of electric energy from 20 units of it (TIHDP, Brazil). The Koyna power plant was started in 1954 and generates 1960 MW of energy. Hydropower has played a major role in combating Europe's climate change and increased energy demand. As one of the futuristic energy systems, Europe is focusing on the pumped storage of hydropower electricity, thus providing an additional energy supply. More such European projects are being planned to harness the increased penetration of various renewable resources over the continent (Lehner et al. 2005). Hydropower has been proved to be a major key role player in controlling the climate change owing to its low-carbon footprint by generating low-carbon electricity. Analyzing the overall impact of greenhouse gas emissions, it has been discovered that hydropower plants generate minimum carbon footprints (Kaunda et al. 2012). Energy researchers are further planning to evolve recent and innovative methods to maximize the production of hydroelectric plants without much exploitation to the environment, thereby attaining sustainable goals (Ardizzon et al. 2014).

8.2.2 Wind Energy

Wind energy is being used for many years, as seen at around 3000 years ago in Egypt, when windmills were used to pump water, whereas, in China, the farmers used the wind wheels to irrigate their rice fields using a vertical axis (Leung and Yang 2012). The electricity is produced using wind as a prime source wherein the kinetic energy is converted into electricity via mechanical energy with the help of wind turbines and other systems (Bharucha 2005). The turbines can be seen at both offshore and onshore depending upon their location on land and water respectively (Bhoi et al. 2020). The energy production depends upon the diameter of the turbine and size of its blades, depending on this, the capacity ranges from several kilowatts (kW) to Megawatts (MW). The generation of energy is dependent on two factors, namely aerodynamic lift and aerodynamic drag (Chakraborty et al. 2011). In the current scenario where most of the energy demands are being fulfilled by fossil fuels, renewable energy sources have provided a path where the energy requirements can be reduced (TEDA). The Tamil Nadu Development Agency (TEDA) was laid to cater the energy demands by the Government of Tamil Nadu in 1985. The Muppandal windfarm located in Tamil Nadu is the largest wind power plant in India which generates 1500 MW. Gansu wind farm located in China is the largest electricity producer using wind energy in the world. Other countries such as the United States and Germany are also among the top producers of wind-based electricity of 47 and 29 GW after China (Abolhosseini et al. 2014). Many companies and several researchers are working onto establish accurate mechanism and way to efficiently use the wind as a source of energy by manipulating turbine length, size and motor capacity (Alabdali et al. 2020).

8.2.3 Geothermal Energy

Geothermal energy is extracted and harnessed to produce energy from the interior of the earth, and this reservoir of energy is due to the friction produced from the collision of continental plates and the radioactive decay of the elements (Herzog et al. 2001; Bharucha 2005; Halkos and Gkampoura 2020). The energy that is produced by the dust and gas remained inside for over four billion years and the process of radioactivity due to rocks occurring inside the earth's crust (Abolhosseini et al. 2014). Some of the various techniques used in the extraction process include steam and water along with direct heating, heating pumps and enhanced geothermal system (EGS) which allows the transfer of the geo-energy (Bhaucha 2005; Halkos and Gkampoura 2020). It has been estimated that around 1% of geothermal heat present from about 10 km beneath is equal to 500 times of all the energy obtained from gas and oil resources (Herzog et al. 2001). The geothermal energy gradient in France is about 4 °C/100 m ranging from 2 °C/100 m to 10 °C/100 m in Pyrenees mountains to Alsace region respectively while due to volcanic eruptions in Iceland, the energy reaches up to 30 °C/100 m (Abolhosseini et al. 2014). There is a gradual increase in the temperature from the surface to crust of the earth from 17 to 30 °C while the mantle contains the viscous and molten rocks with temperatures ranging from 650 to 1250 °C. The primary forms of geothermal energy include (i) magma, (ii) hot dry rocks, (iii) hydrothermal and (iv) geo-pressurized brines. For many years, people have benefited from hot springs and steam vents, using it for varied purposes such as bathing, cooking and heating. With the advancement of technology, the ability to locate and drill into hydrothermal reservoirs, directing the steam through the pipes or hot water to the surface, and use the heat directly in space heating, industrial and aquaculture processes can be used for the conversion of heat into electricity (Herzog et al. 2001). Geysers, located in California, United States, is the largest producer of geothermal energy of the world while the Puga geothermal power plant in Jammu and Kashmir is the largest in India. Many power stations, across the world, that are constructed to harness the geothermal energy which can overcome the problems of growing environmental issues and energy crisis include Argentina (Copahue), Australia (Birdsville and Cooper Basin), Austria (Bad Blumau) and many more where innovative ways are being developed to extract energy (Huttrer 2020).

8.2.4 Solar Energy

The sun's heat energy is utilized to produce electricity (Bharucha 2005). According to the World Energy Council (2013), 'the total energy from solar radiation falling on the earth was more than 7500 times the world's total annual primary energy consumption of 450 EJ' (Owusu and Asumadu-Sarkodie 2016). It is the most abundant form of energy, and only 0.04% of it is used by humans for various purposes with the use of photovoltaic panels. As the setup cost related to the photovoltaic (PV) panels is very high, therefore nowadays, organic photovoltaic materials are being used to enhance energy productivity (Mohtasham 2015). The function of this PV is to convert the energy obtained from the sun into electricity (direct current, DC). The PV module is of size from 60 to 170 W, and the individual units are arranged in series while the PV panel is the collection of more than one PV module. The solar energy can be used in developing technology such as solar heating, solar thermal energy, artificial photosynthesis and solar architecture (Shaikh et al. 2017). Like wind energy, solar energy is also dependent on season, the fluctuations in the weather directly affect the functioning. More innovative technological advancements are being done to improve the efficiency to capture and convert solar energy. Some of the limiting factors include unskilled workers, improper infrastructure and high processing costs while some of these were overcome by the application of PV cells (Durganjali et al. 2020).

8.2.5 Ocean Energy

The stored energy in the ocean is used to harness energy. The classes include tidal energy, tidal current, ocean current, ocean thermal energy, salinity gradient and ocean wave. Most common ocean energy that is used is obtained in the form of wave energy wherein a large wave produced from the ocean is used to generate electricity (Bharucha 2005). Waves are created when wind passes over the surface of the ocean. When the speed of the wind is faster, then wind remains for a longer time which is directly proportional to the wind distance, the height of the wave, and thus more energy will be produced (Bhoi et al. 2020).

8.3 Environmental Impacts of Renewable Resources

Certain management tools are being now applied to attain sustainable goals for environmental assessment. EIA (Environmental Impact Assessment) is one such project-oriented planned tool for linking environment with the development making optimum use of natural resources on sustainable basis. It helps in maintaining a transparency for all the current and proposed developmental projects/activities taking into account the environmental impacts of the project on long-term basis.

Environmental impacts of hydropower include the deteriorating quality of water, which in turn affects the nature of nearby water. Another problem associated is the loss of ecosystem and biodiversity. The composition of the water entering and leaving a reservoir is also different which also disturbs the nature, composition and affects the habitat of various aquatic animals, and it also causes problems in their reproduction, growth and sexual maturity processes. Hence, this acts as a barrier in the migration and disruption to the aquatic animals. Another reason for failure includes the emissions of GHGs which occur in three different phases of hydropower plant, namely at the construction, operation and maintenance and the decommissioning of the plant (Mussa et al. 2018).

The biggest advantage of using wind-based energy is that they neither pollute the environment by releasing or emitting GHGs into the surroundings nor create any kind of radioactive waste. The major limiting factor associated with wind energy is the noise they create due to the operation of onshore wind turbines. Mechanical and aerodynamical noises are created by wind turbines. Apart from this, they also create problems because of their large bodies and size to the people living there and nearby. A study conducted by Pedersen (2011) on the relationship of the well-being of people living around the wind turbines showed that there was a strong correlation between them as it caused headache and stress to the people exposed to the sound. Similarly, sleep disturbances and hearing loss were developed by the people due to the exposure to low-frequency aerodynamic noise created by the wind turbines, which was described by Punch et al. (2010) (Punch et al. 2010; Pedersen 2011; Leung and Yang 2012). On the other hand, there were diverse responses for its visual impact due to the contrasting views on the appearance of the wind turbines. Wind turbines also create problems for the birds and animals living there, as sometimes they get killed by the movement of the turbines (Leung and Yang 2012).

Geothermal energy is a stable source of energy which can be used to produce electricity continuously throughout a year. However, there are several problems which are associated with geothermal energy production. To produce energy, the land has to be drilled for varied purposes such as for laying roads, pipelines, waste dumps, digging rigs and disposal pits etc. but at the same time, they also leave an impact on the environment. Some of the major issues include land disturbance, increase in the noise levels, loss and disturbance to vegetation, wildlife, ecosystem, health issues to the workers, water and air pollution. The negative impacts of geothermal energy include the generation of loads of solid wastes such as cuttings, cement residues and other leftovers after drilling, chemical sediments and other mechanical and machine-related wastes (Bošnjaković et al. 2019).

8.4 Mitigation Strategies and Sustainable Development of Renewable Resources

There are many effective methods and strategies that are being followed for encouraging the use of the renewable resources on a larger scale. Some of the mitigation ways include the complete removal of dams from the places having more disturbances and losses within/to the environment. Secondly, one can also regulate the flow of water to control (minimizing etc.) the impacts of it. Thirdly, sediments can be separated from the flowing water using various kinds of traps and sedimentation methods and finally relocating is the best possible way for maintaining and caring the biodiversity (Mussa et al. 2018). For a better future and growing environmental concerns, sustainable developments play an important role in building a secure and healthier society. Supply of energy should be sustainable, and at the same time, it should be available at an affordable price. The relevance and importance of renewable energy consumption in today's context include that they do not cause any kind of severe harm to the environment, cannot be depleted (inexhaustible resources) and flexible (can be used anywhere as per the requirement). The deciding factors which govern or help in planning for a successful developed sustainable society are public awareness, information, education and training on environment, adoption of innovative energy strategies, promotion of the use of renewable resources, financing and monitoring the system (Dincer 2000).

8.5 Biomass and Microorganisms-Derived Energy

Fuels which are obtained with the help of microorganisms and biomass are termed as biofuels (Shuba and Kifle 2018). Plant (herbaceous and woody) residues such as starch, oils, fats of animal, wastes and those from non-toxic algae are used to obtain the biomass (Shuba and Kifle 2018; Sanderson et al. 2020). The products which are produced include esters, ethers, alcohols and other derivatives. The first-generation biofuels were obtained from arable crops while those of second-generation biofuels are derived from organic matter (biomass) (Sanderson et al. 2020). The amount of energy which is obtained using biomass is very high (Khan et al. 2015). Many factors govern the use of type of microorganisms for the production of biofuels which include the amount of lipids (high and constant), growth and development of the microorganisms (should not be affected by the environment), must remain continuous, easily extractable and have simpler harvesting methods producing minimal waste (Shuba and Kifle 2018). Commonly used microorganisms include *Microcystis aeruginosa*, a prokaryotic cyanobacterium and *Scenedesmus obliquus*, a eukaryotic green alga.

Table 8.1 summarizes the various advantages and disadvantages related to different types of renewable energy sources.

Table 8.1 A comparative account of the different renewable resources of energy.

S. no	Type of energy	Merits	Demerits	References
1	Hydropower	• Clean and environmental-friendly • High-level reliability • Low maintenance costs • High efficiency • A proven method to generate electricity • Also helps in controlling floods, provides water and other recreational benefits	• Availability of water depends on rainfall; no manual control • Initial setup cost is very high • Causes displacement of people to make dams etc. • Disrupts the flora and fauna	Bharucha (2005), Mohtasham (2015)
2	Wind	• Clean and environmental-friendly • Reduced/less GHG emissions • High efficiency • A proven method to generate electricity • It is free, sustainable and abundant	• The wire interferes with TV and radio signals • Sometimes kills birds (migratory also) • It cannot be placed in restricted areas and forests (limited locations) • Initial setup cost is very high • Sometimes creates noise and unattractive in look	Bharucha (2005), Mohtasham (2015)
3	Geothermal	• Reliable and sustainable • It is an eco-friendly and efficient and high capacity	• It releases some amount of GHGs in the atmosphere • It is an expensive technique • Sometimes finding a suitable location is also a tedious task • It provides less energy than others	Bharucha (2005), Mohtasham (2015)
4	Solar energy	• Pollution-free, emission less and does not create waste	• Setting up of PV panels is expensive • Maintenance is less	Mohtasham (2015)
5	Ocean energy	• Pollution-free, emission less • Abundant	• Expensive technologically	Bharucha (2005)

Figures 8.2 and 8.3 were developed using a tool/data available at the International Renewable Energy Agency (IRENA). Figure 8.2 represents the generation of electricity using different energy sources from 2010 to 2018. Water or hydro energy is the major source for the generation of energy over the years which is followed by both wind and biofuels.

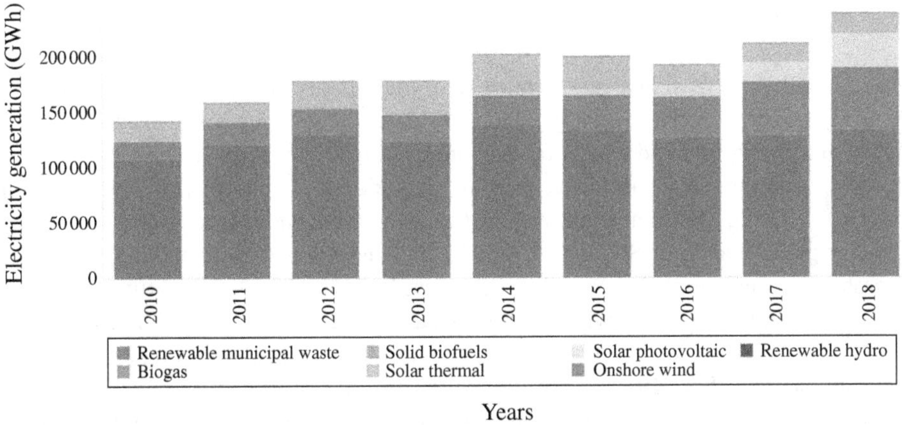

Figure 8.2 The graph shows the production of electricity using different types of renewable resources in India. *Source:* IRENA; https://www.irena.org/Statistics/View-Data-by-Topic/Capacity-and-Generation/Statistics-Time-Series.

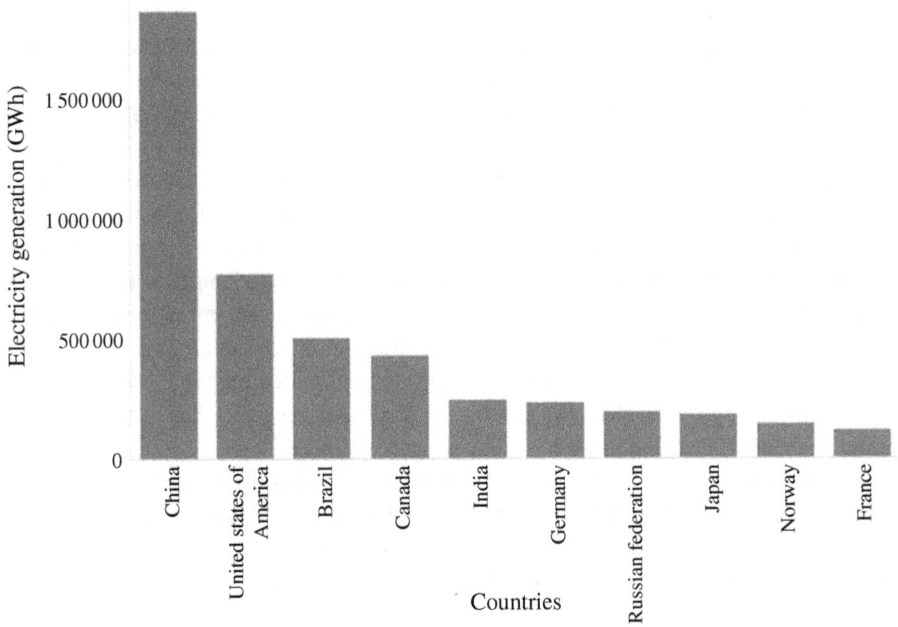

Figure 8.3 The graph shows the production of energy in 2018 by top 10 countries using renewable resources. *Source:* IRENA; https://www.irena.org/Statistics/View-Data-by-Topic/Capacity-and-Generation/Country-Rankings.

Figure 8.3 shows the production of energy using renewable source by different countries. China (1 844 073 GWh) being the highest energy producer in 2018 followed by the United States and Brazil and India (around 240 671 GWh, according to the data available) at fifth place.

8.6 Alternative Energy Resources

The population is increasing at an alarming rate, and it has compelled the development of new or alternative sources of energy which can meet the growing energy demands. It is believed that energy demand will increase by more than 50% by 2030. The consequences of using fossil fuels are very harmful as they are degrading the quality of the environment rapidly by emitting GHGs and alleviating the levels of carbon dioxide (CO_2), thereby enhancing global warming (Shuba and Kifle 2018). Therefore, to overcome the challenges and problems faced due to the excessive use of fossil fuels and depleting natural resources such as water and biomass etc. which thereby causes a reduction in reserves of fossil fuels, exploitation of natural resources, changing climate due to global warming and political issues (Shuba and Kifle 2018), an urge to explore new alternative sources of energy is very much needed. Some of the potential sources which would act as the most promising energy resources include biofuels, hydrogen, synthetic fuels and microalgae-based fuels.

8.6.1 Biodiesel from Bioengineered Fungi

Biodiesel is emerging as a new and alternative source for renewable energy. It is extracted using biological materials such as algae, bacteria, fungi and higher plants with the help of the transesterification process. There are various advantages of using biodiesel over the fossil fuels which include less toxicity, renewable nature and rapid degradation in comparison with diesel. Further, no specialized engine is required to use the biodiesel as it can be used in/for the existing machines. It can be obtained or extracted from either edible or non-edible oil-yielding crops such as soybean, rapeseed and jatropha, tobacco and neem etc. respectively. Some commonly used fungi in the production of biodiesel include *Yarrowia lipolytica*, *Trichosporon pullulan*, *Cryptococcus albidus*, *Rhodotorula graminis* and *Rhodosporidium toruloides*. The biomass and oil obtained from fungi have less generation rate, more productivity, and their cultivation is not dependent on the environmental factors as compared with the plants used for the extraction (Tiwari et al. 2020).

Genetic engineering is used to manipulate the increment or decrement of certain metabolites for varied purposes of desired products. It has been reported that when the gene encoding enzyme glycerol-3-phosphate dehydrogenase (GUT2) was removed from the *Yarrowia lipolytica* (a yeast), it was observed that the synthesis of lipid increased. Similarly, in addition to this effect, when GUT2 was removed; it resulted in more production of lipids as the production of the enzyme glycerol-3-phosphate dehydrogenase 1 (GPD1) was increased due to inhibition of the conversion of glycerol-3-phosphate into dihydrogen phosphate (Tiwari et al. 2020).

Recent studies were carried out on the fungus, *Rhodosporidium toruloides*, an oleaginous yeast which is a potential candidate for the production of many bioproducts such as biodiesel, polyols, alcohols, along with renewable lignocellulosic biomass etc. Phytoene synthase, β-isopropylmalate dehydrogenase (LEU2) and phytoene desaturase were removed using the techniques of CRISPR-cas 9, and it thereby increased the yield of biodiesel (Rao 2020).

Recent metabolic and genetic engineering methods are being regularly used to improve oil and biodiesel production. Mostly essential enzymes, oil-yielding microbes are genetically modified to enhance the production and cloning and modification of desired proteins. For instance, to synthesis, the ester derivatives of fatty acids such as wax esters have been manipulated using a recombinant method in *Pseudomonas citronellolis*, fatty acid butyl esters (FABEs) in *Escherichia coli*, fatty acid methyl esters (FAMEs) in *Saccharomyces cerevisiae*. Similarly, an unspecified acyltransferase from *Acinetobacter baylyi* and pyruvate decarboxylase and alcohol dehydrogenase from *Zymomonas mobilis* were expressed in *E. coli* and it was observed that there was an increase in fatty acid ethyl esters (FAEEs) (Meng et al. 2009).

Therefore, with the application of genetic engineering, these microbes can act as potable agents for extracting biodiesel or for fuel production and thus can become a substitute over other materials due to various advantages (Tiwari et al. 2020).

8.6.2 Microbial Fuel Cells (MFCS)

The mechanism of this device is based on the action and performance of microorganisms and their ability to generate electrogenic impulses via enzymatic reactions. In this process, the microorganisms are being used as an electron donor and these generated electrons are captured and subsequently released using cathode–anode electrodes (Allen and Bennetto 1993). Now, as the energy requirements are increasing day by day, the usage of MFCs is also increasing accordingly. The functioning of MFC depends on the factors such as good design model of MFC, type of microorganism selected, electrodes and on optimization methods (Choudhury et al. 2017). Some microorganisms such as *Shewanella* and *Geothrix* which can produce electrons of their own (electron-donating species) are being used in the process of obtaining fuels using MFC systems. The MFCs are used to generate electricity from sediments of both aquatic (Benthic Unattended Generator, BUGs) and marine organic matter. These BUGs supply electricity to remote locations. Crisis of water is always a major problem all over the world. Water shortage, polluted water and energy production are some of the major hindering issues that are faced globally. This problem can be overcome by using species such as *Shewanella* and *Geobacter* which can use wastewater, marine and lake sediments and natural water to produce electricity (Choudhury et al. 2017). Recent studies have also revealed the usage of MFC in the treatment of pollution with the application of sulfate-reducing bacteria (SRB) and electrogens (electron-emitting molecules). This SRB-MFC complex can remove the sulfate and carbon dioxide from the wastewater and release them (Lee et al. 2015).

8.6.3 Waste-to-Energy Technology

The **Waste to Energy** is an emerging concept in the current scenario which is applied or used on waste materials to extract either fuel, heat or electricity using various methods. The key features of this technology are: they are economical, eco-friendly and follow the principle of sustainable management. The livestock and agriculture wastes are being generated in huge amounts daily which are causing environmental pollution and health issues. Therefore, they can be used to harness energy to meet the increasing energy demand of the overgrowing population. For instance, from the biomass, biochar is obtained which contains an equal amount of energy to that of fossil fuels and can be used for generation of electricity and heat (Moustakas et al. 2020). Municipal wastewater treatment plants contain about 60% of the organic matter derived from the main wastewater treatment (Kor-Bicakci and Eskicioglu 2019; Moustakas et al. 2020). The treatment processes include primary and secondary treatment, while the secondary treatment of the waste is difficult to handle, as due to the production of a microbial cell from the biological material, they are resistant to the biodegradation process. The conventional degradation of the sludge is limited to 30–45% using anaerobic digestion (Moustakas et al. 2020). Various studies have been carried out to extract energy from the wastes such as lignocellulosic biomass and agricultural residues, thereby obtaining alkaline hydrogen peroxide for biofuel production and produced lactic acid and polylactic acid polymers respectively (Djukić-Vuković et al. 2019; Ho et al. 2019; Moustakas et al. 2020).

Organic matter which is obtained from plants is called biomass. Biomass can be obtained from varied sources such as sewage (biosolids from municipal), solid waste from municipal, manures, agricultural residues and other wastes from industries etc. In India, the sources of biomass waste are rice straw and husk, cotton stalks, bagasse, molasses, coconut shells and jute stalks etc. The biomass obtained can be used to produce solid, liquid and gaseous fuels and heat energy with the help of various conversion techniques such as pyrolysis, direct combustion, fermentation, gasification etc. (Khan et al. 2015).

8.6.4 Hydrogen as a Fuel

Hydrogen is one of the good, clean and environmentally friendly fuel (emission-free), and after combustion, it produces water rather than gases which contribute to greenhouse (Kothari et al. 2010). In addition, hydrogen is the most abundant and lightest of all the known and available elements. It produces more energy (2–3 times) than gasoline. It has been reported that due to low energy density of hydrogen, the amount of energy produced by 1 gal of gasoline is equal to 1 kg of hydrogen, and therefore, it can be used in spacecraft and aircraft by NASA. The main problem lies in the extraction of pure hydrogen gas and there are many methods or processes which are used such as electrolysis of water, hydrocarbon stream reforming and autothermal methods that help to produce hydrogen gas. Hydrogen is present in water, methane and other organic matter (Pareek et al. 2020). Some other sources which include cellulose, starch, food and agricultural industrial wastes, *Enterobacter*, *Bacillus*, *Clostridium* that are rich in carbohydrates are also used to extract hydrogen gas (Kothari et al. 2010; Pareek et al. 2020). Some of the merits associated with hydrogen fuel are attributed owing to its significance on the overall scenario of energy needs and crisis, global environmental issues, local quality conditions, security and supply of energy issues, technological innovations, thus promoting the hydrogen fuel as a clean and potential future energy source. On the other hand, the durability and cost of a fuel cell, technical assistance required, improvements to be done in vehicles and dominating nature of fossil fuels are some of the disadvantages which are associated with the commercialization and broad usage of hydrogen fuels (Kothari et al. 2010).

8.6.5 Fuel Cell

Fuel cell refers to an electrochemical cell that produces heat, electricity and water as products using hydrogen and oxygen as raw materials (Figure 8.4). Sir William Grove used the process of reverse electrolysis of water and generated electricity in 1839. The term 'fuel cell' was coined by Charles Langer and Ludwig Mond in 1889 using coal gas (contained a mixture of hydrogen, carbon monoxide, methane etc.). The structure involves two electrodes, namely cathode and anode which are positively and negatively charged respectively along with an electrolyte (solid or liquid). The electrodes produce electricity etc. with the help of inputs (hydrogen and oxygen), sometimes they require catalyst such as palladium or platinum which accelerates the reactions. There are many electrolytes which include potassium hydroxide in water as an alkaline solution, phosphoric acid, molten carbonate consisting of sodium, lithium or magnesium carbonate or potassium etc. The advantageous features of a fuel cell include: the parts are not required to move, the silent machine with no chance of causing noise pollution, safe instrument and eco-friendly nature (Merewether 2003).

8.6.6 Radiant Energy

Radiant energy is a type of active energy (one of the forms of kinetic energy) that is derived from particles, atoms or waves etc. as a result of changes in the electronic configuration (Azeem et al. 2019). In 1930, Nikola Tesla gave the concept of radiant energy that is widely being used in the fields of radiometry, heating and lighting and telecommunication etc.

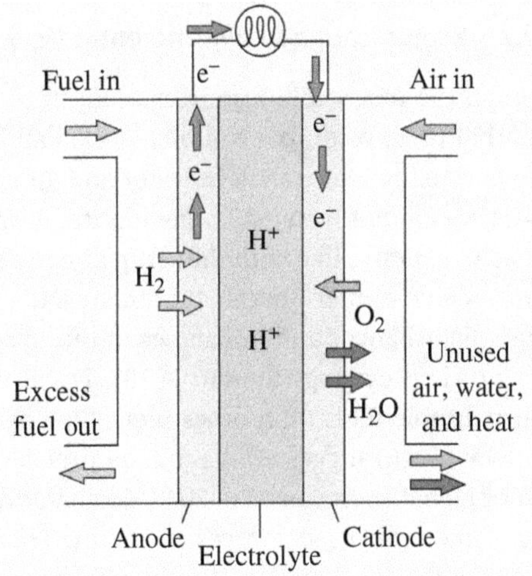

Figure 8.4 A typical fuel cell. *Source:* 'File: Proton Exchange Fuel Cell Diagram.svg' by Mattuci. Licensed under CC BY 1.0. http://creativecommons.org/publicdomain/zero/1.0/deed.en; https://creativecommons.org.

Some other applications include its usage in the forms such as a microwave, space heaters, campfire and hot sidewalk etc. Light and heat are two important components of radiant energy, and this energy moves in a straight line and can be absorbed, reflected and transmitted. The merit of using this method is that it is cheap and able to produce energy in small quantities while the demerits include that this technique is only useful to generate a small amount of energy and cannot be used on a wider scale (Azeem et al. 2019).

8.7 Challenges: Implementation to the Usage of Renewable Energy

Renewable resources have various benefits over other traditional energy sources such as fossil fuels. There are many issues related to the usability of renewable resources as a source of energy, some of which are commercialization, scalability, requirements (input), geographical locations or land usage, technological limitations, ecological and environment-related and social issues (Suman 2018; Seetharaman et al. 2019).

8.7.1 Social Barriers

As the usage of renewable energy is not much popular among people, there may be many conflicts and issues from the point of implementation to consumption (Seetharaman et al. 2019). This is due to lack of proper awareness about the benefits associated with the usage of renewable energy. Though everyone wants their life to be free of difficulties, at the same time they are reluctant to implement these technologies due to various reasons such as (i) improper and insufficient information regarding financial and other aspects (ecological and environmental), (ii) lack of knowledge and awareness about the renewable energy technologies (RET) and their benefits and (iii) issues related with the setup and implementation costs (Seetharaman et al. 2019).

8.7.2 Ecological and Environmental Issues

Though the renewable resources seem to be inexhaustible, it is not the case always. It is said that these resources can be recycled and non-degradable, but in fact, these resources also face many issues such as water and air are prone to get contaminated and with pollution, they cannot be used for generation of energy (Suman 2018). Some changes are irreversible in nature like with the setup of wind turbines, solar panels etc. and construction of dams, many animals lost their habitats and many plants were ruined. Such project setups cause disturbances and imbalance in the ecosystem, thereby disrupting the flow of energy and nutrient cycling (Suman 2018). Fossil fuels always remain the most preferred and indispensable fuels than other fuels. One of the reports of the EIA International Energy Outlook (2016) suggested that oil, natural gas and coal (fossil fuels) are expected to supply 78% of global energy used in 2040. As coal is the cheapest and easy to use, it still remains as one of the most used sources of energy. Therefore, it is difficult for renewable sources to take over the existing ones (Seetharaman et al. 2019).

8.7.3 Commercialization and Scalability

The system with the ability to incorporate the fluctuations in production or number of users without a penalty in cost, reliability, performance and functionality is known as scalability. Therefore, it is an important attribute for pushing the renewable resources to be used as a power source (Suman 2018).

8.7.4 Material Requirement

Demerits of using MFCs include high operational and installation costs as platinum cathodes (recently being replaced by biocathodes) and mediators used for enhancing the efficiency of electron transfer. Although it is a green fuel, the output is very low as compared with chemical MFC ($10^4 W/m^2$) (Seetharaman et al. 2019). Infrastructure is an important factor to set up the power station at any location. For instance, the initial investment of capital is high to construct dams, turbines and wind turbines etc. In the development of fuel cells, elements such as palladium, platinum are used while indium and gallium are required for the PV panels. Therefore, these elements and resources are expensive and require general maintenance to manage them in proper shape for efficient functioning (Suman 2018).

8.8 Conclusion

Sustainable development is a process that integrates all including social, political, economic and environmental aspects together to ensure the current and future potential to meet human needs. Both the quality of human life and ecosystems indicate that the resources should be used judiciously and sustainably on a long-term basis. Long life, increase in both income and knowledge would indicate or are the indicators of sustainable development. As the usage of fossil fuels has tremendously increased over the years, the quality of life and condition of both the ecosystem and environment has deteriorated at a much faster rate, leading to the exploration of the new alternative renewable energy sources which can uplift the status of the environment. The production of fuel and energy from biomass is very economical, efficient, accessible and easy to use. The concept of a 'Greener and Cleaner Planet' can be successfully achieved by using these alternative and traditional sources of renewable energy that will surely result in the future elimination or reduction in pollution levels caused by the major culprits such as fossil fuels.

References

Abolhosseini, S., Heshmati, A., and Altmann, J. (2014). A review of renewable energy supply and energy efficiency technologies. *IZA Discussion Paper No. 8145.*

Alabdali, Q. A., Bajawi, A. M., Fatani, A. M., & Nahhas, A. M. (2020). Review of recent advances of wind energy. *Sustainable Energy*, 8(1), 12–19.

Allen, R. M., & Bennetto, H. P. (1993). Microbial fuel-cells. *Applied Biochemistry and Biotechnology*, 39(1), 27–40.

Ardizzon, G., Cavazzini, G., and Pavesi, G. (2014). A new generation of small hydro and pumped-hydro power plants: advances and future challenges. *Renewable and Sustainable Energy Reviews*, 31, 746–761.

Azeem, S. A., Fatema, Q., & Sharma, S. K. (2019). Ether & radiant energy: a vital source of energy & allied resources. *IRACST – International Journal of Computer Networks and Wireless Communications (IJCNWC)*, 9(2) 22–27.

Bharucha, E. (2005). *Textbook of Environmental Studies for Undergraduate Courses.* Universities Press.

Bhoi, A. K., Sherpa, K. S., Kalam, A., & Chae, G.-S. (Eds.). (2020). Advances in greener energy technologies. *Green Energy and Technology* doi:https://doi.org/10.1007/978-981-15-4246-6.

Bošnjaković, M., Stojkov, M., & Jurjević, M. (2019). Environmental impact of geothermal power plants. *Tehnički vjesnik*, 26(5), 1515–1522.

Chakraborty, S., Sinha, A., Dutta, S., and Biswas, N. (2011). A review paper of wind energy. *National Conference on Advances and Research in Electrical System Technology*, Arya Institute of Engineering and Technology, Jaipur, India (23–24 April 2011), pp. 202–206.

Choudhury, P., Prasad Uday, U. S., Bandyopadhyay, T. K., Ray, R. N., & Bhunia, B. (2017). Performance improvement of microbial fuel cell (MFC) using suitable electrode and Bioengineered organisms: a review. *Bioengineered*, 8(5), 471–487.

Demirbas, A. (2005). Potential applications of renewable energy sources, biomass combustion problems in boiler power systems and combustion related environmental issues. *Progress in Energy and Combustion Science*, 31(2), 171–192.

Dincer, I. (2000). Renewable energy and sustainable development: a crucial review. *Renewable and Sustainable Energy Reviews*, 4(2), 157–175.

Djukić-Vuković, A., Mladenović, D., Ivanović, J., Pejin, J., & Mojović, L. (2019). Towards sustainability of lactic acid and poly-lactic acid polymers production. *Renewable and Sustainable Energy Reviews*, 108, 238–252.

Durganjali, C. S., Bethanabhotla, S., Kasina, S., & Radhika, S. (2020). Recent developments and future advancements in solar panels technology. *Journal of Physics: Conference Series*, 1495(1), 012018).

Halkos, G. E., & Gkampoura, E. C. (2020). Reviewing usage, potentials, and limitations of renewable energy sources. *Energies*, 13(11), 2906.

Herzog, A. V., Lipman, T. E., & Kammen, D. M. (2001). Renewable energy sources. Encyclopedia of life support systems (EOLSS). *Perspectives and Overview of Life Support Systems and Sustainable Development*, 76. http://www.eolss.net/sample-chapters/c08/E3-03-27.pdf

Ho, M. C., Ong, V. Z., & Wu, T. Y. (2019). Potential use of alkaline hydrogen peroxide in lignocellulosic biomass pretreatment and valorization–a review. *Renewable and Sustainable Energy Reviews*, 112, 75–86.

Huttrer, G.W. (2020). Geothermal power generation in the world 2015-2020 update report. *World Geothermal Congress*, International Geothermal Association, Reykjavik, Iceland.

International Energy Outlook (2016). https://www.eia.gov/outlooks/ieo/pdf/0484(2016).pdf

Kaunda, Chiyembekezo S., Kimambo, Cuthbert Z., and Nielsen, T. K. (2012). Hydropower in the context of sustainable energy supply: a review of technologies and challenges. *ISRN Renewable Energy* 2012 1–15.

Khan, A. A. (2018). Why would sea-level rise for global warming and polar ice-melt. *Geoscience Frontiers* doi:10.1016/j.gsf.2018.01.008.

Khan, S., Paliwal, V., Pandey, V. V., & Kumar, V. (2015). Biomass as renewable energy. *International Advanced Research Journal in Science, Engineering and Technology (IARJSET)*, 2(1), 301–304.

Kor-Bicakci, G., & Eskicioglu, C. (2019). Recent developments on thermal municipal sludge pretreatment technologies for enhanced anaerobic digestion. *Renewable and Sustainable Energy Reviews*, 110, 423–443.

Kothari, R., Tyagi, V. V., & Pathak, A. (2010). Waste-to-energy: a way from renewable energy sources to sustainable development. *Renewable and Sustainable Energy Reviews*, 14(9), 3164–3170.

Lee, D. J., Chang, J. S., & Lai, J. Y. (2015). Microalgae–microbial fuel cell: a mini review. *Bioresource Technology*, 198, 891–895.

Lehner, Bernhard, Czisch, G., and Vassolo, S. (2005). The impact of global change on the hydropower potential of Europe: a model-based analysis. *Energy Policy*, 33(7), 839–855.

Leung, D. Y., & Yang, Y. (2012). Wind energy development and its environmental impact: a review. *Renewable and Sustainable Energy Reviews*, 16(1), 1031–1039.

Meng, X., Yang, J., Xu, X., Zhang, L., Nie, Q., & Xian, M. (2009). Biodiesel production from oleaginous microorganisms. *Renewable Energy*, 34(1), 1–5.

Merewether, E. A. (2003). *Alternative Sources of Energy: An Introduction to Fuel Cells.* US Department of the Interior, US Geological Survey.

Mohtasham, J. (2015). Renewable energies. *Energy Procedia*, 74, 1289–1297.

Moustakas, K., Loizidou, M., Rehan, M., and Nizami, A.S. (2020). A review of recent developments in renewable and sustainable energy systems: key challenges and future perspective. doi: 10.1016/j.rser.2019.109418.

Mussa, M., Teka, H. and Ayicho, H. (2018). Environmental impacts of hydropower and alternative mitigation measures. *Current Investigations in Agriculture and Current Research* 2(2). doi: 10.32474/CIACR.2018.02.000133

Østergaard, P. A., Duic, N., Noorollahi, Y., Mikulcic, H., & Kalogirou, S. (2020). Sustainable development using renewable energy technology. *Renewable Energy*,146. 2430 – 2437.

Owusu, P. A., & Asumadu-Sarkodie, S. (2016). A review of renewable energy sources, sustainability issues and climate change mitigation. *Cogent Engineering*, 3(1), 1167990.

Panwar, N. L., Kaushik, S. C., & Kothari, S. (2011). Role of renewable energy sources in environmental protection: a review. *Renewable and Sustainable Energy Reviews*, 15(3), 1513–1524.

Pareek, A., Dom, R., Gupta, J., Chandran, J., Adepu, V., & Borse, P. H. (2020). Insights into renewable hydrogen energy: recent advances and prospects. *Materials Science for Energy Technologies*, 3, 319–327.

Pedersen E. (2011). Health aspects associated with wind turbine noise. Results from three field studies. *Noise Control Engineering Journal*, 59(1), 47–53.

Punch J, James R, Pabst D. (2010). Wind-turbine noise: what audiologists should know. *Audiology Today*, 22(4), 20–31.

Rao, C.V. (2020). Exploring oleaginous yeast *Rhodosporidium toruloides* as a platform organism for production of chemicals and fuels. https://genomicscience.energy.gov/pubs/2020abstracts/abstracts/Zhao_Schultz.pdf

Sanderson, M. A., Adler, P., & Martin, N. P. (2020). *Biomass, Energy, and Industrial Uses of Forages*. Forages: The Science

K. Seetharaman, Moorthy, N., Patwa, N., A. Saravanan & Gupta, Y. (2019). Breaking barriers in deployment of renewable energy. *Heliyon*, 5(1), e01166.

Shaikh, M. R. S., Waghmare, S. B., Labade, S. S., Fuke, P. V. and Tekale, A. (2017). A review paper on electricity generation from solar energy. *International Journal for Research in Applied Science & Engineering Technology (IJRASET)* 5(9).1884–1889.

Shuba, E. S., & Kifle, D. (2018). Microalgae to biofuels: 'promising' alternative and renewable energy, review. *Renewable and Sustainable Energy Reviews*, 81, 743–755.

Suman, S. (2018). Hybrid nuclear-renewable energy systems: a review. *Journal of Cleaner Production*, 181, 166–177.

Tester J. W. (2005). *Sustainable Energy: Choosing Among Options.* London: MIT Press.

Tester, J. W., Drake, E. M., Driscoll, M. J., Golay, M. W., & Peters, W. A. (2012). *Sustainable Energy: Choosing Among Options*. MIT Press.

Tiwari S., Jadhav S.K., Sharma G.D., Beliya E. (2020) Fungal bioengineering in biodiesel production. In: Hesham AL., Upadhyay R., Sharma G., Manoharachary C., Gupta V. (eds) *Fungal Biostechnology and Bioengineering. Fungal Biology*. Springer, Cham.

World Energy Council (2013). World Energy Resources: Hydro. https://www.worldenergy.org/wp-content/uploads/2013/10/WER_2013_5_Hydro.pdf

Xu, Jiuping, Ni, T., and Zheng, B. (2015). Hydropower development trends from a technological paradigm perspective. *Energy Conversion and Management*, 90, 195–206.

Suggested Readings

Abdulkareem-Alsultan, G., Asikin-Mijan, N., Lee, H. V., & Taufiq-Yap, Y. H. (2020). Biofuels: past, present, future. In *Innovations in Sustainable Energy and Cleaner Environment.* Springer, Singapore. 489–504.

Edenhofer, O., Seyboth, K., Creutzig, F., & Schlömer, S. (2013). On the sustainability of renewable energy sources. *Annual Review of Environment and Resources*, 38, 169–200.

Esteban, M., Gasparatos, A., & Doll, C. N. (2017). Recent developments in ocean energy and offshore wind: financial challenges and environmental misconceptions. In *Sustainability Through Innovation in Product Life Cycle Design*, Springer, Singapore. 735–746.

Rittmann, B. E. (2008). Opportunities for renewable bioenergy using microorganisms. *Biotechnology and Bioengineering*, 100(2), 203–212.

Saidur, R., Rahim, N. A., Islam, M. R., & Solangi, K. H. (2011). Environmental impact of wind energy. *Renewable and Sustainable Energy Reviews*, 15(5), 2423–2430.

Sen, S., Ganguly, S., Das, A., Sen, J., & Dey, S. (2016). Renewable energy scenario in India: opportunities and challenges. *Journal of African Earth Sciences*, 122, 25–31.

9

Sustainable Energy Policies of India to Address Air Pollution and Climate Change

Prem Lata Meena[1], *Vinay*[2], *and Anirudh Sehrawat*[2]

[1]*Department of Polymer Science, Bhaskaracharya College of Applied Sciences, University of Delhi, Dwarka, New Delhi, India*
[2]*University School of Environment Management, Guru Gobind Singh Indraprastha University, Dwarka, New Delhi, India*

9.1 Introduction

The power sector of India is one of the most diversified sectors in the world (Kumar et al. 2014) which includes different sources ranging from different conventional sources to non-conventional sources (Prasanthrajan and Keerthika 2020). But in India, about 75% of energy still comes from burning fossil fuels making this sector a world's dirtiest electricity system (CarbonBrief analysis 2019). This sector emits half of the CO_2 in India. The electricity demand is increasing exponentially, and it is also further expected to grow in the years ahead. Therefore, there is a requirement of massive addition to the installed generating capacity to satisfy the increasing energy demand of the country for electricity.

Climate change is changing in long-term weather patterns which is directly or indirectly attributed to anthropogenic activities that alter the global atmospheric composition. This leads to change in temperature, precipitation, sea level and, the severity and frequency of extreme events such as heat strokes, drought, flood, hurricane, and storms, etc. These changes will be likely to affect energy production, delivery and consumption adversely leading to an energy crisis.

Energy production and use are interconnected with other factors such as consumption of water, use of goods and services, land use, population, transportation and economic growth. An increase in temperature and more heat strokes will lead to more demand for energy. During the warmer climate, the efficiency of the production of energy reduces for existing fossil fuel and nuclear plants that use water for cooling purposes. Change in precipitation and more drought lead to less water availability and adversely influence energy production. More evaporation due to climate change leads to more demand for energy for drinking and agricultural purposes as desalination and pumping of water are energy-intensive technologies. More sudden and frequent events such as hurricanes, floods and storms can damage the coastal power plants, their platforms and powerlines, which ultimately affect the electricity distribution.

Energy: Crises, Challenges and Solutions, First Edition. Edited by Pardeep Singh, Suruchi Singh, Gaurav Kumar, and Pooja Baweja.

Depleting conventional resources is forcing countries to develop effective energy mix strategies that require a novel approach for planning and sponsoring research on different energy resources. Therefore, this chapter first discusses the energy sector in India, followed by national strategies and financial instruments launched by the Indian Government for encouraging renewable energy (RE) use.

9.2 Energy Sector of India

9.2.1 Energy Reserves

9.2.1.1 Coal and Lignite

India's eastern and south-central parts are rich in coal deposits which majorly include Jharkhand, West Bengal, Chhattisgarh, Odisha, Maharashtra, Telangana and Madhya Pradesh which is approximately equal to 98.26% of the total coal reserves in India. Jharkhand had the maximum share in the overall coal reserves (26.06%) followed by Odisha (24.86%) as of March 2018. The estimated coal reserves as of March 2018 were 319.04 billion tons, whereas lignite was 45.66 billion tons in India. During the year 2017–2018, there has been an increase of 1.23%, i.e. 3.88 billion tons in the coal reserves with Odisha contributing a maximum increase of 2.6% (MOSPI 2019).

9.2.1.2 Petroleum and Natural Gas

The western offshore (40%) is the largest reserves of crude oil followed by Assam (27%), as per the regional distribution of crude oil. The crude oil reserves in India were 594.49 million tons as of March 2018 compared to 604.10 million tons as of March 2017. The projected production of crude oil declined by 1.59% in 2017–2018 relative to the previous year. The estimated reserves of natural gas in India amounted to 1339.57 billion cubic meters (BCM) as of March 2018 compared to 1289.70 billion cubic meters as of March 2017. The reserves of natural gas have risen by 3.87% as compared to the past year (MOSPI 2019).

9.2.1.3 Renewable Energy Sources

India has a huge capacity for producing renewable electricity from a range of sources, such as solar, small hydropower, wind, biomass and bagasse cogeneration. The total capacity to generate RE in the country is projected at 1 096 081 MW as of March 2018, which includes 68.33% of solar power capacity (748 990 MW), 27.58% of wind power potential (302 251 MW) at a hub height of 100 m, 1.80% of small hydropower capacity (19 749 MW), 1.60% of biomass power (17 536 MW), 0.46% from bagasse-based cogeneration in sugar mills (5000 MW) and 0.23% from waste to energy (2554 MW) (MOSPI 2019, Figure 9.1).

9.2.2 Production of Energy

The commercial electricity generation suggested that the coal was the primary source of energy during 2017–2018 which nearly accounted for almost 70.69% of total production followed by crude oil (10.34%), and in the end, natural gas (8.7%). During 2017–2018, the

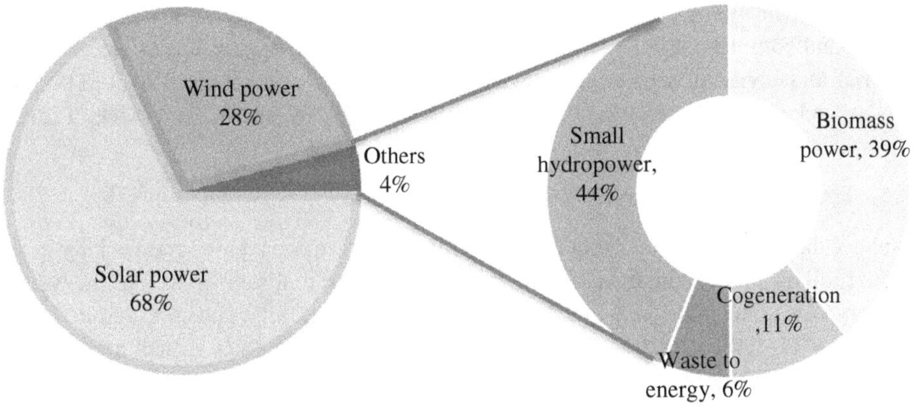

Figure 9.1 The renewable power potential in India source wise (as of March 2018) (MOSPI 2019).

contribution of RE sources and lignite was just 6.63 and 3.64%, respectively. Thus, India's energy sector is focused exclusively on fossil fuels (MOSPI 2019).

9.2.3 Consumption of Fossil Fuel and Electricity

9.2.3.1 Coal and Lignite

The total raw coal consumption was 549.57 million tons in 2008–2009 by industry which increased to 896.34 million tons in 2017–2018 with a compound annual growth rate of 5.01%. The lignite consumption was 31.85 million tons in 2008–2009 which has also been increased to 45.82 million tons in 2017–2018 with a CAGR of 3.70%. The lignite consumption is the largest in the energy generation, approximately 83.7% of overall lignite consumption. The primary usage of raw coal is in electricity production, followed by the steel industry (MOSPI 2019).

9.2.3.2 Crude Oil and Natural Gas

The crude oil consumption increased gradually from 160.77 million metric tons (MMT) in 2008–2009 to 251.93 MMT in 2017–2018, with a compound growth of 4.59%. The highest utilization of natural gas is in the fertilizer sector (27.78%), followed by energy production (22.77%) and natural gas used for domestic transport (16.25%). Natural gas is used for both energy and non-energy purposes which is 60.68 and 39.32%, respectively, in India (MOSPI 2019).

9.2.3.3 Petroleum Products Consumption

In 2017–2018, high-speed diesel contributed to 39.3% of total consumption excluding refinery fuel and losses of all types of petrochemicals, accompanied by 12.7% Petrol, 12.4% Pet Coke, 11.3% LPG and 6.1% Naphtha (MOSPI 2019).

9.2.3.4 Consumption of Electricity

The electricity demand which was 553 995 GW per hour in 2008–2009 had been increased to 1 130 244 GW per hour in 2017–2018, with a CAGR of 7.39%. Of the overall electricity usage, the industry held the largest share (41.48%), followed by domestic (24.20%), agricultural (18.08%) and commercial (8.51%) sectors in 2017–2018. The consumption of electricity in industrial and domestic sectors escalated at a much faster rate between 2008–2009 and 2017–2018 relative to other sectors, with a CAGR of 8.39 and 7.58%, respectively (MOSPI 2019).

9.2.4 Energy Sector and Greenhouse Gases Emission

The above data highlight that India is still relying on conventional sources especially fossil fuels to fulfil its energy demands making it the third largest emitter of GHGs in the world, although per capita emission is low. In the Paris Agreement, India vowed to reduce GHG emission by 33–35% of 2005 levels by 2030. Although due to the key contribution of the energy sector to GHGs, several policy measures have been undertaken by the government to promote renewable energies and, in particular, solar energy.

9.3 India's Potential and Policies to Exploit Renewable Sources

The country has set an objective of creating 175 GW of RE capacity by 2022, comprising solar power (100 GW), wind (60 GW), bioenergy (10 GW) and small hydropower (SHP) (5 GW) (Digambar et al. 2019).

9.3.1 Solar Energy

India possesses immense potential solar energy capacity, exceeds annual energy consumption and can make India a global leader in solar power generation. Approximately 5000 trillion kWh of solar energy incident over an Indian land area per year, which mostly receives 4–7 kWh/m^2 daily (Chandra et al. 2019).

International Solar Alliance, launched in November 2015, is an alliance of 121 solar-rich countries which are located partly or entirely between the two tropics – Tropic of Cancer and the Tropic of Capricorn with a purpose of promoting solar energy among its members and increasing investment by $1 trillion by 2030 (Shidore and Busby 2019). The **Jawaharlal Nehru National Solar Mission** launched with the goal of 20 GW grid capacity and 2 GW off-grid installations by 2022 by Indian Government in 2010. Then, The Government of India (GoI) increased this target to 100 GW under the Union budget in 2015 (Kashyap et al. 2020).

The United Nations Environment Program (UNEP) supported the **Indian Solar Loan Programme** through which the GoI developed a loan platform for solar home power systems. This programme has also received the renowned Energy Globe World Award for sustainability. Over three years, 2000 bank branches sponsored more than 16 000 solar home energy systems (Cleantech n.d.).

9.3.2 Wind Energy

India was one of the early starters in wind power and the fourth largest in global wind power generation as compared to other developing nations. In India, the wind power's installed capacity was 34 293 MW as of March 2018, accounting for 10% of India's total installed capacity. In India, as of 2018, the wind power generation is highest in Tamil Nadu which is 7269.50 MW, followed by 4100.40 MW by Maharashtra, then 3454.30 MW by Gujarat, then 2784.90 MW from Rajasthan, then 2318.20 MW by Karnataka, 746.20 MW by Andhra Pradesh and 423.40 MW by Madhya Pradesh. India has the potential of about 60 GW to generate electricity from wind power which is also India's **ambitious target to generate wind energy of 60 000 MW** by 2022. But, it would go up substantially when low capacity old wind power stations could be replaced with the wind turbines which have higher capacity.

The government initially had the **National Wind Power Policy,** but subsequently, two wind policies were developed one for onshore and other offshore. In 2015, the **National Offshore Wind Energy Policy** framework came intending to develop the offshore wind energy along the Indian coastline. The Indian Government allocated Rs. 10 000 crores for the offshore wind energy sector, which is collected from Coal Cess, as the initial seed money from clean energy fund (CEF).

In May 2018, the MNRE of the GoI announced a new hybrid policy on wind and solar energy to develop a framework for the promotion of large grid-connected solar photovoltaic hybrid systems. This **wind-solar hybrid policy** ultimately promotes the effective and productive use of wind and solar power together, transmission networks, and land (Kuriakose et al. 2019).

9.3.3 Hydropower

India has a hydro-electric potential of about 145 000 MW and stands at fifth position in the world in terms of exploitable hydro-potential. India's installed hydropower capacity was 12.35% (45 699 MW) as of March 2018. The decline in the share of total hydropower from 50.36% in the 1960s to about 13% in 2018–2019 was reported.

The MNRE offers opportunities for these hydropower ventures and intends to make it more competitive in the marketplace by funding 75% of the costs. In addition to that, other Indian states have introduced low-interest loans, special tariffs and other financial frameworks to promote technological innovations. Small hydropower, which can generate power up to 25 MW, falls within the scope of the MNRE, whereas large-scale hydropower, which can generate more than 25 MW, falls within the scope of the Ministry of Power. India also has a vast scope for pumped hydroelectric storage that can efficiently transform non-dispatchable RE.

9.3.4 Biomass Energy

With its tropical location, plentiful sunlight and precipitation, India has a suitable environment for biomass production. The enormous agricultural potential of the nation provides massive volumes of agro-residues that can be used to fulfil energy demands. India has an estimated potential to generate 16 000 MW of biomass energy and an additional 3500 MW of bagasse cogeneration.

The **Unnat Chulah Abhiyan** (UCA) is a programme initiated by the MNRE in 2014. The programme aims (i) to improve biomass cooking chulha to provide cleaner cooking energy solutions; (ii) to alleviate drudgery by women and children using conventional chulha for cooking and (iii) to minimize black carbon and other emissions from burning biomass for food, therefore, to mitigate adverse impacts of climate change. As of March 2018, 36 940 family-type and 849 community-type improved cooker stoves were delivered (PIB 2018).

National Biogas and Manure Management Programme launched by GoI to build household types of biogas plants to meet the need for cooking energy in rural and semi-urban areas. According to MNRE data, there was built up of approximately 49.6 lakh family-type biogas plants in the country as of March 2018 (MNRE 2019).

National Biofuel Policy was launched by Indian Government in 2018 with the target to blend 20% of biodiesel in diesel and 20% of ethanol in petrol. The objectives are to strengthen the ongoing supply of ethanol/biodiesel by expanding domestic production, to establish second-generation (2G) bio-refineries, to introduce new biofuels feedstocks and to develop new technologies for conversion to biofuels (Shukla and Rathore 2020). The benefits of this policy include: (i) reduced fossil-fuels import dependency; (ii) provides cleaner Environment and health benefits; (iii) provides a way for municipal solid waste management; (iv) generates rural livelihood; (v) additional income to farmers and (vi) provides cheaper alternatives to consumers. The GoI plans to set up 12 new biofuel refineries in the country to save Rs. 12 000 crores in the next four years by the use of ethanol. The government is indeed planning to develop a new policy and framework for the application of biofuel in the aviation industry which will enable the utilization of excess food grains, non-edible oilseeds, short-lived crops and recycled cooking oil.

9.4 National Strategies to Promote Renewable Energy: Policy Framework with Their Objectives

In 1981, India started its drive for the quest of RE with the commission establishment to exploit an inexhaustible supply of clean energy sources including solar, wind and biomass. In 1992, the commission transformed into the Ministry of Non-Conventional Energy Sources. And again in 2006, this ministry transformed into the Ministry of New and Renewable Energy (MNRE). A new agency Indian Renewable Energy Department Agency (IREDA) was established in 1987 (Burke et al. 2019) to fund Renewable Energy initiatives. But today, the MNRE and its state affiliates are enriching the clean energy sector through encouraging renewables and by active participation in the corporate sector.

The objective of MNRE is to make certain (i) energy security by reducing reliance on oil imports and developing alternative energy sources to minimize the difference between oil supply and demand; (ii) the energy access and availability; (iii) an increase in the RE share; (iv) cost-effective, safe, convenient, efficient and reliable energy supply alternatives and (v) by sustainable and diverse fuel mix.

9.4.1 India's Electricity Act

In 2003, Indian Electricity Act that came into force is said to have strengthened and promoted India's quest for alternative sources of energies to fossil fuels, thus substituting all previous electricity legislation. There are 185 sections in this Act, some of which are linked to RE, such as-(i) Part 2, section 3, which outlines the 'National Electricity Policy' and emphasizes on the development of a 'National Electricity Policy', new policy on tariffs and a plan for national electricity that maximizes the usage of all resources, including thermal, nuclear, hydro and primarily RE; (ii) section 4, which supports a national policy on stand-alone systems for rural areas with an emphasis on RE; (iii) Section 61(h) which uses RE in the electricity production by means of tariff legislation and (iv) Section 86(1(e)) which encourages the electricity production from renewable sources to link it to Indian power grids and to permit the sale of electricity to any individual (IEA 2003).

9.4.2 National Electricity Policy (NEP), 2005

In 2005, India formulated its National Electricity Policy (NEP). The objectives of the policy are to boost access to power for all by 2010; to ensure energy supply by meeting energy needs by 2012; to provide competitively and appropriately cost, reliable electricity; to increase per-capita electricity by more than 1000 units by 2012 and to enhance economic stability and long-term demand. The NEP also promotes conditions for increasing the proportion of RE by granting authority to the State Electricity Regulatory Commissions to agree on a tariff for energy generated from renewable sources to promote competition in the energy market (Bardhan et al. 2019).

9.4.3 NAPCC-National Action Plan on Climate Change, 2008

The NAPCC was launched in 2008 to strengthen environmental and economic ambitions of India by achieving a sustainable development path. This climate action plan is driven by a set of principles, some of which are (i) to implement a sustainable development approach to minimize hunger, vulnerability and adverse impacts of climate change; (ii) to achieve India's development goals through a transition in energy policies that foster sustainability, resulting in a reduction in GHGs; (iii) to design and deploy fast-paced technologies that enable adaptation and mitigation of GHGs and (iv) to establish new marketplaces and compliance frameworks to promote sustainable development. It encompasses the following eight missions,

1) **National Solar Mission:** The mission is for the promotion of the utilization of solar energy power generation, establishment of a Solar Research Centre, enhancing international cooperation on technological development, boosting domestic production and manufacturing capability which is also involved in this mission.
2) **National Mission for Enhanced Energy Efficiency:** This advocates significant cuts in energy utilization in large energy-extensive enterprises, with a scheme allowing businesses to exchange energy-saving certificates, renewable subsidies, including reducing taxes on energy-efficient appliances and funding public–private collaborations to minimize energy use through demand-side control systems in various sectors.
3) **National Mission on Sustainable Habitat:** The objective of the mission is seeking to promote energy sustainability as a central aspect of urban development through expansion of the current Energy Conservation Building Code, by introducing pricing policies to facilitate the procurement of energy-efficient vehicles, by improving the

implementation of automotive fuel economy regulations and providing incentives for the use of public transport. The mission also emphasizes waste reduction, recycling and management.

4) **National Water Mission:** This mission is for enhancing water use efficiency by 20% through pricing and other interventions.
5) **National Mission for Sustaining the Himalayan Ecosystem:** This mission aims to resist Himalayan glacier's melting and conservation of biodiversity in the Himalayan region.
6) **Green India Mission:** Green India Mission aims to afforest six million hectares of degraded forest land with enhancing the forest cover from 23 to 33% in the Indian territory.
7) **National Mission for Sustainable Agriculture:** The mission intends to strengthen climate resilience and adaptation in the agriculture sector through climate-resistant crops cultivation, the implementation of weather insurance schemes and smart agricultural practices.
8) **National Mission on Strategic Knowledge for Climate Change:** The mission includes a dedicated Climate Science Research Fund, advanced climate models and enhanced international cooperation to develop understanding of climate change and its challenges. This mission also promotes efforts by the corporate sector to establish adaptation and mitigation technological solutions by venture capital funds (Bansal et al. 2019).

9.4.4 Copenhagen Accord

Under the Copenhagen Accord, India, in 2009, has pledged that it will commit to lessen the emission intensity by 20–25% of GDP (excluding agriculture) by 2020 which was a voluntary goal (Saveyn et al. 2012).

9.4.5 India's Intended Nationally Determined Contribution (INDC)

Different countries submitted their INDCs which specifies their commitments to reduce emissions by 2025 or 2030 in Conference of Parties 21. India also submitted its INDC commitments on 1 October 2015, with significant commitments for the period from 2021 to 2030, as shown in Table 9.1 (UNFCCC 2015; Shukla et al. 2017).

9.5 Financial Instruments to Promote Renewable Sources in India

9.5.1 Coal Tax

The carbon tax is one of the popular financial instruments which is used to reduce GHG emissions. On 1 July 2010, the GoI introduced a nationwide coal tax as a Clean Energy Tax. This tax was initially fixed at ₹50 per ton of coal of both domestic and imported to set a price on the carbon content of goods and services to discourage their use. By this, India has

Table 9.1 Intended nationally determined goals of India.

S. no.	Category	Targets	Target year
Renewable energy targets			
a	Solar power capacity	100 GW installed capacity	By 2022
b	Small hydropower capacity	5 GW installed capacity	By 2022
c	Wind power capacity	60 GW installed capacity	By 2022
d	Biomass power capacity	10 GW installed capacity	By 2022
e	Non-fossil fuel-based electric power capacity	40% total electric power installed capacity from non-fossil fuel	By 2030
f	Nuclear power capacity	63 GW installed capacity	By 2032
Carbon intensity and energy efficiency			
a	Carbon intensity	33–35% lessen from 2005 level	By 2030
b	Energy efficiency	10% lessen in the current energy consumption	By 2019
Afforestation			
a	Afforestation	Increase to 5 million hectares (MHa) in forest/tree cover and 100 million tonnes CO_2 equivalent carbon sequestration annually	—
Industrial sector			
a	Industry	4–5% reduction in specific energy consumption compared to 2012 level	—
Transportation sector			
a	Transportation	i) Blending of biofuels. Both for biodiesel and bioethanol-20%, by 2017 ii) Switching from Bharat Stage IV (BS IV) fuels to Bharat Stage VI (BS VI) across the country iii) 6–7 million hybrid and electric vehicles (HEV) per year by 2020	—

Source: Adopted from Shukla et al. (2017; UNFCCC 2015).

shifted from the regime of carbon subsidization to the regime of carbon taxation. This has significantly increased petrol and diesel prices with an intent to reduce CO_2 emissions.

The tax increased four times from ₹50 to ₹200 per ton of coal in 2015. The funding of the National Clean Energy and Environment Fund (Bhat and Mishra 2020) which comes from coal tax is used to invest in different RE initiatives and engineering sciences. Recently, the coal tax doubled from ₹200 to ₹400 per ton in 2016 enabling renewable energies to become more attractive in the energy marketplace and helping to fund more clean energy ventures.

The coal tax has lots of benefits such as (i) the tax is Pigouvian in nature assessed against peoples who are engaged in private businesses that create adverse side effects for society, thus to balance the marginal social costs; (ii) high prices will cause a break towards the greener alternatives; (iii) more tax revenue for other clean energy projects and (iv) more employment generation due to high employment elasticity in RE forms.

9.5.2 Subsidy Cuts on Fossil Fuels

There was a decline of 70% in fossil fuel incentives from ₹73 330 crores in the financial year-14 to ₹52 980 crores in the financial year-17 in India as a result of amendments to subsidies for cooking gas, diesel, petrol and kerosene use; and lower global oil prices. Moreover, there was a substantial increase of six times in subsidies for RE since 2014. This includes the biggest increase of ₹5770 crores from Financial Year-16 to Financial Year-17.

There are lots of benefits for subsidy cuts such as (i) diverting towards clean energy ultimately cuts harmful emissions; (ii) creating a field for alternative sources of energy; (iii) fossil fuel subsidies cut make it costly, and its natural offshoot would be conservation of fossil fuels; (iv) saved money can be diverted for the development of other sectors such as rural development, education, etc. and (v) consumption would decrease ultimately reduces country's dependence on import of such fossil fuel. However, according to the report, even still in the financial year 17, subsidies to fossil fuels were found to be more than three times the subsidies for renewables and electric vehicles in India. Coal alone (₹15 990 crores) surpassed that for clean energy (₹15 000 crores) by Financial Year-17.

9.5.3 Renewable Energy Certificates (RECs)

The MNRE established the RECs in 2010 to endorse RE in India with an intent to promote sustainable development. The different names of RECs are Tradable Renewable Certificates (TRCs), Green tag, Renewable Energy Credits or Renewable Electricity Certificate. In March 2011, the government approved RE certificates trading to address the difference between the availability of RE and the requirements of the companies to comply with their Renewable Purchase Obligation. This system was initiated by the National Authority-Central Electricity Regulatory Commission (CERC) which provides the producer with two alternatives, the first of which is to encourage them to sell the produced energy at different rates and, second, providing power with sustainable virtues that can be exchanged as Renewable Energy Certificates. The key factors that impede the programme's ability to meet its full potential are lack of long-term objectives, absence of compliance and lack of potential pricing signals for shareholders (Shrimali and Tirumalachetty 2013).

9.5.4 Perform, Achieve and Trade Scheme

The PAT scheme is to improve energy efficiency in energy-intensive industries through energy trading mechanisms. The first cycle of this scheme was launched for three years between 2012 and 2015 with a target to reduce energy consumption by 4.1% more than 400 energy-intensive enterprises. But its proper implementation managed to reduce the energy consumption by 5.3%, which was above the target. All sectors except thermal power generation exceeded their targets in the first cycle. Thermal power generation had the maximum target of the eight sub sectors unable to meet its target due to the absence of new efficient natural gas power plants (Sharma et al. 2019).

The second PAT Cycle was notified from 1 April 2016, for three years up to March 2019 covering 11 sectors. The reduction target of energy consumption given to 621 units is

8.869 MT of Oil Equivalent (MTOE). Then again, the third PAT cycle was notified from 1 April 2017, for three years including 116 new units with a reduction target of 1.06 MTOE (BEE 2020).

9.5.5 Other Government Policies, Their Budget and Status

- Allocation of ₹15 875 crores to the Ministry of Power and ₹5500 crores to ambitious policy **Deen Dayal Upadhyay Gram Jyoti Yojana**. Under this programme, cent percent of village electrification was achieved as of 28 April 2018 (Heynen et al. 2019).
- The government aims to create a **500 GW renewable energy** capacity by 2030.
- Introduction of '**Draft amendment to Electricity Act, 2003**' in September 2018 which discussed direct benefit transfer (DBT) of subsidy, the introduction of smart meters and prepaid meters, obligation of 24 × 7 power supply.
- **Capping of solar power tariff** – Capping at the rate of ₹2.50 and ₹2.68 units for developers using domestic solar cells and modules (without safeguard duties), and imported solar cells (with safeguard duties), respectively, by MNRE in 2018.
- **FAME-India: Faster Adoption and Manufacturing of (Hybrid &) Electric Vehicles** – The government launched Phase-1 of the FAME India Scheme for two years in March 2015, and the scheme was extended time to time in different phases. The scheme aims to promote the adoption of electric vehicles to reduce dependence on fossil fuels and to address concerns related to vehicular emissions (Preetha and Poornachandran 2019). The government has started providing incentives to the consumers on the purchase of electric vehicles under the FAME-India Scheme Phase-II.

9.6 Conclusion

India's heavy dependence on fossil fuels to meet its energy needs, particularly coal, leads to the generation of harmful and toxic emissions that adversely affects human and environmental health. Dramatic increase in energy demand is expected in the years ahead due to population growth, a rise in the use of per capita electricity and rapid industrialization. Climate change such as an increase in temperature and more heat strokes will lead to more demand for energy; change in precipitation and more drought lead to less water availability and adversely influence energy production; more evaporation leads to more demand for energy for drinking and agricultural purposes as desalination and pumping of water are energy-intensive technologies, more sudden and frequent events such as hurricanes, floods and storms can damage the coastal power plants, their platforms and powerlines, which ultimately affect the electricity distribution. As a result, a change from conventional energy to RE is required. The country has an immense capacity for RE, including solar, wind, hydropower and biomass.

The government has launched various programmes, strategies and financial instruments to promote renewable clean energy as discussed above. But, more diversification in energy sources at a faster pace is required. Supercritical technologies should be encouraged in thermal power plants by tax incentives, enforcement of power purchase agreements and supply of high-quality coal. According to an IPCC report, staying below 1.5° in 2100 requires zero emissions by 2030, which is moving towards decarbonization. Despite high global sensitivity towards the use of nuclear energy, it can reduce the cost of achieving zero emissions for India. The more focus should be given on energy efficiency with the view of a 24×7 power supply target. Environmental policies and acts should be integrated with energy policies in view to reduce energy intensity, to reduce CO_2 emission and to mitigate climate change.

References

Bansal, N., Srivastava, V. K., & Kheraluwala, J. (2019). Renewable energy in India: policies to reduce greenhouse gas emissions. In *Greenhouse Gas Emissions* (pp. 161–178). Springer, Singapore.

Bardhan, R., Debnath, R., & Jana, A. (2019). Evolution of sustainable energy policies in India since 1947: a review. *Wiley Interdisciplinary Reviews: Energy and Environment*, 8(5), e340.

BEE, PAT Cycle. (2020). Retrieved 29 August 2020 from https://beeindia.gov.in/content/pat-cycle.

Bhat, A. A., & Mishra, P. P. (2020). Evaluating the performance of carbon tax on green technology: evidence from India. *Environmental Science and Pollution Research*, 27(2), 2226–2237.

Burke, P. J., Widnyana, J., Anjum, Z., Aisbett, E., Resosudarmo, B., & Baldwin, K. G. (2019). Overcoming barriers to solar and wind energy adoption in two Asian giants: India and Indonesia. *Energy Policy*, 132, 1216–1228.

CarbonBrief Analysis. (2019). Analysis: India's CO_2 emissions growth poised to slow sharply in 2019. https://www.carbonbrief.org/analysis-indias-co2-emissions-growth-poised-to-slow-sharply-in-2019 (accessed 1 August 2020).

Chandra, Y. P., Singh, A., Kannojiya, V., & Kesari, J. P. (2019). Solar energy a path to India's prosperity. *Journal of The Institution of Engineers (India): Series C*, 100(3), 539–546.

Cleantech. (2020). http://www.nishithdesai.com/information/areas-of-service/industry/cleantech.html. (accessed 1 August 2020).

Digambar Singh, A., Yog Raj Sood, B., & Deepak, C. (2019). Recent techno-economic potential and development of solar energy sector in India. *IETE Technical Review*, Volume 37, Issue 3 1–12.

Heynen, A. P., Lant, P. A., Smart, S., Sridharan, S., & Greig, C. (2019). Off-grid opportunities and threats in the wake of India's electrification push. *Energy, Sustainability and Society*, 9(1), 1–10.

IEA. (2003). https://powermin.nic.in/sites/default/files/uploads/The%20Electricity%20Act_2003.pdf (accessed on 1 August 2020).

Kashyap, K., Nayim, S. T. I., Thakur, R., Kumar, A., & Thapa, S. (2020). A study of solar energy in India; Utility, status, and procurement. In *Intelligent Communication, Control and Devices* (pp. 435–443). Springer, Singapore.

Kumar, S., Bhattacharyya, B., & Gupta, V. K. (2014). Present and future energy scenario in India. *Journal of The Institution of Engineers (India): Series B*, 95(3), 247–254.

Kuriakose, K. M., Pinkymol, K. P., Koshy, D. E., & Gopi, A. (2019). Design & development of PV-wind hybrid grid connected system. In *2019 3rd International Conference on Electronics, Communication and Aerospace Technology (ICECA)* (pp. 1012–1017). IEEE.

MNRE. (2019). https://mnre.gov.in/img/documents/uploads/file_f-1580897894906.pdf (accessed on 1 August 2020).

MOSPI. (2019). *Energy Statistics 2019*. MOSPI

PIB. (2018). https://pib.gov.in/PressReleaseIframePage.aspx?PRID=1525934 (accessed 1 August 2020).

Prasanthrajan, M., & Keerthika, A. (2020). *Unit-10 Energy Systems*. Indira Gandhi National Open University, New Delhi.

Preetha, P. K., & Poornachandran, P. (2019). Electric vehicle scenario in India: roadmap, challenges and opportunities. In *2019 IEEE International Conference on Electrical, Computer and Communication Technologies (ICECCT)* (pp. 1–7). IEEE.

Saveyn, B., Paroussos, L., & Ciscar, J. C. (2012). Economic analysis of a low carbon path to 2050: a case for China, India and Japan. *Energy Economics*, 34, S451–S458.

Sharma, A., Roy, H., & Dalei, N. N. (2019). Estimation of energy intensity in Indian iron and steel sector: a panel data analysis. *Statistics*, 107 pp 107–121.

Shidore, S., & Busby, J. W. (2019). One more try: the International Solar Alliance and India's search for geopolitical influence. *Energy Strategy Reviews*, 26, 100385.

Shrimali, G. and Tirumalachetty, S. (2013). Renewable energy certificate market in India—a review. *Renewable and Sustainable Energy Reviews*, 26, 702–716.

Shukla, S. K., & Rathore, P. K. S. (2020). Production of biodiesel and its application in engines. In *Waste Biorefinery* (pp. 379–389). Elsevier.

Shukla P.R., Mittal S., Liu JY., Fujimori S., Dai H., Zhang R. (2017) India INDC assessment: emission gap between pledged target and 2 °C target. In: Fujimori S., Kainuma M., Masui T. (eds) *Post-2020 Climate Action*. Springer, Singapore. https://doi.org/10.1007/978-981-10-3869-3_7.

UNFCCC. (2015). *India's Intended Nationally Determined Contribution: Working Towards Climate Justice United Nations Framework Convention Climate Change*. UNFCCC

10

A Regime Complex and Technological Innovation in Energy System

A Brazilian Experience

Pooja Sharma

Daulat Ram College, University of Delhi, New Delhi, India

10.1 Introduction

Brazil, a Latin American, is the fifth largest country in the world, while biggest in South America (IBGE) 'The Brazilian Institute of Geography and Statistics'. It is the most prominent country in terms of international trade in South America. The institutional regime of Brazil facilitated Brazil to accomplish carbon-intensive growth. In Brazil, around 45% of primary energy demand is achieved by renewable energy (IEA 2012). 'It has laid down ambitious targets of Bio-power 19.3 GW, small – scale hydropower 7.8 GW and wind power 15.6 GW by 2021' (Renewable Energy Policy Brief BRAZIL+ 2015).

Brazil is instrumental in promoting and advocating the consumption of biofuel, and it exhibits one of the greatest proportions of renewable sources. Consequently, Brazil began substituting oil for ethanol in the 1970s. 'Biofuel has been a prior concern of Brazilian foreign policy and ethanol diplomacy from 2006 to 2010'. 'Between 1960 and 1990 the federal government worked in the arena of national and regional interest between 1960 and 1990, while states limited their scope for generation, transmission and distribution of electricity' (Roehrkasten 2015).

This phenomenal success has been an outcome of a considerable investment in Research and Development and innovation in renewable technology in Brazil. The country has provided a well-coordinated environment of institutional structure, financing system and political structure to foster the innovation and make it commercially viable. Brazil's stance in the international environmental regime, mainly the Kyoto Protocol, has evolved with time, from opposition to supporting the mechanism.

Brazil demonstrates a catalyst role in bridging the gap between developing and developed nations. Brazil assisted the European Union (EU) in opposing carbon sinks, ultimately reaching the final agreement in Bonn (2001) and Marrakech (2001). Brazil played a leading role in attaining the ratification of Protocol in 2001 (Viola 2004). Brazil has evolved in its stance concerning environment regime over time; it becomes highly significant to examine

Energy: Crises, Challenges and Solutions, First Edition. Edited by Pardeep Singh, Suruchi Singh, Gaurav Kumar, and Pooja Baweja.

the impact of the energy transition on energy regime. Its supremacy in technological innovation, research and development is well reflected in the filing of patents during that period.

In this backdrop, this chapter attempts to examine the energy regime of Brazil and the challenges it faces related to global renewable energy governance. The energy regime of Brazil evolves because of the reverse impact of Brazil's growing stance in the global environmental regime. Section 10.2 describes the changing role in global governance in Brazil, followed by the discussion on the regime complex in Brazil in Section 10.3. The role of climate change on regime complex is described in Section 10.4. Section 10.5 describes the shift in energy regime due to technological change in the energy sector followed by the conclusion in Section 10.6.

10.2 Brazil: Its Changing Role in Global Governance

Brazil has gradually moved from a nationalist form to a more liberal and globalist form of governance. Brazilian stance in the negotiation of the Kyoto Protocol has evolved to a great extent over the last decade. Brazil extended the assistance in bridging the gap between the developed and developing countries and offered an alliance with United States intending to transform the Clean Development Fund (Stevenson 2013).

Brazil facilitated and supported the EU and supported strongly in the inclusion of carbon sinks. Brazil strongly opposed all the voluntary commitments by the emerging and developing countries to any binding legal regulations. Brazil is also not in agreement with the eligibility of primary forest protection for pursuing Clean Development Mechanism (Lahsen 2009). Brazil also criticized withdrawal of United States in March 2001. Brazil also played an instrumental role in supporting the final agreement in Bonn (July 2001) and Marrakech (November 2001). It was also critical in the Protocol's ratification in 2002 (Boyd and Schipper 2002).

However, it is crucial to highlight a gradual shift in Brazilian stance during UNCED negotiations (1990–1992). Brazil played a leadership role in the Biodiversity Convention and supported Agenda 21. In the context of carbon emissions, Brazil has three significant advantages of being an intermediate country; one is that it is out of the obligatory commitments for the carbon emissions reduction, another to sustain an energy matrix with a higher weightage of hydroelectricity. However, the biggest drawback which Brazil has is a considerable amount of carbon emissions contributed by agricultural burning coupled with Amazon deforestation.

'Brazil produces about 2.5% of world emissions, contributed more by the modern economy than the traditional agriculture' (Siqueira et al. 2009). About 80% of the Brazilian population is engaged in carbon-free productive activities. In comparison, 20% is tied up to traditional agriculture, conversion in agricultural frontier and land use, inefficient timber industry all of them responsible for higher per capita carbon emission than the average of developed and developing countries (Siqueira et al. 2009).

The Multilateral Treaties being signed by Brazil, the treaties coloured in green reflect the related environmental treaties across nations. In contrast, the ones in purple indicate treaties related to intellectual property rights signed by Brazil WIPO (World Intellectual Property Organization). This reconfirms that Brazil has played a dominant role in shaping international relations through international institutional regimes.

Brazil underwent many regional treaties to administer intellectual property rights effectively, and at the same time look forward to trade and economic cooperation within the region WIPO (World Intellectual Property Organization). This further strengthened the role and bargaining power of Brazil in the international institution regime, maintaining the regime complex in favour of Brazil.

The Kyoto negotiation process from 1996 to 2003 was guided by national interest and commitment to global governance by Brazil: First, to ensure the right to development as a fundamental component of world order in compliance with Brazilian foreign policy. Second, to advocate the vision of sustainable development guided by public awareness on environment in Brazil (United Nations Conference on Environment & Development). Further to promote funding for climate mitigation related projects. Third, 'to ensure the leadership role of Brazil in the international prestige of the country during the Cardoso administration' (United Nations Conference on Environment & Development). Finally, to block international regulation on the use of forests (United Nations Conference on Environment & Development).

10.3 Brazilian Energy: A Regime Complex

The international regime is constructed based on interest by states. The state's interest reflects the prominent constituencies that would exert pressure over the state leaders (Beyers et al. 2008). The asymmetrical distribution reflects complex interdependence where the bargaining power will be determined by each other's decision on each other (Rana 2015).

Some of the components which constitute a regime system are interests, power, information and beliefs which tend to change over time. 'It is observed that interest and power vary among different states; therefore, governments form "clubs" and intend to create club goods' (Institute for International Law and Justice, New York University). These are those goods which limit the benefits to the states that do not share the interest or try to act as free riders (Keohane and Nye Jr 2001). The international regime, which is based on the foundation of interest and beliefs, tend to change depending on its membership. Eventually, the international regime is extremely vulnerable to changes and is not an outcome of one international conference but evolves through the process of coinciding interests and bargaining (Young 1997). The international regime is finally path-dependent and is shaped and framed historically. An agreement on any regulatory arrangement or an international regime would be an outcome of a continuum.

The international regime can take many forms. One kind of extreme regime would be the one which is highly integrated and imposes regulations based on hierarchical rules. The other category could be a disintegrated institution without any identifiable core, such are the arrangements of loose variety with relatively narrow regime, referred to as 'regime complex' (Keohane and Victor 2011).

There are cases of comprehensive regimes where powerful actors all similar in terms of interest and information and broad issue area pertain to a substantial amount of credibility, compliance costs also yielding a single entity of institution with no potential rival, example GATT encouraged the formulation of one integrated regime since the benefits accrued to all the stakeholders.

10.3.1 Role of Brazil and Regime Complex for Climate Change

Climate change is now a global concern involving the interest, motivation and belief of similar kind among nations, thereby bringing them together under one umbrella (King et al. 2012). As a result, the climate change issue is also governed under 'a regime complex'. This regime has to be distinguished from a comprehensive regime, fragmented or disintegrated regime. One of the climate change regimes is UNFCCC which has almost a universal membership Kyoto protocol that was held basically to enhance the comprehensiveness of UNFCCC, explicit from the fact that Kyoto placed no obligation on developing countries proved to be non-binding.

Many nations have tried to form small groups coming together to cooperate on climate change issues. A group was created by the United States to abandon 'the Kyoto treaty' named as APP (Asia Pacific Partnership). It was created to deploy new carbon technologies; a total of six countries joined this club. Soon after APP, the central 'Emitters Forum' was created in 2007 consisting of sixteen states formulating their own rules to reduce emissions. Further, 'G8 club' continued to resolve climate-related issues. Finally, the 'G20 forum' was created by finance minister replaced G8 in non-security matters.

'Brazil became the first country to sign UNFCCC and hosted the UN conference 1992'. In 1994, Brazil strongly proposed that it is the responsibility of each country to mitigate carbon emissions. Therefore, the historical account of the contribution to global warming should be considered. 'It further implies that accumulated stock of GHG emissions should be considered instead of the instantaneous change in emission levels'.

This was primarily proposed with the aim of institutionalizing the collective and historical responsibility of all nations in the UNFCCC process. However, the Kyoto Protocol failed to adopt this proposal in Brazil. This proposal was primarily endorsed by developing countries. Brazil supports the top-down policy of the Kyoto protocol and considers it as the most appropriate legal instrument globally to reduce climate change.

Brazil has adopted the strategy of quantified emission instrument for reducing emission reduction, unlike India and China (King et al. 2012). Therefore, Brazil mainly advocates that it is essential to set clear targets and commitments for funding. Moreover, it supports a direct funding mechanism (King et al. 2012). 'In case of direct fund mechanism funds are transferred directly to the National Amazon Fund, mainly to support National framework to encounter deforestation'.

'Brazil was pushed to adopt global Sustainable Development Goals (SDGs) in the Rio+20 Summit in 2012' (G20-SDG Report, OECD). 'Later, in the UNFCCC seventh conference, Brazil confronted European Union in negotiating the 2010–2020 commitments for developing countries, also received support from US (Viola 2004)'.

The developing and emerging countries favoured Kyoto as long as it did not imply binding obligations. On one side, formation of the alliance was seen in countries against Kyoto comprising Russia, Australia and United States and emerging nations, while on the other side, there existed a pro-Kyoto alliance between the European Union and Japan. Finally, in the Milan conference 2003, Russia exited from ratifying Kyoto based on business and geopolitics (UNFCCC, Kyoto Protocol).

'The Lula administration in 2003 established a great divide in Brazilian governance owing to the complexities of federative structure' (Visentini and Silva 2010). This administration advocated disengagement from the Kyoto Protocol. 'The outcome of this administration was increased disengagement from Kyoto Protocol, but the Ministry of Environment tried to revive the Brazilian involvement with the Treaty and took a stand that it would not be legally binding, at the same time it looked for alternatives to implement Clean Development Mechanism' (Eckersley 2007).

10.4 Implications of Climate Regime on Brazilian Energy Regime

> Brazil is among the topmost five nations with potential to mitigate emissions, playing an instrumental role in mitigating emission from LULUCF and also fostering emission reduction in forestry sector. (King et al. 2012)

'Besides mitigation of emission, Brazil aims at increasing energy efficiency, specifying ambitious targets' (International Emissions Trading Association 2013). 'Solar water heating and developing the markets for public housing are some of the policies implemented to attain these targets' (Centre for Development and Enterprise 2012).

With the implementation of the National Alcohol Programme in 1975, Brazil emerged as the most advanced biofuel market, promoting the blending of gasoline with ethanol produced from sugarcane. 'Brazil stimulated the market for ethanol by collaborating with international market of ethanol to enhance the business opportunities' (Koizumi 2003). Aiming to support the biodiesel industry, 'The National Biodiesel Programme was initiated in 2003, mandating that diesel be mixed with biodiesel, advocating varied proportion of the blend' (Transport Policy, Brazil).

'Brazil adopted the policies related to climate change by synchronizing the sectoral plans' (King et al. 2012). Prevention and control of fires in Cerrado were the two action plans that motivated and advocated the emission-free agriculture sector. 'Emission mitigation plans were also incorporated in other sectors in PNMC' (Climate Policy Database, Brazil).

It is observed that one can deduce that as a consequence of various treaties signed by Brazil, several laws and decrees were issued under the treaty approvals. Some of the laws enacted by the Brazilian government were strictly on environmental aspects, and some were only based on intellectual property rights. The ones in green are the environmental-related laws, while the ones in purple are laws related to property rights.

'Approximately 18% of total Brazilian energy supply is biogas, with a substantial proportion of flex-fuel as light vehicles, the outcome of blending gasoline and ethanol' (Rapid Transition Alliance 2018). 'Brazil produces paper from planted forests instead of old-growth forests and also mandates a substantial proportion of charcoal in iron and steel industry' (Asher 2019). Consequently, Brazil becomes the third country among all the countries implementing CDM projects by promoting planted forests for mitigating emissions.

'Vale and selected Brazilian companies presented a representation, insisting the government to play an instrumental role in REDD schemes' (King et al. 2012). Approximately 132 industries comprise 150 000 companies instigated and instructed government about the responsibility that lies with all the developing countries. This culminated into lobbying of business class. 'Later, with the empowerment of such lobbying of agriculture business, the controversial 2011 Forest law was passes' (BBC 2012).

10.5 A Shift in Energy Regime: Technological Innovations in Energy Sector

The emerging need to curtail emissions owing to the global concerns of climate change has led many nations across the globe to undertake innovations in clean technologies. Innovation, therefore, becomes an essential factor guiding social and economic change; ability to innovate becomes fundamental to sustain competitive advantage. The possibility of depletion of crucial natural resources has made both developing and developed nations to seek new ways to produce and consume. In this context, the substitution or adaptation of technological methods would lead a country towards the path of sustainable development. Technological innovations, specifically in the energy sector, are deemed to change the energy regime.

Large firms can deal with risky projects and activities related to Research and Development, and some investments provide economies of scope because of diversified nature. Sometimes, the market power gives the firms incentives to invest in innovation since it reduces rivalry and uncertainty associated with the innovation process (Lucchesi et al. 2014). Some of the ways to gather market power are the patents and intellectual property rights.

Generally, innovation is the activities associated with developed countries, while developing countries focus on the diffusion of the new technologies embedded in capital goods purchased from advanced countries. Multinational companies transfer their technologies to their subsidiaries; firms can import technology embodied in capital goods or inputs. In such a case, developing countries are seen as buyers or borrowers of technology from developed countries.

In the context of environmental innovations, both scarcities of natural resources and environmental regulation lead to more expensive environmental inputs, therefore changing the relative prices or shadow prices. In Brazil, the control of pollution is decentralized to states through environmental state agencies backed up by the federal legislation. Firms face two different types of legal sanctions: one is administrative fines imposed by environmental state agencies and redemption and cleans up legal sanctions imposed by judiciary. Given the complexities of Brazilian environmental legislation at the state level, other crucial determinants influence the firm's innovative activities, the firm's size and physical intensity. Apart from this, foreign-owned firms are more likely to innovate green technology.

The intellectual property rights for patents etc., emerging out of various laws and treaties being signed by Brazil, indicates that Brazil started deliberate and successful efforts towards innovation in new technologies. The graph in Figure 10.A.1 shows that the number of patents filed by Brazil has been consistently rising since 1990, reaching its peak in 2012 and declined sharply till 2014. In 2011, INPI Resolution No.263/2011 advocated the quality management system of the Trademark Directorate.

With the formation of this Regulatory authority, the number of patents started falling. The Resolution PR.No. 100/2013 in 2013 approves the version 2 of the Quality Manual of the National Institute of Industrial Property (INPI) issued in August 2012. It can easily be visualized in Figure 10.A.2 that the patents filed under lighting and heating are higher than the patents filed under electricity as percentage of total payments, and all the patents irrespective of lighting, heating or electricity fall around 2011. The graph of patents in Figure 10.A.3 shows a smaller number of patents being filed in solar energy and tidal sources. The data further indicate an evident rise in patents in 2011 owing to second phase reforms in heating industry through a network of policy framework involving institutional and legal regime, ensuring a stable funding structure.

10.6 Conclusion

Brazil, which is a Latin American country, is most prominent in South America, fifth in the world. It is the most prominent country in terms of international trade in South America. Brazil's institutional regime, policy framework and choices have made it possible to achieve successful carbon-intensive growth. This phenomenal success has been an outcome of a considerable amount of investment in Research and Development and innovation in renewable technology in Brazil. The country has provided a well-coordinated environment of institutional structure, financing system and political structure to foster the innovation and make it commercially viable.

From 1996 to 2003, Brazilian energy transition policies were focussed on national interest and commitment to global governance. Some of the specific objectives of the Brazilian energy transition are first, to ensure the right to development as basic right for all developing countries as a significant determinant of balancing world order. This became the fundamental of foreign policy advocated by Brazil. Another is to promote a world vision of sustainable development guided by the motive of public awareness on environment in Brazil. Finally, to encourage the flow of funding from the developed countries as a part of their responsibility for the global issue and concern of climate mitigation and all the projects related to research and implementation and deployment of clean energy technologies in all the developing nations.

Brazil promotes and supports the top-down policy framework of the Kyoto protocol as the most effective legal instrument since the bottom-up approach is not sufficient enough for meeting sustainable development goals. As a consequence of various treaties signed by Brazil, several laws and decrees were issued under the treaty approvals. Some of the laws enacted by the Brazilian government were strictly on environmental aspects, and some were only based on intellectual property rights. It is directly revealed that in case of Brazil, there is considerable impact of the International Regime on Brazilian regime.

Given the complexities of Brazilian environmental legislation at the state level, other crucial determinants influence the innovative activities of firms, the firm's size and physical intensity. Apart from this, foreign-owned firms are more likely to innovate green technology. It can easily be visualized that the patents filed under lighting and heating are higher than the patents filed under electricity as a percentage of total payments, and all the patents irrespective of lighting, heating or electricity fall around 2011. Through the various regulations and laws in compliance with the global energy scenario, Brazil has undoubtedly achieved a remarkable success in technological breakthrough transforming energy regime and final move on the path of sustainable growth.

References

Asher, C. (2019). Brazil soy trade linked to widespread deforestation, carbon emissions.

BBC (2012). https://www.bbc.com/news/world-latin-america-17851237

Beyers, J., Eising, R., & Maloney, W. (2008). Researching interest group politics in Europe and elsewhere: much we study, little we know?. *West European Politics*, 31(6), 1103–1128.

Boyd, E., & Schipper, E. L. (2002). The Marrakech Accord—at the crossroad to ratification: seventh conference of the Parties to the United Nations Framework Convention on Climate Change. *The Journal of Environment & Development*, 11(2), 184–190.

Centre for Development and Enterprise (2012). https://www.cde.org.za/our-work/our-publications/page/12/

Eckersley, R. (2007). Ambushed: the Kyoto Protocol, the Bush administration's climate policy and the erosion of legitimacy. *International Politics*, 44(2–3), 306–324.

IEA (2012). https://www.iea.org/reports/world-energy-outlook-2012

International Emissions Trading Association (2013). https://www.ieta.org/Reports

Keohane, R., and Nye Jr, J.S. (2001). Between centralization and fragmentation: the club model of multilateral cooperation and problems of democratic legitimacy. Available at SSRN 262175.

Keohane, R. O., & Victor, D. G. (2011). The regime complex for climate change. *Perspectives on Politics*, 7–23.

King, D., Cole, M., Tyldesley, S., & Hogarth, R. (2012). The Response of China, *India and Brazil to Climate Change: A Perspective for South Africa*. Centre for Development and Enterprise (CDE).

Koizumi, T. (2003). The Brazilian ethanol programme: impacts on world ethanol and sugar markets. *Agriculture*, 113.

Lahsen, M. (2009). A science–policy interface in the global south: the politics of carbon sinks and science in Brazil. *Climatic Change*, 97(3–4), 339.

Lucchesi, A., Cole, M.A., Elliot, R.J., and Menezesfilho, N.A. (2014). Determinants of environmental innovation in Brazilian manufacturing industries. *Proceedings of the 42th Encontro Nacional de Economia da ANPEC*, pp. 9–12.

Rana, W. (2015). Theory of complex interdependence: a comparative analysis of realist and neoliberal thoughts. *International Journal of Business and Social Science*, 6(2) 290.

Rapid Transition Alliance 2018. https://www.rapidtransition.org/events/

Renewable Energy Policy, Brazil (2015). https://www.irena.org/-/media/Files/IRENA/Agency/Publication/2015/IRENA_RE_Latin_America_Policies/IRENA_RE_Latin_America_Policies_2015_Country_Brazil.pdf

Roehrkasten, S. (2015). Global governance on renewable energy. In *Global Governance on Renewable Energy* (pp. 73–116). Springer VS, Wiesbaden.

Viola, E. (2004). Brazil in the context of global governance politics and climate change, 1989-2003. *Ambiente and Sociedade*, 7(1), 27–46.

Visentini, P. G. F., & Silva, A. L. R. D. (2010). Brazil and the economic, political, and environmental multilateralism: the Lula years (2003–2010). *Revista brasileira de política internacional*, 53, 54–72.

Young, J. T. (1997). *Economics as a Moral Science*. Books.

Websites

https://www.climatechangenews.com/2019/10/11/brazil-fights-attempt-cancel-old-carbon-credits
www.iilj.org
https://unfccc.int/kyoto_protocol
https://www.edf.org/sites/default/files/EDF_IETA_Brazil_Case_Study
https://issuu.com/cdesouthafrica/docs/the_response_of_china_india_and_bra
https://www.transportpolicy.net/standard
http://www.climatepolicydatabase.org/index.php/Brazil_CDlinks_project
https://www.bbc.com/news/world-latin-america-17851237

Appendix A

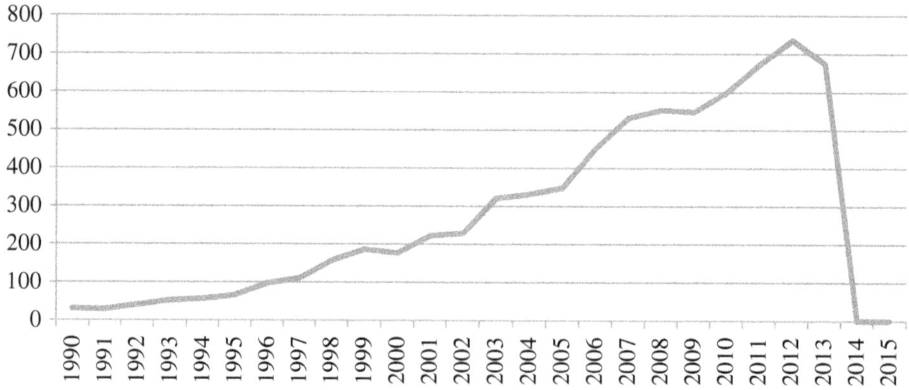

Figure 10.A.1 Number of patents in Brazil. *Source:* Author's compilation, OECD.stat.

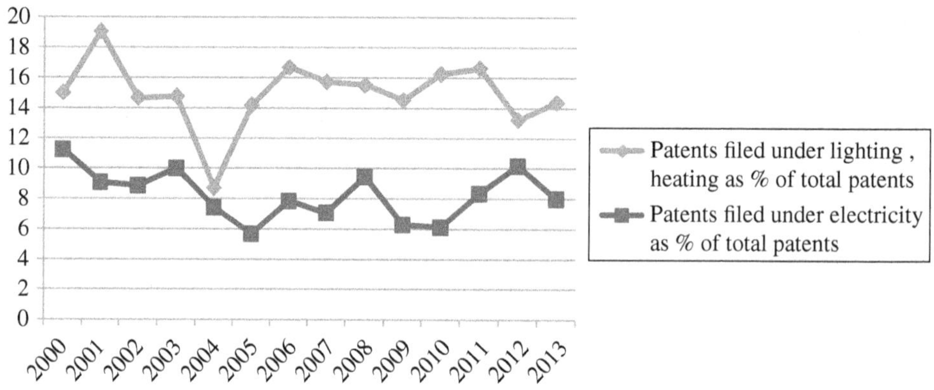

Figure 10.A.2 Patents filed under various energy uses. *Source:* Author's compilation, OECD.

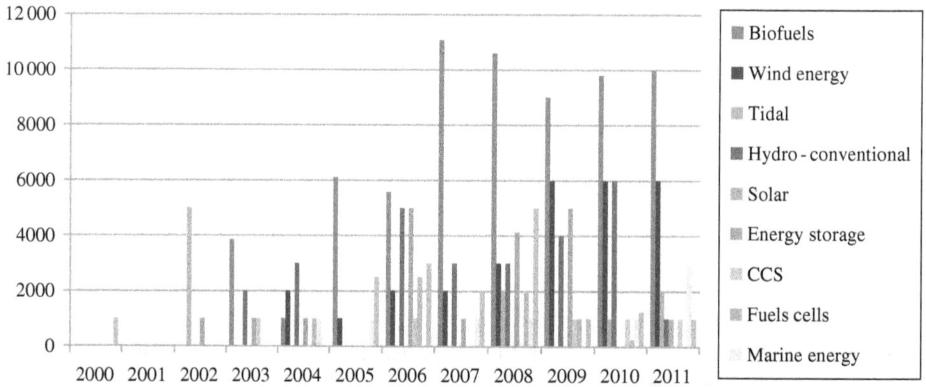

Figure 10.A.3 Patents in various renewable energy sources in Brazil. *Source:* Author's compilation, OECD.

11

Opportunities in the Living Lights

Special Reference to Bioluminescent Fungi

Pramod Kumar Mahish[1], *Nagendra Kumar Chandrawanshi*[2], *Shriram Kunjam*[3], *and S.K. Jadhav*[2]

[1]*Department of Biotechnology, Govt. Digvijay Autonomous P.G. College Rajnandgaon (C.G.), Rajnandgaon, Chhattisgarh, India*
[2]*School of Studies in Biotechnology, Pt. Ravishankar Shukla University Raipur (C.G.), Raipur, Chhattisgarh, India*
[3]*Department of Botany, Govt. V. Y. T. Autonomous P.G. College Durg (C.G.), Durg, Chhattisgarh, India*

11.1 Introduction

Life on earth is so diverse. Bioluminescent organisms are present ubiquitously in the world. It comprises anglerfish in the deepness, and the outside flashing dinoflagellates can be seen. The diversity of bioluminescent organisms is so interesting because many mushrooms are self glowing, but plants cannot; similarly in animals, various fishes and insects are luminescent but frog, spiders, birds and mammals cannot (Wilson and Hastings 2013). Bioluminescence is a beautiful natural occurrence that stimulates the imagination of people and increases the curiosity of the scientific world to explore it more. Bioluminescence provides a certain advantage to the living organisms for their survival in an ecological context, like visual communication, attack prey, etc. (Fleiss and Sarkisyan 2019). Bioluminescent organisms are mainly found in the ocean. These organisms are dispersed in 16 major phyla and nearly about 700 genera (Waldenmaier et al. 2012).

Bioluminescence is a light produced by living organisms due to chemical reactions. It is a type of chemiluminescence. Bioluminescence is sometimes referred to as phosphorescence, but it is erroneous. In phosphorescence, light production keeps it up for a restricted stage of occasion following the elimination of the excitation cause (Widder 1999). Bioluminescence is a cold light that generates lesser thermal radiation (Pandey and Sharon 2017). Bioluminescence requires energy, produced from a reaction system known as the bioluminescence system. These systems include molecules that act as enzymes, substrates, energy storage, etc. The most important is the small luciferins that act as substrates which are identified from various luminescent organisms. Luciferins emit lights upon oxidation. The chemical process is catalyzed by luciferase (Kaskova et al. 2016).

The advantage of luminescence is that it is a visible indicator of the reaction system and can also be numerically converted using biophysical devices like Luminometer and other advanced

Energy: Crises, Challenges and Solutions, First Edition. Edited by Pardeep Singh, Suruchi Singh, Gaurav Kumar, and Pooja Baweja.

light-sensitive devices. Therefore, recently, it is being utilized for various applications in biomedical (Sharifian et al. 2018), cancer study (Alsawaftah et al. 2020), food safety (Doh et al. 2019), environmental monitoring (Wan et al. 2014), viral study (Perdiguero et al. 2019), bioassay and therapy (Yan et al. 2019).

Fungi are very unique among the organisms because of their cell wall characteristics. Fungi are saprophytes, plant pathogens, human pathogen and decomposers. Various fungi have been already proven their utilization in bio-fertilizer, production of alcohols, antibiotics, acids, enzymes and bioactive secondary metabolites of different importance (Kour et al. 2019; Devi et al. 2020). However, bioluminescence in fungi is not so far investigated. Only about 81 species of bioluminescent fungi have been recognized, and little is known about their applications too. Therefore, the present chapter aims to explore the bioluminescent fungi, their diversity and opportunities in the different areas.

11.2 History of Bioluminescence

The characteristics of the bioluminescent organism were earlier reported by Romans and Greeks. The first written record of the bioluminescent organism was presented by Aristotle (384–322 BCE). He reported that the luminescence of organisms was not acquired by other sources. Aristotle explored glowworms, fireflies, dead fish and flesh. In this ancient time, the first detailed and specific study of the bioluminescent organism was done by Pliny the Elder (23–79 CE). The Natural History (Latin: *Naturalis Historia*) was written by him that contain details about fireflies, glowworms, luminous mollusk (*Pholas dactylus*), purple jellyfish (*Pelagia noctiluca*), bioluminescent birds, etc. (Lee 2008).

A German monk Albertus Magnus (1206–1280 CE) compiled some luminescence living organisms that were published in the book 'De Animalibus' (1478) after the use of printing miracle. Luminescence in the Red Sea was seen by Don Joao de Castro (1500–1548) a Portuguese naval officer, and the tropical fireflies were also observed during these days by Sir Francis Drake (1540–1596). A book written by Conrad Gesner (1516–1565) was fully focused on the luminous animals, plants and luminous stones. The book was published at Tiguri (Zurich) in 1555 referred to as *De Lunariis* (Harvey 1957).

The seventeenth century was based on scientific knowledge. Francis Bacon (1561–1626) is credited for building up the logical technique. The luminance of glowworm was mentioned by Shakespeare (1564–1616) in Hamlet. *De Luce Animalium* (1647) was a book written by Bartholin (1616–1680) who was a Danish general practitioner. He presented the phenomenon of luminescence and its difference between friction and reflection. Robert Boyle (1627–1691) was another personality who had contributed through his scientific experiments. His experiment was known as Boyle's law which was based on the work in the air. He observed the effect of air in luminescence too. The effect of applying and re-applying air in various objects like flam, coal, iron and live mouse was an interesting experiment done by Robert Boyle (Roda 2010). The Eighteen century was full of experiments with heat, light and electricity. Francis Hauksbee was the first who studied electroluminescence in the laboratory. Several new bioluminescent organisms were discovered these days which includes Meriam's (1705) observation of light from the lanternfly, Fulgora. French mathematician, Jacques Ozanam (1640–1717) has written a book that was published in several volumes that mentioned the natural and artificial phosphorus about luminescence. Volume 4 of the Paris edition was published after his death (1735) contained 240 pages related to natural phosphorus and bioluminescence by several living organisms like fireflies, fishes and glowworms (Harvey 1957; Lee 2008).

HMS Challenger was the breakthrough of bioluminescence history in the nineteenth century. HMS Challenger was a steam-based Royal Navy corvette, and the challenger expenditure was a grand tour (1872–1876) of the world organized by the Royal Society. The tour makes it possible to explore bioluminescent organisms of the ocean. Raphael Dubois was a French physiologist (1849–1929) who demonstrates luciferin and luciferase, as luminous molecules in Pyrophorus and Pholas. He concluded that the production of light by plants and animals was a biological process that is controlled by enzymes. E. Newton Harvey (1887–1959) was a leading worker in the field of bioluminescence in the twentieth century. He was an American zoologist. He devoted himself to work on bioluminescent Ostracod *Vargula hilgendorfii*. He discovered that the luciferins and luciferases are species-specific and not interchangeable (Johnson 1967).

11.3 Bioluminescence in Terrestrial Organisms

Most luminescent organisms are found in the deep-sea where bioluminescence is the main cause of illumination. Luminescence is almost deficient in freshwater. On land, fireflies are the mainly noticeable examples, microorganisms like bacteria; insects like flies and springtails. Mushrooms, earthworms, snail, centipedes, millipedes and beetles are some other terrestrial bioluminescent organisms. The constant ecological situation with an extensive continual evolutionary account favors the origin of luminescence in deep-sea. Optically obvious contrast, stable darkness and communications among huge living organisms are also very important factors responsible for the existence of bioluminescence in the ocean (Haddock et al. 2010).

There are few reasons suggested by most investigators that why bioluminescence is not common in freshwater; freshwater habitat is a short period, the lesser benefit of bioluminescence as the availability of sunlight, lesser diversity and interaction among organisms, lesser requirement of adaptation. The use of luminescence by terrestrial organisms is mainly for attracting prey, intraspecies communication and escape from a predator (Pandey and Sharon 2017). The terrestrial organism emits green bioluminescence color comparing blue from marine habitats. The reason behind this is the photic environment due to sunlight. The terrestrial bioluminescent organism includes bacteria, fungi, mollusks, annelids and arthropods (Viviani 2007). A bioluminescent *Xenorhabdus luminescence* bacterium was isolated from a human leg wound clinical specimen. This discovery further leads to find some other bioluminescent bacteria such as *X. nematophilus* which is also a pathogen for some insects like *Galleria*. The bioluminescent bacteria discovered were very poor luminescence (Farmer III 1989). Terrestrial mollusks also emit greenish luminescence. *Dakya striata* (snail) was discovered from grasses in Malaysia, and limpet *Latia neritoides* were found from freshwater of North Island New Zealand. *Latia neritoides* are the only fluorescence producing organism found in freshwater (Rochow and Moore 1988; Viviani 2007).

Clitellata is the only class of earthworm that emits luminescence in the terrestrial environment. There are five families in *Clitellata* class. *Fridericia* and *Michaelseniella* are the genera of *the Enchytraeidae* family that are found in organic-rich soil (Verdes and Gruber 2017). The members of *the Lumbricidae* family are related to the biogeochemical cycle (Edward and Bohlen 1995). *Ancanthodrilidae, Octochaetidae* and *Megascolecidae* are other families that include around 30 luminescent species (Rota et al. 2003). *Ancanthodrilidae, Octochaetidae* and *Megascolecidae* families are the largest luminescent earthworm in *Clitellates.* Millipedes are also terrestrial bioluminescent organisms. Bioluminescence in Millipedes provides an ecological role against predators by emitting visual signals as they are blind. Their distribution is restricted in California (Marek et al. 2011). The millipedes emit greenish-blue light that belongs to the genus of motyxia (Hastings and Davenport 1957). *Xystocheir bistipita* luminescent Millipedes rediscovered from California confirmed its evolutionary sister group motyxia (Marek and Moore 2015). Lights from motyxia evolve from the exoskeleton by a photoprotein as its functional group (Shimoura 1984).

Many terrestrial arthropods are also luminescent that includes a wide range of insect species (Herring 1987). Coleoptera beetles are among well known. Elateroidea is the superfamily that includes click beetles (*Elateridae*), fireflies (*Lampyridae*), and railroad worms (*phengodiae*) (Longkumer and Kumar 2018). Click beetles emit green light luminescence, but during flying, it emits orange light. *Pyrophorus Augustus* was recently observed from the Amazon forest in Brazil that can produce orange luminescence (Amaral et al. 2016). Railroad worms are widely distributed in nature. Railroads of South-American species emit yellow-green light. Brazilian *Pyrearinuster mililluminans* demonstrate the occurrence of glowing termite mounds (Bechara 1988; Bechara and Stevani 2018). Fireflies are the well known and most studied luminescent organism. Fireflies emit green-yellow luminescence (Viviani 2001). Diptera is an insect widely distributed in nature. In the Australian and New Zealand region, *Arachnocampa* species were identified which showed luminescence characteristics. *Arachnocampa* produces blue-green luminescence to prey flying insects (Meyer-Rochow 1990). Sinopyrophorinae is a new subfamily of *Elateridae* Arthropods. Luminescence click beetles of *Sinopyrophorinae* were first recorded from China (Bi et al. 2019).

11.4 Bioluminescence Molecules

Forty bioluminescence systems have been reported on the earth in which nine luciferins and seven luciferase gene families were identified (Haddock et al. 2010; Oba et al. 2017). Bioluminescence has resulted from the chemical process which produces energy. Luciferin is a molecule that emits light after oxidation. This oxidation reaction is catalyzed by the luciferase enzyme that also controls the reaction rate of luciferin. The emitted light is dependent on ATP and NADH. The emitted light may differ in color, localized in the cell, catalyze rate, etc. (Kaskova et al. 2016). There are various bioluminescence reaction systems in the sea and terrestrial organisms. But, there are only a few examples of the light-emitting process where luciferin and luciferase both are recognized. So, many light systems are yet to discover; therefore, these are the focused area of present-day researchers.

Long-chain myristic aldehyde, oxygen, flavin mononucleotide (FMN) and nicotinamide adenine dinucleotide (NADH) are the key molecules involved in the bacterial bioluminescence system. The bacterial bioluminescence emitted blue color of 490 nm which is encoded by the *lux* operon gene. Oxidized aldehyde is known as luciferin, and FMN plays a role in the light source (Meighen 1991). Click beetles, fireflies and railroad worms utilize similar luminescence reactions which are D-luciferin based (Wood 1995). Orange, yellow and sometimes red color light is emitted by the D-luciferin (Fallon et al. 2018). Benzothiazoylthiazole is the luciferin of fireflies. Condensation of luciferin takes place by the luciferase in the existence of ATP and Mg^{2+}. Dioxetanone is the intermediate in this reaction system, and its breakdown releases energy to excite oxyluciferin. Oxidized luciferin thus produces photons (McElroy and DeLuca 1985). The chemical structure of D-luciferin is presented in Figure 11.1a.

The mechanism of bioluminescence in marine organisms is based on the coelenterazine luciferin-dependent system. Imidazolopyrazine a type of coelenterazine is also found in marine organisms that do not show luminous properties (Shimomura 1987; Thomson et al. 1997). The organisms which are based on the coelenterazine luciferin substrate emit blue light of 450–500 nm where luciferases are generally independently evolved (Fleiss and Sarkisyan 2019). Mnemiopsis, jellyfish Aequorea, sea pansy Renilla and hydroid Obelia are some marine coelenterates and ctenophores which emit fluorescence based on a coelenterazine luciferin-dependent system. Coelenterazine luciferin is shown in Figure 11.1b.

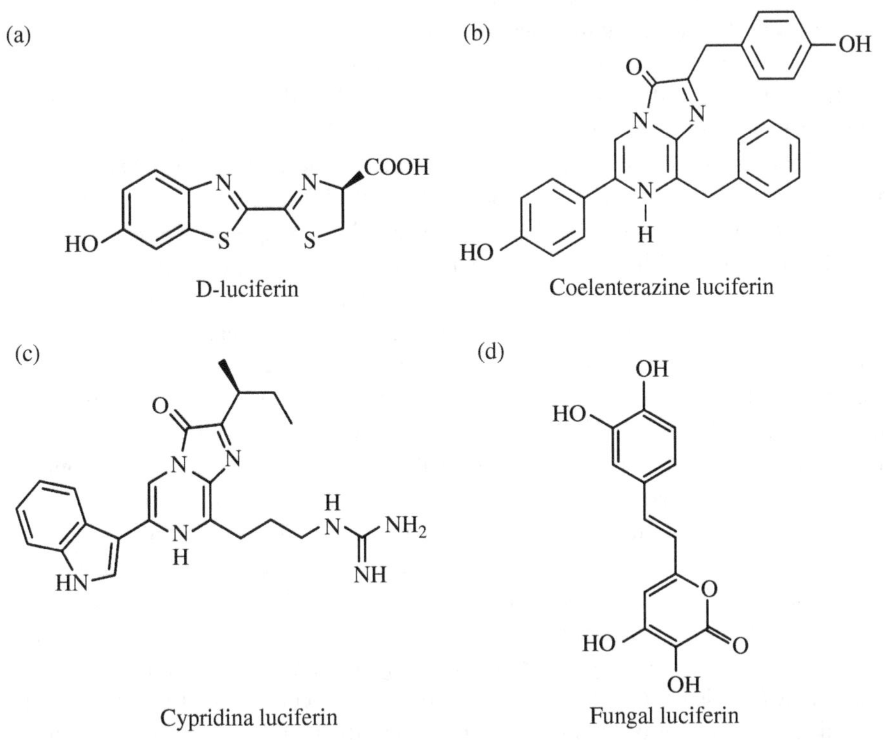

Figure 11.1 Molecular structure of luciferins found in the various bioluminescent organisms.

Another bioluminescence reaction system is cypridina luciferin-based system. This system is mainly found in midshipman fish and cypridinid ostracods. *Pempheris* and *Parapriacanthus* like fishes also follow the cypridina luciferin-based fluorescence system (Haddock et al. 2010). The synthesis of ostracods luciferin and its detailed steps are still unclear, but the study based on amino acid isotopes suggests that the luciferin may synthesize from various amino acids like tryptophan, isoleucine and arginine (Kato et al. 2007). The organization of Cypridina luciferin is presented in Figure 11.1c.

11.5 Bioluminescent Fungi

11.5.1 Diversity

Hawksworth estimated that there are 1.5 million fungal counts in nature (Hawksworth 2001, 1991). Till the Hawksworth record, molecular methods were not frequently in use. Now the recent study expects 3.5–5.1 million fungal species (O'Brien et al. 2005; Blackwell 2011) which are based on the most advanced tools frequently in use nowadays. Of the known fungi, bioluminescent species belong to four lineages. Oliveira et al. (2012) suggested that these four lineages (Omphalotus, Armillaria, Mycenoid, and Lucentipes) show a similar evolutionary history. But, the most recent evolutionary model proposed by Ke et al. (2020) suggests an alteration in the luciferase cluster cross the bioluminescent fungi. Genome sequencing of the *Mycena* species uncovers that the luciferase cluster consequent the phenomenon of duplication and translocation.

There are about 81 known bioluminescent fungi worldwide (Chew et al. 2015). These fungi are saprophytic, mushrooms and sometimes plant pathogens, which are mostly found in the temperate and tropical climate where growth and reproduction are favoured by the relatively high temperature (Deheyn and Latz (2007). Oliveira et al. (2012) confirm that Omphalotus lineage contains 12 species, Armillaria lineage contains 10 species and the mycenoid lineage contains more than 50 species. Oliveira et al. (2012) also suggested a newly identified lineage Lucentipes. This lineage contains two species. These are distributed worldwide, Mycenoid have been recorded from Asia, Europe, the Americas, Africa, Caribbean, Australia, and the Pacific Islands, Armillaria are natives of South Asia, Omphalotus were obtained from Asia, Europe, the Americas, Caribbean, and Australia and Lucentipes which includes *Gerronema viridilucens* and *Mycena lucentipes* were investigated from Brazil (Oliveira et al. 2012). Large luminous mushrooms belong to the Omphalotus lineage. The fungi of Omphalotus lineage belong to the genera of *Omphalotus* and *Neonothopanus.* These fungi have been identified by morphological, chemotaxonomical and advanced molecular methods based on ribosomal DNA sequencing (Kirchmair et al. 2002, 2004). Omphalotus lineage includes *Omphalotus olivascens* known as western Jack-o-Lantern mushroom of western North America; *O. japonicas* the Moon Night mushroom or Tsukiyotake of Japan; *O. olearius* and *O. illudens* Jack-o-Lantern mushrooms of Europe and eastern North America and *O. nidiformis* the Ghost Fungus of Australia. The Saprophytic, occasionally pathogen, plant root fungi belong to the Armillaria lineage. These are edible and also known as Honey Mushroom. Guyot (1927) preliminary observed the bioluminescence property of honey mushrooms in their rhizomorphs and mycelium. Armillaria is widely distributed in the forest of California. Armillaria is a wood decomposer and plays a significant function in nutrient recycling (Baumgartner and Rizzo 2001). Few species have been seen in Australia and New Zealand (Coetzee et al. 2001). Morphological and molecular methods were applied to identify these fungi (Anderson and Stasovski 1992; Pegler 2000). Few species of Armillaria are unique living organisms of our earth. *Armillaria bulbosa* present name *Armillaria gallica* accounts for 1500 years old, 15 ha occupied area and 10 000 kg in weight (Smith et al. 1992).

Mycenoid lineages fungi are saprotrophic and white-rot decomposers. They are described in the genus of Mycena. Mycenoid is derived from the Mycena due to its dominance in the lineage. These fungi are gills forming and having the spore-bearing plane. Mycena fungi are having huge diversity with 60 sections and about 500 species. Out of which, 35 species of 17 sections belong to the bioluminescent fungi (Desjardin et al. 2008). Most recently, six new species of bioluminescent fungi were isolated from the forest of Mexico (Cortes-Perez et al. 2019). The luminescent organ of Mycena fungi is much complex. In *M. lamprospora* and *M. pruinoso-viscida var. raboulensis,* spores are observed to fluorescence, while in *M. irritans, M. lamprospora* and *M. sublucens,* basidiome is luminescent (Horak 1978). Pilei, lamellae and stipes are also luminescent in some fungi (Desjardin et al. 2007). *Mycena lucentipes* and *Gerronema viridilucens* belong to a new lineage known as Lucentipes (Oliveira et al. 2012). Six bioluminescent species including *Mycena lucentipes* and *Gerronema viridilucens* were obtained from the Atlantic green area of Brazil (Desjardin et al. 2007).

11.5.2 Mechanism of Bioluminescence in Fungi

In the fungal bioluminescence system, following are the important molecules that accompany the reaction system. Luciferin serves as a substrate that gives rise to chemical change due to enzyme action. Luciferase acts as a biocatalyst that controls and accelerates the reaction system along with the presence of the oxygen molecule. Finally, energy pack photon is released from the reaction system, and the Adenosine triphosphate (ATP) role is storing the energy (Perry 2007). Luciferases direct the chemical reaction in which substrate luciferin combines with oxygen. The reaction of luciferases and luciferin increases the level of energy of luminescence material. Thereafter, photon emissions that take place resulted in the form of cold light (Ilondu and Okiti 2016). Airth and Foerster (1962) proposed equation is given in Figure 11.2, in which reductase acts as a soluble enzyme and luciferase as an insoluble enzyme. Airth and Foerster's work was successfully verified employing hot and cold sources of luminescent fungi. It is observed that fungal bioluminescence is certainly based on NAD(P)H-dependent reductase and a luciferase (Oliveira and Stevani 2009).

Step ahead, Purtov et al. (2015) successfully identified the precursor hispidin, its *cis* isomer and true fungal luciferin. The true fungal luciferin holds one extra oxygen in contrast to the hispidin; therefore, it is known as 3-hydroxyhispidin. Further, Purtov and co-workers suggested two-step fungal bioluminescence mechanisms. In the first step, luciferin precursor and NADPH were fully utilized as substrate and give rise to true luciferin. And in the second step, true luciferin produces light in the presence of Luciferase the second enzyme. The structure of true fungal luciferin is presented in Figure 11.1d. Fungi utilize an α-pyrone 3-hydroxyhispidin which is oxidized by an unsolvable luciferase in a chemical process that simply needed oxygen and gives rise to green light fluorescence (~520 nm). Recently, Kaskova et al. (2017) described the structure of fungal oxyluciferin in much detailed and

$$L + NAD(P)H + H^{-} \xrightarrow{\text{Reductase}} LH_2 + NAD(P)^{-}$$

$$LH_2 + O_2 \xrightarrow{\text{Luciferase}} LO + H_2O + hv$$

where
L = Luciferin;
LH_2 = Reduced luciferin and
LO = Oxyluciferin

Figure 11.2 Two-step enzyme-dependent fungal bioluminescence mechanisms proposed by Airth and Foerster (1962).

multicolor enzymatic luminescence of a-pyrones luciferins. They also proposed a high-energy endoperoxide intermediate between luciferin the oxyluciferin.

Circadian rhythm plays an important role in bioluminescence which is observed in various luminescent organisms (Hastings 2013; Goya et al. 2016). A current investigation uncovers the existence of circadian rhythm in fungi too. Oliveira et al. (2015) investigated the circadian rhythm in *Neonothopanus gardneri* which is a bioluminescent mushroom found in Amazonas Forest to the dry Caatinga Forest, Brazil. During the study, it is observed that the repeating light intensity oscillations occur at 25 °C around 22 hours when the mushrooms were kept at various temperatures with a 12/12 hour light/dark cycle. From regulatory circadian rhythm, it is now understood that the fungal bioluminescence system prevents loss of energy during the day and maximum light intensity during the night. It has also a great advantage to collect the active and rich quantity of luminescence dependable biomolecules from organisms.

11.5.3 Significance

Apart from fluorescence, the bioluminescence system also plays some important significance in fungi. However, some studies also suggest that biochemical reaction releases bioluminescence as a byproduct; therefore, it has no biological role (Lingle 1993). But many previous investigators also find that there are various ecological roles of bioluminescence in fungi. It is observed that bioluminescence works as predators of fungivores. Dispersal of fungal spores is an important ecological phenomenon. It is found that bioluminescence plays a major position to attract insects so that the spore may spread to a large region. The presence of bioluminescence in the spore area and gills also support this presumption (Bermudes et al. 1992; Desjardin et al. 2008). The fruiting body of mushrooms has been used to lightening in the forest for the finding of the pathway during the Vietnam War (Dao 2006).

11.6 Opportunities in Fungal Bioluminescence

11.6.1 Glowing Tree

Future lighting is the major scope of bioluminescence. The bioluminescence pathway of fungi is now fully understood, and the fungal luciferins are similar in temperament to the plant biology; hence, it is much possible to synthesis luciferins by the plant. The technology may develop to replace traditional street lighting, road signs with the glowing tree. Similar to the fungal fluorescence system, the lightning is possible at night only and it may off in a day or invisible due to the sunlight (Kushwaha and Hajirnis 2016). Living light replacing traditional light pole is a dream of urban society. Ongoing, many recent scientific and in-depth investigations creating hope to come to this dream may true. Fungal bioluminescence enzyme system based on luciferin metabolic pathway is a simple two-step process. Recently, this fungal bioluminescence system was genetically encoded and successfully introduced into the yeast cell (Kotlobay et al. 2018). The resulted new stain was found autonomously bioluminescent. It was a great finding because a eukaryotic bioluminescence system was effectively expressed in the other eukaryotic organism; on the other hand, bacteria and another prokaryotic bioluminescence system have poor significance in higher organisms. *Neonothopanus nambi* luciferase gene, two *E. coli* 4-hydroxyphenylacetate 3-monooxygenase components and *Rhodobacter capsulatus* tyrosine ammonia-lyase were introduced into the *P. pastoris* yeast which carries npgA, hisps, h3h and luz genes. This novel work gives rise to a new eukaryotic yeast organism that was self luminescent when developed in a suitable growth condition.

Mitiouchkina et al. (2019) have developed an engineered self glowing and self-sustained tobacco plant with the use of expressing enzymes of *Neonothopanus nambi* fungus. This enzyme is responsible for the caffeic acid cycle. Various enzyme systems accompany the caffeic acid cycle results in the formation of light. The lighting was observed during all developmental stages of the tobacco plant. This development requires no external subtract and is visible to the naked eye. This most recent achievement can play a milestone in the further development of a self glowing plant because the seeds of the transgenic tobacco are also able to fluoresce upon germination. Roots and the tips of the cotyledons have been seen as a source of light. The other advantages were noticed that the imported genes related to the caffeic acid cycle do not affect the plant growth and the phenotype of the plant remains the same as wild-type.

11.6.2 Bioassay of Toxicity

A rapid, cost-effective and advanced than previous, bioassay technology is always a subject of investigation in the scientific community. *Mycena citricolor, Gerronema viridilucens* and *Armillaria mellea* were investigated for the heavy metal toxicity test (Horsewell et al. 2005; Mendes and Stevani 2010). A very low concentration of mercury (Hg) was successfully detected from the use of luminescent mushrooms (Wan et al. 2014). Noxious properties of 3,5-dichlorophenol (3,5-DCP), pentachlorophenol (PCP), copper and zinc were analyzed using bioluminescence-based bioassay. Bioluminescent *Armillaria mellea* and *Mycena citricolor* fungi were used for the toxicity testing, and bioassay takes only 60 minutes exposure period to give results (Weitz et al. 2002). This bioassay-based technique has a lot of advantages over the chemical-based environmental analysis process. The chemical process needs extraction steps that may affect the target receptor, along with human risk. But, the bioassay-based sensor technology overcomes these problems as well as to detect some pollutant which could not be determined chemically (Weitz 2007).

11.6.3 In-Vivo Imaging

Bioluminescence imaging is one of the very important tools looking for future biological investigation. Bioluminescence imaging depends on the uncovering of visible light emitted from a chemical reaction (Papon et al. 2014). Biological monitoring with bioluminescence imaging is already made in the field of immunology, cancer biology, bacteriology, neurobiology, virology, etc. (Badr and Tannous 2011). Bioluminescence imaging senses luminescence from cells and tissues of living organisms (Sato et al. 2004).

This tool has a great opportunity to find pathogenic infections in animal in vivo (Avci et al. 2018). Without harming of animals, monitoring at the same time, progression observation, spatial and temporal resolutions are some advantages of bioluminescence imaging (Hutchens and Luker 2007). The fungal infection in humans has been raised in the last few decades, but its diagnosis is largely based on conventional technology like plate culture method, colony-forming units, historical investigations, etc. Bioluminescence imaging is a noninvasive imaging technique used for the monitoring of fungal infection in vivo (Dyck et al. 2020). On-time monitoring of fungal infection could be achieved with the detection of the fungal photon by a sensitive detector. Monitoring of *Candida glabrata* and *Aspergillus terreus* was successfully done with the help of this technique (Slesiona et al. 2012; Persyn et al. 2019). The virulence of pathogenic *Candida albicans* was investigated with the aid of bioluminescence imaging in vivo (Doyle et al. 2006). Brock et al. (2008) used *Aspergillus fumigatus* genes for the construction of bioluminescence imagining for studying the progression of Aspergillosis. The Mechanism of bioluminescence imaging is presented in Figure 11.3.

The study also finds successful observation in the efficiency of drugs against Aspergillosis using the bioluminescence imagining tool. The bioluminescence imaging is also used for the study of *Cryptococcus neoformans* in the mouse model. It causes serious infections in the central nervous system and brain. Spatial and temporal study of infection was studied, and infection pack was observed in respective organs (Vanherp et al. 2019). There are so many advanced tools for the monitoring of our organs. The lung is mainly diagnosed by the three-dimensional anatomical information from micro-computed tomography (μCT), and the

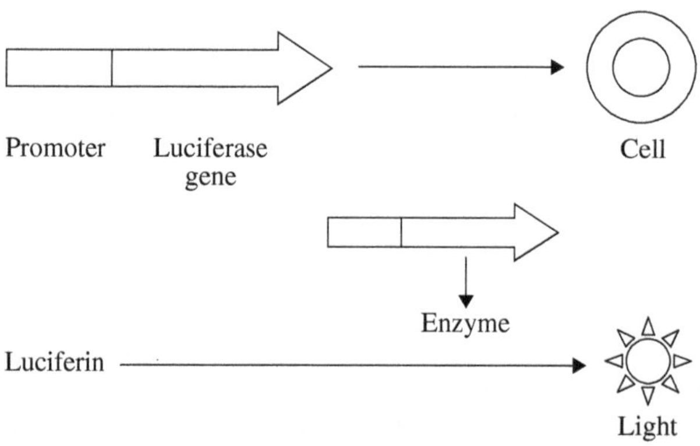

Figure 11.3 Mechanism of bioluminescence imaging. Transfer of promoter and Luciferase gene in the choice of the cell. When luciferin is added, the gene synthesizes the Luciferase enzyme. It leads to the utilization of substrate and the production of photons in the presence of oxygen.

brain is monitored by magnetic resonance imaging (MRI), but nowadays, bioluminescence imaging has been established as a complement in the field. These findings aim to generate better preventive and therapeutic measures for the fungal infection. Apart from *Candida albicans, Aspergillus fumigatus*, and *Cryptococcus neoformans,* there is a huge count of human pathogens. Therefore, there are great opportunities to develop bioluminescence imaging for the other pathogen too.

11.6.4 Animal Model Study

Animal models play an important role to investigate fungal pathogen development and antifungal target identification in vivo. These studies traditionally need a fungal biofilm which may cause various difficulties related to the resistance in antifungal drugs. Bioluminescence imaging-based biofilms are now a hope to overcome this limitation. Bioluminescence imaging-based *Candida albicans* biofilm was successfully implanted into the mice with the aid of a catheter. Now, this practice allows in vivo monitoring of biofilm development, follow-up monitoring and testing the antifungal drug. Bioluminescence imaging-based bio-films have other advantages like it is also suitable for testing in small animals as well as dropping the quantity of animal models in experimental batch (Velde et al. 2014). Apart from a few known biofilm, there is a need to develop biofilms based on other fungi that may support us in the battle against pathogens.

11.6.5 Bioactive Secondary Metabolites

A bioluminescent *Neonothopanus nambi* was also investigated for the production of Bioactive Secondary Metabolites. These biomolecules include nambinones A-C, 1-epi-nambinone B and nambinone D (Mehmet and Gulsen 2015). These molecules are found to be antimalarial against *Plasmodium falciparum* (nambinones A; nambinones C), antitubercular against *Mycobacterium tuberculosis* (nambinones A-C, 1-epi-nambinone B and nambinone D) and cytotoxic (nambinones A-C, 1-epi-nambinone B and nambinone D). Nambinones C was found toxic to the NCI-H187 cancer cell line. Aurisin A and aurisin K were also obtained from *Neonothopanus nambi* and found cytotoxic against the cholangiocarcinoma cell lines (Kanokmedhakul et al. 2012). Apart from *N. nambi*, there is a long list of bioluminescent fungi that synthesize bioactive compounds. Sesquiterpenes like – isoomphadione, illudiolone, illudosone hemiacetal and illudin S – were identified from *Omphalotus illudens* (McMorris et al. 2002; Lehmann et al. 2003). Illudin S was found antiviral. *Omphalotus japonicas* were also found with various Sesququiterpenes metabolites (Aoki et al. 2020). Cytotoxic activity was recorded in one of them. Therefore apart from lighting significance, bioluminescent fungi are a good source of bioactive molecules and it needs to be explored more.

11.7 Conclusion

Light illumination is very essential in modern living. Various live organisms are self-illuminating due to nature made enzymatic phenomenon. It is very exciting that many new luminescent organisms have been discovered in the last few decades and the luminescence molecules and mechanisms are understood. Step ahead, researchers are looking to use this luminescence system for the welfare of human society. The fungal luminescent system has been genetically encoded, and a self-glowing plant is produced. In vivo imaging, real-time disease monitoring, environmental assay, bioassay and biosensor development are the various areas where the fungal bioluminescence system was greatly applied and successful outcomes were achieved. The number of identified bioluminescent fungi is still little, so the finding of the new one is an opportunity for us apart from this more is needed in the heavy metal and xenobiotics monitoring and bioassay of some more pathogens. Replacing traditional street lights to the glowing tree is also yet not in practice. Therefore, some things are done, and some things are needed to be done.

References

Airth, R.L. and Foerster, G.E. (1962) The isolation of catalytic components required for cell-free fungal bioluminescence. *Archives of Biochemistry and Biophysics*, 97, 567–573.

Alsawaftah, N., Farooq, A., Dhou S., and Majdalawieh, F. (2020). Bioluminescence imaging application in cancer: a comprehensive review. *IEEE Reviews in Biomedical Engineering*. https://dx.dog.in.org/10.1109/RBME.2020.2995124

Amaral, D.T., Oliveira, G., Silva, J.R. and Viviani, V.R. (2016) A new orange emitting luciferase from the southern-Amazon *Pyrophorous angustus* (Coleoptera: Elateridae) Click-beetle: structure color relationship, evolutional and ecological considerations. *Photochemical and Photological Science*, 15, 1148–1154. doi:https://doi.org/10.1039/C6PP00165C

Anderson, J.B. and Stasovski, E. (1992) Molecular phylogeny of Northern Hemisphere species of *Armillaria*. *Mycologia*, 84, 505.

Aoki, S., Abosi, T., shiono, Y., Kimura, K., Murata, T., Arai, d., Iizuka, Y. and Murayama, T. (2020) Constituents of the fruiting body of poisonous mushroom *Omphalotus japonicas*. *Chemical and Pharmaceutical Bulletin*, 68(5), 436–442. doi:https://doi.org/10.1248/cpb.c19-01009

Avci, P., Karimi, M., Sadasivam, M., Antunes-Melo, W.C., Carrasco, E. and Hamblin, M.R. (2018) *In-vivo* monitoring of infectious diseases in living animals using bioluminescence imaging. *Virulence*, 9, 28–63. doi:https://doi.org/10.1080/21505594.2017.1371897

Badr, C.E. and Tannous, B.A. (2011) Bioluminescence imaging: progress and applications. *Trends in Biotechnology*, 29(12), 624–633. doi:https://doi.org/10.1016/j.tibtech.2011.06.010

Baumgartner, K. and Rizzo, D.M. (2001) Distribution of *Armillaria* species in California. *Mycologia*, 93(5), 821–830.

Bechara, E.J.H. (1988) Luminescent elaterid beetles: biochemical, biological and ecological aspects. *Advanced Oxidation Processes*, 1, 123–178.

Bechara, E.J.H. and Stevani, C.V. (2018) Brazilian bioluminescent beetles: reflactions on catching glimpses of light in the Atlantic forest and Cerrado. *Anais da Academia Brasileira de Ciências*, 90(1), Sup. 1 doi:https://doi.org/10.1590/0001-3765201820170504.

Bermudes, D., Peterson, R.H. and Neason, K.N. (1992) Low level bioluminescence detected in *Mycena haematopus* basidiocarps. *Mycologia*, 84, 799–802.

Bi, W.X., He, J.W., Chen, C.C., Kurendra, R. and Li, X.Y. (2019) Sinopyrophorinae, a new subfamily of Elateridae (Coleoptera, Clateroidea) with the first record of a luminous click beetle in Asia and evidence of multiple origins of bioluminescence in Elateridae. *Zookeys*, 864, 79–97.

Blackwell, M. (2011) The fungi: 1, 2, 3 . . .5.1 million species? *American Journal of Botany*, 98 (3), 426–438.

Brock, M., Jouvion, G., Droin-Bergère, S., Dussurget, O., Nicola, M.A. and Ibrahim-Granet, O. (2008) Bioluminescent *Aspergillus fumigatus*, a new tool for drug efficiency testing and *in vivo* monitoring of invasive Aspergillosis. *Applied and Environmental Microbiology*, 74(22), 7023–7035.

Chew, A.L.C., Desjardin, D.E., Tan, Y. et al. (2015) Bioluminescent fungi from Peninsular Malaysia – a taxonomic and phylogenetic overview. *Fungal Diversity*, 70, 149–187. doi:https://doi.org/10.1007/s13225-014-0302-9

Coetzee, M.P.A., Wingfield, B.D., Bloomer, P., Ridley, G.S., Kile G.A. and Wingfield, M.J. (2001) Phylogenetic relationships of Australian and New Zealand *Armillaria* species. *Mycologia*, 93, 887.

Cortes-Perez, A., Desjardin, D.E., Perry, B.A., Ramirez-Cruz, V., Guillen, F.R., Arambula, A.R.V. and Rockefeller, A. (2019) New species and records of bioluminescent *Mycena* from Mexico. *Mycologia*, 111(2), 319–338. doi:https://doi.org/10.1080/00275514.1554172

Dao, T.V. (2006) *Bioluminescence – The Mystery of the Nature. Research on Cultivation of Luminescent Mushroom*. Springbrook Research Center, Queensland, Australia.

Deheyn, D.D. and Latz, M.I. (2007) Bioluminescence characteristics of a tropical terrestrial fungus (Basidiomycetes). *Luminescence*, 22, 462–467.

Desjardin, D.E., Capelari, M. and Stevani, C.V. (2007) Bioluminescent *Mycena* species from Sao Paulo, Brazil. *Mycologia*, 99, 317. DOI: https://doi.org/10.3852/mycologia.99.2.317

Desjardin, D.E., Oliveira, A.G. and Stevani, C.V. (2008). Fungi bioluminescence revisited. *Photochemical & Photobiological Sciences*, 7, 170–182.

Devi, R., Kaur, T., Guleria, G., et al. (2020). Fungal secondary metabolites and their biotechnological applications for human health. In: *New and Future Developments in Microbial Biotechnology and Bioenginnering, Trends of Microbial Biotechnology for Sustainable Agriculture and Biomedicine Systems: Perspective for Human Health*. doi:10.1016/B978-0-12-820528-0.00010-7

Doh, I.J., Min, H.J., Aguilera, C.C., et al. (2019). Portable bioluminescence detection for food safety: smartphone vs silicon photomultiplier. *Proceedings of the SPIE1106, Sensing for Agriculture and Food quality and safety XI*, 11016E. doi:10.1117/12.2521452

Doyle, T.C., Nawotka, K.A., Kawahara, C.B., Francis, K.P. and Contag, P.R. (2006) Visualizing fungal infections in living mice using bioluminescent pathogenic *Candida albicans* strains transformed with the firefly luciferase gene. *Microbial Pathogenesis*, 40, 82–90.

Dyck, V.K., Rogiers, O., Vande, G.V. and Dijck, P.V. (2020) Let's shine a light on fungal infections: a noninvasive imaging toolbox. *PLoS Pathogens*, 16(3), e1008257. doi:https://doi.org/10.1371/journal.ppat.1008257

Edward, C.A. and Bohlen, P.J. (1995) *Biology and Ecology of Earthworms*, Springer, London.

Fallon, T.R., Lower, S.E., Chang, C.H., Bessho-Uehara, M., Martin, G.J., Bewick, A.J., Behringer, M. et al. (2018) Firefly genomes illuminate parallel origins of bioluminescence in beetles. *eLife*. https://doi.org/10.7554/eLife.36495

Farmer III, J.J., Jorgensen, J.H., Grimont, P.A.D., Akhurst, R.J., Poinar Jr., G.O., Ageron, E., Pierce, G.V., Smith, J.A., Carter, G.P., Wilson, K.L. and Brenner, F.W.H. (1989) *Xenorhabdus luminescens* (DNA Hybridization Group 5) from human clinical specimens. *Journal of Clinical Microbiology*, 27(7), 1594–1600.

Fleiss, A. and Sarkisyan, K.S. (2019) A brief review of bioluminescent systems (2019) *Current Genetics*, 65, 877–882. doi:https://doi.org/10.1007/s00294-019-00951-5.

Goya, M.E., Romanowski, A., Caldart, C.S. Benard, C.Y. and Golombek, D.A. (2016) Circadian rhythms identified in Caenorhabditis elegans by in vivo long term monitoring of a bioluminescent reporter. *Neurobiology*, doi:https://doi.org/10.1073/pnas.1605769113

Guyot, R. (1927) Myc′elium lumineux de l'Armillaire. *Comptes Rendus des Seances de l Academie des Sciences*, 96, 114.

Haddock, S.H.D., Moline, M.A. and Case, J.F. (2010) Bioluminescence in the sea. *Annual Review of Marine Science*, 2, 443–493.

Harvey, E.N. (1957) *A History of Luminescence from the Earliest Times Until 1900*. American Philosophical Society, Philadelphia.

Hastings, J.W. (2013) Circadian rhythms in Dinoflagellates: what is the purpose of synthesis and destruction of proteins? *Microorganisms*, 1(1), 26–32.

Hastings, J.W. and Davenport, D. (1957) The luminescence of the Millipede, *Luminodesmus sequoia. The Biological Bulletin*, 113, 120–128.

Hawksworth, D.L. (1991) The fungal dimension of diversity: magnitute, significance, and conservation. *Mycological Research*, 95, 641–655.

Hawksworth, D. L. (2001) The magnitude of fungal diversity: the 1.5 million species estimated revisited. *Mycological Research*, 105, 1422–1432.

Herring, P.J. (1987) Systematic distribution of bioluminescence in living organisms. *Journal of Bioluminescence and Chemiluminescence*, 1, 147–163.

Horak, E. (1978) *Mycena rorida* (Fr.) Quel. and related species from the Southern Hemisphere. *Berichte der Schweizerischen Botanischen Gesellschaft*, 88(1-2), 20–29.

Horsewell, J. Percival, H.J. and Speir, T.W. (2005) Impact of heavy metal amended sewage sludge on forest soils as assessed by bacterial and fungal biosensors. *Biological Fertile Soils*, 42, 569–576.

Hutchens, M. and Luker, G.D. (2007) Applications of bioluminescence imaging to the study of infectious diseases. *Cellular Microbiology*, 9, 2315–2322

Ilondu, E.M. and Okiti, A.A. (2016) Bioluminescence in mushroom and its application potentials. *Nigerian Journal of Science and Environment*, 14(1), 132–139.

Johnson, F.H. (1967) "*Edmund Newton Harvey (1887-1959) – A Biographical Memoir*, National Academy of Sciences, Washington, DC.

Kanokmedhakul, S., Lekphroma, R., Kanokmedhakul, K., Hahnvajanawong, C., Bua-art, S., Saksirirat, W., Prabpai, S. and Kongsaeree, P. (2012) Cytotoxic sesquiterpenes from luminescent mushroom *Neonothopanus nambi*. *Tentrahedron*, 68, 8261–8266.

Kaskova, Z.M., Tsarkova, A.S. and Yampolsky, I.V. (2016) 1001 Lights: luciferins, luciferases, their mechanisms of action and applications in chemical analysis, biology and medicine. *Chemical Society Reviews*, 45(21), 6048–6077.

Kaskova, Z.M., Felipe, A.D., Valentin, N.P., Konstantin, V.P., Aleksandra, S.T., Natalja S. R. et al. (2017) Mechanism and color modulation of fungal bioluminescence. *Science Advances*, 3, e1602847. http://advances.sciencemag.org/content/3/4/e1602847#BIBL

Kato, S.I., Oba, Y., Ojika, M., Inouye, S. (2007) Biosynthesis of cypridina luciferin in *Cypridina noctiluca*. *Heterocycles*, 72, 673–76.

Ke, H.M., Lee, H.H., Lin, C.I., et al. (2020). *Mycena* genomes resolve the evolution of fungal bioluminescence. bioRxiv preprint. doi:10.1101/2020.05.06.079921.

Kirchmair, M., Poder R., Huber, C.G. and Miller, O.K.J. (2002) Chemotaxonomical and morphological observations in the genus *Omphalotus Fayod* (*Omphalotaceae*), *Persoonia –Molecular Phylogeny and Evolution of Fungi*, 17(4), 583–600.

Kirchmair, M., Morandell, S., Stolz, D., Poder R. and Sturmbauer, C. (2004) Phylogeny of the genus *Omphalotus* based on nuclear ribosomal DNA sequences. *Mycologia*, 96, 1253–1260.

Kotlobay, A.A., Sarkisyan, K.S., Mokrushina, Y.A., Marcet-Houben, M., Serebrovskaya, E.O., Markina, N.M., Somermeyer, L.G. et al. (2018) Genetically encodable bioluminescent system from fungi. *PNAS*, 115(50), 12728–12732.

Kour, D., Rana, K.L., Yadav, N., Yadav, A.N., Singh, J., Rastegari, A.A. and Saxena, A.K. (2019) Agriculture and industrially important fungi: current development and potential biotechnological applications. In *Recent Advancement in White Biotechnology Through Fungi, Fungal Biology*, Springer, Cham. doi:10.1007/978-3-030-14846-1_1

Kushwaha, V. and Hajirnis, S. (2016) A review on bioluminescent fungi: a torch of curiosity. *International Journal of Life Sciences*, A7, 107–110.

Lee, J. (2008) Bioluminescence: the first 3000 year (review). *Journal of Siberian Federal University. Biology* 3, 194–205.

Lehmann, V.K., Huang, A., Ibanez-Calero, S., Wilson G.R. and Rinehart, K.L. (2003) Illudin S, the sole antiviral compound in mature fruiting bodies of Omphalotus illudens. *Journal of Natural Products*, 66(9), 1257–1258.

Lingle, W.L. (1993) Bioluminescence and ligninolysis during secondary metabolism in the fungus *panellus stypticus*. *Journal of Bioluminescence and Chemiluminescence*, 8 100.

Longkumer, I.Y. and Kumar, R. (2018) Bioluminescence in insect. *International Journal of Current Micrbiology and Applied Sciences*, 7(3), 187–193.

Marek, P. E. and Moore, W. (2015) Discovery of glowing Millipedes in California and the gradual evolution of bioluminescence in *Diplopoda*. *PNAS*, 112(20), 6419–6424.

Marek, P., Papaj, D., Yeager, J., Molina, S. and Moore, W. (2011) Bioluminescent sposematism in millipedes. *Current Biology*, 21(18)R, 680–681.

McElroy, W.D. and DeLuca, M. 1985. Firefly luminescence. In *Chemi- and Bioluminescence*, ed. JG Burr, New York, Dekker.

McMorris, T.C., Kashinatham, A., Lira, R., Rundgren, H., Gantzel, P., Kelner, M. and Dawe, R. (2002) Sesquiterpenes from *Omphalotus illudens*. *Phytochemistry*, 61(4), 395–398.

Mehmet, E.D. and Gulsen, T.C. (2015). Biologically active terpenoids from mushrooms origin: a review. *Records of Natural Products*, 9(4), 456–483.

Meighen, E.A. 1991. Molecular biology of bacterial bioluminescence. *Microbiological Reviews*, 55,123–142

Mendes, L.F. and Stevani, C.V. (2010) Evaluation of metal toxicity by a modified method based on the fungus *Gerronema viridilucens* bioluminescence in agar medium. *Environmental Toxicological Chemistry*, 29, 320–326.

Meyer-Rochow, V.B. (1990) The New Zealand Glow worm. Waitomo Cave Museum society, Waitomo Caves, New Zealand. *Proceedings of the National Academy of Sciences of the USA*, 115(50), 12728–12732.

Mitiouchkina, T., Mishin, A.S., Somermeyer, L.G., et al. (2019). Plants with self-sustained luminescence. Biorxiv.org 809376.

O'Brien, H.E., Parrent, J.L., Jackson, J.A., Moncalvo, J.M. and Vilgalys, R. (2005) Fungal community analysis by large-scale sequencing of environmental samples. *Applied and Environmental Microbiology*, 71(9), 5544–5550.

Oba, Y., Stevani C.V., Oliveira, A.G., Tarkova, A.S., Chepurnykh, T.V. and Yampolsky, I.V. (2017) Selected least studied but not forgotten bioluminescent systems. *Photochemistry and Photobiology*, 93, 405–415. doi:https://doi.org/10.1111/php.12704

Oliveira, A.G. and Stevani, C.V. (2009) The enzymatic nature of fungal bioluminescence. *Photochemical & Photobiological Sciences*, 8, 1416–1421.

Oliveira, A.G., Desjardin, D.E., Perry, B.A. and Stevani, C.V. (2012) Evidence that a single bioluminescent fungal lineages. *Photochemical and Photobiological Sciences*, 11, 848–852. doi:https://doi.org/10.1039/C2PP25032B

Oliveira, A.G., Stevani, C.V., Waldenmaier, H.E., Viviani, V., Emerson, J.M., Loros, J.J. and Dunlap J.C. (2015) Circadian control sheda light on fungal bioluminescence. *Current Biology*, 25(7), 964–968.

Pandey, A. and Sharon, M. (2017) Bioluminescent organisms. *BAOJ Chemistry*., 3(2), 029 (1–20).

Papon, N., Courdavault, V., Lanoue, A., Clastre, M. and Brock, M. (2014) Illuminating fungal infections with bioluminescence. *PLoS Pathogens*, 10(7), e1004179. doi:https://doi.org/10.1371/journal.ppat.1004179

Pegler, D.N. (2000) Taxonomy, nomenclature and description of *Armillaria*, in *Armillaria Root Rot: Biology and Control of Honey-Fungus*, ed. R. T. V. Fox, Intercept Ltd., Andover.

Perdiguero, B., Gomez, C.E. and Esteban, M. (2019) Bioluminescence imaging as a tool for Poxvirus biology. Vaccinia virus, In *Methods in Molecular Biology*, 2023 269–285.

Perry, B.A. (2007) Mycodigest: bioluminescence in fungi. *The Mycological Society of San Francisco*, 58(3),1–8.

Persyn, A., Rogiers, O., Brock, M., Vande, V.G., Lamkanfi, M., Jacobsen, I.D., et al. (2019) Monitoring of fluconazole and caspofungin activity against *in vivo Candida glabrata* biofilms by bioluminescence imaging. *Antimicrobial Agents and Chemotherapy*, 63(2), e01555–18. doi:https://doi.org/10.1128/AAC.01555-18 PMID:30420485

Purtov, K.V., Petushkov, V.N., Barnov, M.S., Mineev, K.S., Rodionova, N.S., Kaskova, Z.M. et al. (2015) The chemical basis of fungal bioluminescence. *Angewandte Chemie, International Edition*, 54, 8124–8128.

Rochow, V.B.M. and Moore, S. (1988) Biology of *Latia neritoides* Gray 1850 (*Gastropoda, Pulmonata, Basommatophota*): the only light- producing freshwater Snail in the world. *Internation Revue der gesamten Hydrobiologie und Hydrographie*, 73(1), doi:https://doi.org/10.1002/iroh.19880730104.

Roda, A. (2010) *Chemiluminescence and Bioluminescence: Past, Present and Future*. Royal Society of Chemistry, Cambridge, UK.

Rota, E., Zalesskaja, N.T., Rodionova, N.S., Petushkov, V.N. (2003) Redescription of *Fridericia heliota* (Annelida, Clitellata: Enchytraeidae), a luminous worm from the *Siberian taiga*, with a review of bioluminescence in the Oligochaeta. *Journal of Zoology*, 260, 291–299.

Sato A, Klaunberg, B. and Tolwani, R. (2004) *In vivo* bioluminescence imaging. *Comparative Medicine*, 54(6), 631–634.

Sharifian, S., Homaei, A., Hemmati, R., Luwor, R.B. and Khajen, K. (2018) The emerging use bioluminescence in medical research. *Biomedicine and Pharmacotherapy*, 101, 74–86.

Shimomura, O. (1987) Presence of coelenterazine in non-bioluminescent marine organisms. *Comparative Biochemistry and Physiology*, 86B, 361– 63.

Shimoura, O. (1984) Porphyrin chromophore in Luminodesmus photoprotein. *Comparative Biochemistry and Physiology*, B79, 565–567.

Slesiona, S., Ibrahim-Granet, O., Olias, P., Brock, M. and Jacobsen, I.D. (2012) Murine infection models for *Aspergillus terreus* pulmonary aspergillosis reveal long-term persistence of conidia and liver degeneration. *The Journal of Infectious Diseases*, 205(8), 1268–1277. doi:https://doi.org/10.1093/infdis/jis193

Smith, M., Bruhn, J. and Anderson, J. (1992) The fungus *Armillaria bulbosa* is among the largest and oldest living organism. *Nature*, 356, 428–431. doi:https://doi.org/10.1038/356428a0

Thomson, C.M., Herring, P.J. and Campbell, A.K. 1997. The widespread occurrence and tissue distribution of the imidazolopyrazine luciferins. *Journal of Bioluminescence and Chemiluminescence*, 12, 87–91.

Vanherp, L., Ristani, A., Poelmans, J., Hillen, A., Lagrou, K., Janbon, G., Brock, M., Himmelreich U. and Velde G.V. (2019) Sensitive bioluminescence imaging of fungal dissemination to the brain in mouse models of *cryptococcosis. Disease Models & Mechanisms*, 12, dmm039123, In The Company of Biologists Ltd, doi:10.1242/dmm.039123

Velde G.V., Kucharíková, S., Van, D.P. and Himmelreich, U. (2014) Bioluminescence imaging of fungal biofilm development in live animals. In: Badr C. (eds) *Bioluminescent Imaging. Methods in Molecular Biology (Methods and Protocols)*, vol 1098. Humana Press, Totowa, NJ.

Verdes, A. and Gruber, D.F. (2017) Glowing worms: biological, chemical, and functional diversity of bioluminescent annelids. *Integrative and Comparative Biology*, 57(1), 18–32.

Viviani, V.R. (2001) Fireflies (*Coleptera: Lampyridae*) from southern Brazil: habitats, life history and bioluminescence. *Annals of the Entomological Society of America*, 94, 129–145.

Viviani, V. (2007). Terrestrial Bioluminescence. In: *Photobiological Science Online* (American Society for Photobiology., ed) P1 Online Recourse. McLean, VA: Americal Society of Photobiology.

Waldenmaier, H.E., Oliveira, A.G. and Stevani, C.V. (2012) Thoughts on the diversity of onvergent evolution of bioluminescence on earth. *International Journal of Astrobiology*, 11, 335–343.

Wan, N.A., Jusoh, W. and Wong, L.S. (2014). Explaining the potential of whole cell biosensor: a review in environmental applications. *International Journal of Chemical, Environmental and Biological Sciences*, 2(1), 52–56.

Weitz, H.J. (2007) Bioluminescence based fungal biosensors. In *Exploitation of Fungi, British Mycological Society Symposia*, Cambridge University Press, Cambridge, UK.

Weitz, H.J., Campbell, C.D. and Killham, K. (2002) Development of a novel, bioluminescence-based, fungal bioassay for toxicity testing. *Environmental Microbiology*, 4, 422–429.

Widder, E.A. (1999) Bioluminescence. In: Archer S.N., Djamgoz M.B.A., Loew E.R., Partridge J.C., Vallerga S. (eds) *Adaptive Mechanisms in the Ecology of Vision*. Springer, Dordrecht, doi:https://doi.org/10.1007/978-94-017-0619-3_19

Wilson, T. and Hastings, J.W. (2013) *Bioluminescence: Living Lights, Lights for Living*. Harvard University Press, Cambridge, MA, London.

Wood, K.V. (1995) The chemical mechanism and evolutionary development of beetle bioluminescence. *Photochemistry and Photobiology*, 62, 662– 673.

Yan, Y., Shi, P., Song, W. and Bi, S. (2019) Chemiluminescence and bioluminescence imaging for biosensing and therapy: in vitro; and In vivo perspective. *Theranostics*, 9(14), 4047–4065. Doi:https://doi.org/10.7150/thno.33228.

12

Production of Liquid Biofuels from Lignocellulosic Biomass

Manoj Kumar Singh, Sumit Sahni, and Anita Narang

Department of Botany, Acharya Narendra Dev College, University of Delhi, New Delhi, India

12.1 Introduction

In the past two decades, gradually exhausting fossil fuel reserves and increasing environmental pollution due to excessive use of fossil energy forced many countries to put forward the 'energy strategy' for their sustainable development. This strategy promotes the use of renewable energy resources in place of fossil fuels (Han et al. 2019). According to the 2018 report of the US Energy Information Administration (USEIA), fossil fuels contribute 80% of the total energy consumed worldwide, and this share is going to reduce somewhat (70%) by 2050 (International Energy Outlook 2019), which is clearly unsustainable. Therefore, the production of renewable fuels is urgently required by using renewable resources to replace these non-renewable conventional fuels. At the present time, the main sources of renewable energy include solar energy, wind energy, biomass, hydrogen energy, geothermal energy and ocean energy. Among these renewable energy resources, biomass contributes 70%, the highest of all the available renewable energies (Panwar et al. 2011; Jurasz et al. 2020). The term biomass comprises any material coming from microbes, plants and animals which can be used as an energy source. In the perspective of biomass energy, it is generally called lignocellulosic biomass (LCB), which mainly consists of plant-based materials and plant dry matter. LCB consists of three structural components i.e. cellulose, hemicellulose and lignin with composition varying from 40 to 60, from 20 to 40 and from 10 to 25 wt%, respectively. Cellulose, a linear carbohydrate polymer, consists of 100 to over 10 000 of β-D-glucose units linked through glycosidic linkages. In contrast, hemicellulose is a branched copolymer of pentose and hexose monomer units situated in a plant cell wall along with lignin. Lignin the most complex of all, is an amorphous polymer of phenolic compounds and has high energy density than cellulose and hemicellulose. After depolymerization of lignin, phenolic compounds such as phenol, guaiacol, syringol and other derivatives were obtained (Saidur et al. 2011; Schutyser et al. 2015; Nanda et al. 2016).

Transformation of LCB into liquid fuels can be achieved through many routes such as high-pressure liquefaction (Wang et al. 2008), fast pyrolysis (Wang et al. 2017), hydrolysis and fermentation (Lu et al. 2010) etc. as shown in Figure 12.1. Fast pyrolysis generates

Energy: Crises, Challenges and Solutions, First Edition. Edited by Pardeep Singh, Suruchi Singh, Gaurav Kumar, and Pooja Baweja.

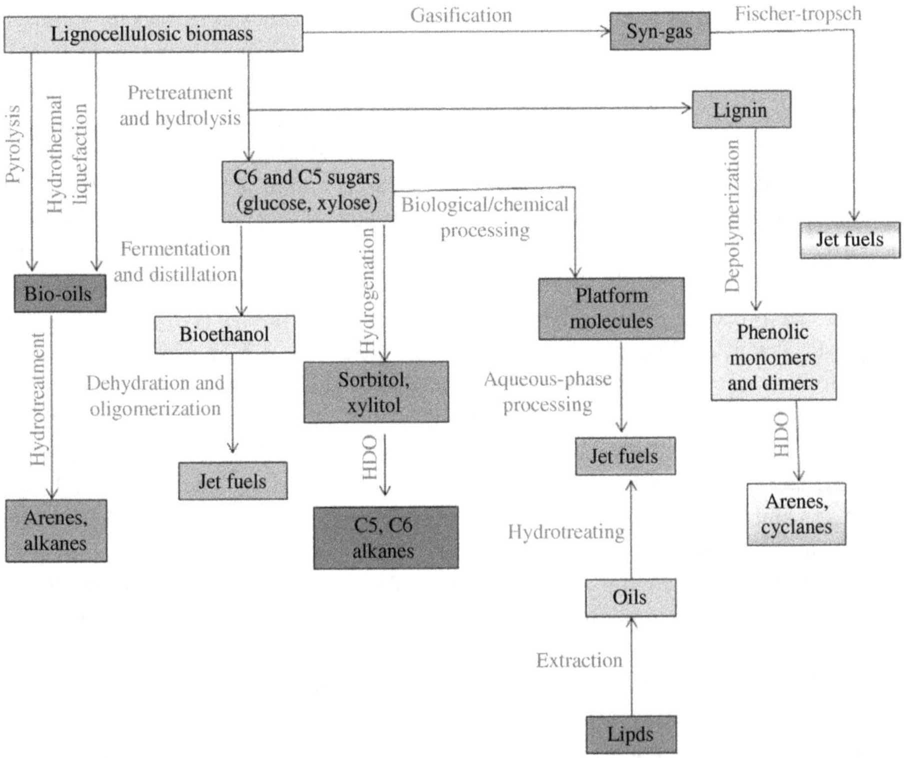

Figure 12.1 Strategies for the conversion of biomass to liquid biofuel.

bio-oil, which is a complex mixture of many oxygenated compounds. Due to high oxygen content of bio-oil, it is immiscible with fossil fuels, which limits its use as an engine fuel (Rahman et al. 2018). The hydrolysis of carbohydrate polymers present in the LCB into sugars followed by its fermentation yields another biofuel i.e. ethanol (Shuai et al. 2010).

The process to recover these carbohydrates from LCB requires a pretreatment step for breaking the interaction between cellulose, hemicellulose and lignin. Pretreatment of LCB can be performed using physical, chemical and biological agents. Physical and chemical pretreatment methods expense a lot of energy as the lignin acts as a barrier for them (Kumari and Singh 2018). Enzymatic method requires lignocellulolytic enzymes (cellulases, hemicellulases and ligninases) which can selectively depolymerize the major lignocellulosic components to produce fermentable sugars (Sindhu et al. 2016). For making this biological pretreatment economically viable, continuous research is going on to minimize the production cost of lignocellulolytic enzymes (Singh et al. 2017). However, as compared with starch and sugar-based ethanol, lignocellulosic bioethanol is still facing certain technical blockades that make the process economically non-competitive (Cardona et al. 2005). The major drawbacks in lignocellulosic bioethanol production are expensive pretreatment process, many steps in the overall process, diversity in complexity of biomass, limited microbial fermentation of pentose sugar and inprocess generation of inhibitors. Despite these techno-economical limitations, research is growing on to make the process economically feasible and bring it closer to industrialization (Zabed et al. 2016).

Because of certain limitations in bioethanol production methods, scientists have started a new route for valorization of lignocellulosic resources by converting them into hydrocarbon fuels.

Recently through a modified refining approach, LCB can also be converted into biogasoline. The process initially involves conversion of cellulose and hemicellulose to five and six carbon sugars by hydrolysis, followed by hydrogenation of sugars to polyols and finally conversion of polyols to C_5/C_6 alkanes via aqueous-phase reforming (APR) (Wang et al. 2020). APR process was first proposed by Huber and Dumesic (2006) as one of the most promising routes for liquid alkane (i.e. C_5–C_6) production. These polyols-derived C_5–C_6 alkanes can directly be used in the gasoline pool. This technology is considered as one of the most efficient methods for producing second-generation (2G) biofuel and can conquer the limitations of ethanol–gasoline blending (Wang et al. 2020).

Worldwide the aviation sector is rapidly growing; still the major energy source in this sector is derived from fossil resources. While in other sectors such as road transport, alternate energy sources are growing to replace the fossil fuels such as bioethanol, biodiesel, bio-hydrogen and hybrid or sole-electric alternatives (Wang et al. 2019). Currently, aviation fuel is a hydrocarbon mixture of *n*-alkanes, cyclo-alkanes, iso-alkanes and aromatics with definite carbon chain lengths and properties (Díaz-Pérez and Serrano-Ruiz 2020). Worldwide the total consumption of fuel in the aviation sector by commercial airlines is 340 million m^3 per year in 2018 and expected a further increase of about 5% annually till 2050. Therefore the development of bio-aviation fuel from lignocellulosic resources should be a promising future strategy (Wang et al. 2019). The bio-aviation fuel can be produced by integration of different technologies such as chemo-catalytic conversion, thermochemical conversions and biochemical conversions. Conversion of biomass into bio-aviation fuel involves many innately linked steps including intermediate state generation, carbon chain length adjustment and hydro-processing. Here we will discuss the production of bio-aviation (jet) fuel from LCB and plant oils (Díaz-Pérez and Serrano-Ruiz 2020).

Lignin is an amorphous phenol-based polymer generated as a by-product after pretreatment of LCB. It can be converted into cycloalkanes and aromatic hydrocarbons by catalytic depolymerization followed by hydrodeoxygenation (HDO) reactions (Han et al. 2019). These different types of hydrocarbons derived from cellulose, hemicellulose and lignin can be used as alternative sustainable fuels in different sectors. In recent years, there is a rapid development in integrated refining technologies, especially the lignocellulose conversion to alkanes in aqueous phase, production of branched long-chain alkanes, cycloalkanes and aromatic hydrocarbons. This review is focused on converting lignocellulosic resources to bioethanol and hydrocarbon fuel via integrated refining methods.

12.2 Ethanol from Lignocellulosic Biomass

Ethanol produced from LCB is most frequently discussed second-generation biofuel. Process of ethanol production from LCB comprises many steps as shown in Figure 12.1. The main steps are pretreatment and/or detoxification, hydrolysis, fermentation and product recovery (Zabed et al. 2016). The collected lignocellulosic raw material first undergoes drying process to optimize moisture level followed by pulverization through grinding.

Before pretreatment the grinded biomass is mixed with appropriate quantity of water to make slurry (Fatma et al. 2018).

12.2.1 Pretreatment of LCB

The utilization of LCB for ethanol production through biochemical processes requires a pretreatment step, because the complicated recalcitrant arrangement of the plant cell wall makes it resistant to enzymatic and microbial attack. The cellulosic component of LCB is usually layered with hemicelluloses forming a cellulose–hemicellulose complex. This complex gets further complicated with lignin to create a recalcitrant structure which prevents the hydrolysis of biomass to produce fermentable sugars (Cavka and Jönsson 2013). Hence, pretreatment process is required to modify the LCB and break the interaction among lignin–cellulose–hemicellulose, due to which the cellulose and hemicellulose components of LCB become available to hydrolytic enzymes that ultimately hydrolysed them into fermentable sugars. After pretreatment the major structural changes of biomass include lignin removal, altered crystal lattice of cellulose, more exposed surface area and perviousness of the LCB (Zheng et al. 2009; Chiaramonti et al. 2012).

There are many studies available on pretreatment methods of a wide range of LCB. Every pretreatment method has its own pros and cons with respect to others in its techno-economic and process conditions. However, a potential pretreatment method should have some general features such as altering the polysaccharides, its efficient release from LCB, maximum recovery of polysaccharides, less degradation of monosaccharides, generating less toxic compounds, high lignin recovery and requiring low investment and operational costs. Broadly pretreatment methods can be categorized into four groups such as physical, chemical, solvent and biological (Kumar and Sharma 2017).

Physical pretreatment techniques alter the biomass by exposing the surface area and pore volume, subsiding the crystallinity and extent of polymerization in cellulose content, hydrolysis of hemicellulose and incomplete degradation of lignin components. It also includes some mechanical size reduction techniques such as milling, grinding, chipping etc. However, these size reduction methods require higher energy input which makes the processes quite expensive (Brodeur et al. 2011; Karimi et al. 2013). Physical pretreatment methods also include steam explosion, irradiation and hot water techniques. Radiation such as microwaves is highly penetrating but could not be used for processing large quantities of LCB because of high power input. Whereas among physical pretreatment techniques, steam explosion and liquid hot water techniques gained more attention in perspective to bioconversion of LCB (Brodeur et al. 2011).

Major chemical pretreatment methods include alkali and/or acid treatment that works on LCB through delignification, declining the degree of polymerization and cellulose crystallinity. Acids which are commonly used for pretreatment of LCB include H_2SO_4, HCl, H_3PO_4 and HNO_3. The most commonly used alkali is NaOH, whereas H_2SO_4 is the most commonly used acid in this process. Acid pretreatment solubilizes the hemicellulosic component of biomass and makes cellulose available to hydrolytic enzymes.

Solvent pretreatment works on the principle of fractionation method that involves aqueous organic solvents with or without catalysts to remove lignin from LCB. Methanol, ethanol, tetrahydro furfuryl alcohol, trimethylene glycol, glycerol, ethylene glycol, acetone, phenol and/or *n*-butanol are the chemicals used in the processes (Brodeur et al. 2011). These solvents extract the lignin as well as solubilize hemicellulose making cellulose available for hydrolysis. Solvent pretreatment also includes Ionic Liquids (ILs) containing salts composed of small anion and large organic cations, which is in liquid state at room temperature having less vapour pressure (Elgharbawy et al. 2016).

Biological pretreatment of LCB requires the action of microorganisms mainly fungi such as white rot, soft rot and brown rot fungi. White rot fungi (WRF) degrade all components of LCB, including lignin, whereas brown rot fungi degrade cellulose and hemicellulose leaving lignin largely intact (Machuca and Ferraz 2001). On the basis of pattern of decay, WRF have been characterized in two groups, one is selective white-rot, and other is simultaneous white-rot. Selective white-rot degrades lignin selectively while simultaneous white-rot degrades a large proportion of cellulose, hemicellulose and lignin simultaneously (Martínez et al. 2005; Kamei et al. 2012). To degrade lignocellulose, WRF secreted a complex set of synergistically acting lignocellulolytic enzymes, which are very specific for action at particular sites.

Since different biomasses have different physicochemical complexity in their structures, a single common pretreatment method is not suitable for every kind of biomass. Therefore, pretreatment method and its conditions vary on the basis of nature of biomass.

12.2.2 Detoxification

Under pretreatment conditions, many lignocellulose-derived by-products were produced, which may prove as inhibitors for enzymes and fermenting microorganisms if they were accumulated in higher concentration (Cavka and Jönsson 2013). By-products generated after the pretreatment process include sugar acids, formic acid, levulinic acid, acetic acid, hydroxymethylfurfural (HMF) and furfural (Jönsson et al. 2013). Most of the by-products are generated after hemicellulose and lignin degradation. These are categorized into three major groups such as furan derivatives, aliphatic acids and phenolic compounds. Many physical, biological and chemical methods have been employed to remove or neutralize the by-products in pretreated biomass. These methods include chemical additives (such as alkali $Ca(OH)_2$, NaOH, NH_4OH and reducing agents such as dithionite, dithiothreitol, sulphite), enzymatic treatment (laccase and peroxidase), heating and vaporization, liquid–liquid extraction (ethyl acetate, trialkylamine), liquid–solid extraction (activated carbon, ion exchange) and microbial treatment (*Coniochaeta ligniaria, Trichoderma reesei* etc.) (Jönsson and Martín 2016). Feedstock selection and their engineering, selection of microorganism, genetic/metabolic engineering of microorganisms, evolutionary engineering and culturing schemes can also be used as alternative measurement strategy to minimize the inhibitor by-product generation after pretreatment (Jönsson et al. 2013).

12.2.3 Hydrolysis

Cellulose and hemicellulose content of LCB are converted into fermentable sugars by the process of hydrolysis. Two fractions were obtained after every pretreatment process. Cellulose was found in water-insoluble solids fraction, whereas hemicellulose is present in the liquid part generated in the process. Depending on the pretreatment method, hemicellulose was either obtained in completely hydrolysed form or in the form of oligosaccharides. Completely hydrolysed sugars were directly fermented to alcohol, whereas oligosaccharides need further hydrolysis before fermentation (Sun and Cheng 2002). Mannose and xylose are the dominant sugars found in hemicelluloses of softwood and hardwood respectively. Small quantities of arabinose and galactose are also present in hemicelluloses. Most of the cellulosic parts are intact and require hydrolysis before fermentation. Acid or enzymatic hydrolysis is used for converting cellulose and hemicellulose into their monomeric sugars (Zabed et al. 2016).

12.2.3.1 Acid Hydrolysis

Two types of acid hydrolysis can be performed: first is treatment with dilute acid at higher temperature and pressure with shorter reaction time and second with concentrated acid at lower temperature. Certain drawbacks of acid hydrolysis include neutralization or recovery of the acids earlier to conducting fermentation and large amounts of wastes production. Sulphuric acid (H_2SO_4) is commonly used for acid hydrolysis, though other inorganic acids for example, hydrochloric acid (HCl), trifluoroacetic acids (TFA), nitric acid (HNO_3), phosphoric acid (H_3PO_4) have also been used for hydrolysis (Gírio et al. 2010). Dilute acid hydrolysis can be used for both the carbohydrates of LCB at different reaction conditions. Hemicellulose segment of the biomass is depolymerized at lower temperature, whereas cellulose hydrolysis requires higher temperature range between 230 and 240 °C. During dilute acid hydrolysis, concentration of H_2SO_4 and temperature ranged between 0.5 and 1.5% and 120 and 160 °C. Concentrated H_2SO_4 (up to 70–90%) can also be used for hydrolysing both the carbohydrates with longer reaction time. Though, concentrated acid offers a rapid and complete hydrolysis of cellulose to β-D-glucose and hemicelluloses to pentose sugars (Balat 2011; Zabed et al. 2016).

12.2.3.2 Enzymatic Hydrolysis

Lignocellulose-degrading microorganisms especially WRF secrete two types of extracellular enzymatic systems: first, the hydrolytic system which hydrolyses cellulose and hemicellulose; second, a distinctive ligninolytic system, which hydrolysed the lignin into its constituent monomers. Therefore, for hydrolysis process, these enzymes can directly be produced by the microorganisms or commercially available enzymes can be used (Thiribhuvanamala et al. 2017). Where the latter is more feasible but to make the production of 2G ethanol economically viable, cost of enzymes needs to be minimized. Substrate specificity of the enzyme, low temperature requirement for enzyme action and minimum generation of inhibitors make it more promising. Three major enzyme systems, required for the hydrolysis of LCB, include cellulases, hemicellulase (xylanase) and ligninases (Sanchez 2009; Sampathkumar et al. 2019).

Cellulase is a mixture of three enzymes, endo-1-4-β-glucanase or carboxymethylcellulases (EC3.2.1.4), exoglucanase or cellobiohydrolase (CBH) (EC3.2.1.91) and β-glucosidase (EC3.2.1.21) that acts synergistically for the hydrolysis of cellulose to β-D-glucose. An endoglucanase is thought to hydrolyse β-1, 4-glycosidic linkage of D-glucan randomly at multiple internal sites within the cellulose fibre which opens up sites for consequent attack by the CBH. CBH removes monomers and dimers from any of the reducing or non-reducing end of the cellulose chain. β-Glucosidases hydrolyse cellobiose to glucose (Singh et al. 2017).

Hydrolysis of hemicellulose requires highly specific and multiple enzyme systems because they contain diversity of linkages, branching and different monomer units. Hemicellulases include endo-1, 4-β-xylanase or endoxylanase (E.C.3.2.1.8), xylan1, 4-β-xylan esterases, *p*-coumaric and ferulic esterases, α-glucuronidase (E.C.3.2.1.139), α-1-arabinofuranosidases, α-arabinofuranosidase (E.C.3.2.1.55), acetyl xylan esterase (E.C.3.1.1.72) and α-4-*O*-methyl glucuronosidases xylosidase (E.C.3.2.1.37). Main xylan chain was hydrolysed by endo-xylanase, whereas xylan1, 4-β-xylan esterase depolymerizes xylooligosaccharides into xylose. Enzymes α-arabinofuranosidase and α-glucuronidase break the xylan backbone and take out arabinose and 4-*o*-methyl glucuronic acid, respectively. The acetyl esterases act on the acetyl substitutions of xylose, whereas feruloyl esterases break the bond between arabinose substitutions and ferulic acid (Zabed et al. 2016; Singh et al. 2017).

12.2.4 Fermentation

After pretreatment and hydrolysis of LCB, C_6 and C_5 sugars are produced, which are then fermented into ethanol with the help of microbial enzymes. It can be done either by solid state or submerged fermentation. Submerged fermentation is also known as liquid-state fermentation as it contains a predefined solid with water. In solid-state fermentation, LCB was incubated in its natural state with appropriate moisture content. Though solid-state fermentation is a cost-effective process but due to techno-economical limitations, it is not used for commercial ethanol production. Whereas submerged fermentation is a most common technique used for most of the commercial ethanol production facilities (Mussatto and Teixeira 2010). Three fermentation techniques, batch, fed-batch and continuous fermentation are commonly used for ethanol production. Most suitable technique is selected on the basis of type of feedstock and kinetics of fermenting microorganism (Kumar et al. 2016; Fatma et al. 2018).

12.2.5 Product Recovery

Fermented broth was subjected to either distillation alone or distillation followed by adsorption for bio-ethanol recovery. Distilling the fermented broth separates ethanol from water. The obtained ethanol concentration is up to 95%, and the remaining liquid contains lignin, non-converted cellulose and hemicellulose, enzymes, microorganisms and other components. These are sent to the waste water treatment. These materials can also be used as fuel or converted to various co-products after concentration (Vane 2008).

12.3 Bio-gasoline from Lignocellulosic Biomass

LCB can be converted into polyols (sorbitol, mannitol and xylitol) through an effective approach of hydrolysis integrated with hydrogenation and hydrogenolysis. Presently, hydrogenation of glucose and fructose is used for the production of sorbitol and mannitol, whereas hydrogenation of xylose produces xylitol (Liu et al. 2014). In an aqueous-phase reforming (APR) reaction under mild conditions, sorbitol, xylitol and other polyols can be tranformed to C_5/C_6 alkanes (Wang et al. 2020). Through the process of hydrolysis, catalytic hydrogenation and APR reaction, the biomass carbohydrates can be converted to light hydrocarbon fuels (Huber et al. 2003).

Earlier we have discussed ethanol production from LCB through pretreatment, hydrolysis, detoxification and fermentation. Compared with ethanol production, light hydrocarbon fuel (alkane) formation has many distinct advantages (Huber et al. 2003). First, their recovery is far easier and cost-efficient than ethanol as they can be easily removed from the reaction system through phase separation method (because of covalent nature of alkanes), whereas recovery of ethanol from the reaction system requires distillation which is an energy-consuming process. Second, they are more efficient in utilizing carbon biomass because pentose sugar via polyols formation can be converted to C_5 alkane, whereas in cellulosic ethanol production, fermentation of C_5 sugar is difficult (Wang et al. 2020).

12.3.1 Hydrolysis to Monosaccharides

As discussed earlier, cellulose is a well-crystalline homopolymer of glucose monomer unit, hemicellulose is a branched heteropolymer of glucose, xylose and other sugars and lignin is a random copolymer of phenylpropane structural unit (Binder and Raines 2010). These biopolymers are so densely interwoven with each other to form a complex recalcitrant structure that must be broken down for the convenience of utilization and consequent transformation into different value-added products. The corresponding monosaccharide components revealed after hydrolysis, cellulose and hemicellulose can further be converted to variety of organic molecules (such as sorbitol, xylitol, levulinic acid, 5-hydroxymethyl furfural etc.) and fuels. Various hydrolysis technologies studied till now include enzymatic hydrolysis, dilute acid hydrolysis, solid acid hydrolysis, alkaline hydrolysis and heteropoly acid hydrolysis etc. (Tian et al. 2010; Amezcua-Allieri et al. 2017). Among these the most commonly used method is dilute acid hydrolysis because of many advantages, namely it generates H^+ ion which can easily trespass lignin without any pretreatment and has faster rate of hydrolysis compared with enzymatic methods. The acid hydrolysis method generally uses H_2SO_4 and HCl at 100–150 °C and concentration of 1–10% by weight. Acid hydrolysis method is widely used in industries; however, it has many inherent disadvantages such as environmental pollution, equipment corrosion and reduced yield of monosaccharide because of its degradation under acidic conditions. Enzymatic hydrolysis is the most promising method because of milder reaction conditions and high substrate selectivity. However, this method requires pretreatment of LCB to enhance the enzymes accessibility. A suitable hydrolysing enzyme cocktail efficiently hydrolyses cellulose and hemicellulose component and enhances the yield of monomeric sugars (Yu et al. 2008; Ying et al. 2018).

However, the potential disadvantage of this method is the high costs associated with both pretreatment and enzymes. Other methods such as alkaline hydrolysis, heteropolyacid hydrolysis and solid acid hydrolysis are not yet commercialized.

12.3.2 Hydrogenation of Monosaccharides to Polyols

Hydrogenation of monosaccharides to produce polyols has been performed in the presence of different catalysts, of which Raney nickel is a most effective and conventionally used one. Many industries use raney nickel for the commercial production of sorbitol from glucose. The Raney nickel catalysts lose Ni and Al under more acidic reaction conditions, because of their corrosion, lead to their deactivation. During hydrogenation of glucose, pH of the reaction mixture decreases, causing the deactivation of the Raney nickel catalysts. Thus, glucose solution pH should be maintained at 8–9 during its hydrogenation with the Raney nickel catalyst (Hoffer et al. 2003). However, under alkaline conditions, glucose is unstable and through isomerization reaction converted into mannose, fructose and mannitol, thus reducing the product selectivity in the glucose hydrogenation reaction. Therefore the hydrogenation of glucose is generally carried out at high temperatures (≥120 °C) while using Raney nickel as a catalyst. Unavoidably, humins (water-insoluble carbon-based macromolecule) and coke were formed in the process, causing a steep decline in the catalyst activity.

Many catalysts have been screened for facilitating the conversion of glucose to produce sorbitol. Ru, Ni, Rh, Pd in decreasing order of their catalytic activity affect the hydrogenation of glucose (Betancourt et al. 1998). Among the different catalysts studied, Ru/C is an excellent hydrogenation catalyst, works under mild reaction conditions with high yield. For hydrogenation of glucose to sorbitol, Ru/C catalyst shows many advantages over other catalysts, continuously used for long hours (up to 312 hours) with very slight loss in activity and works under acidic condition to avoid the isomerization of glucose and increases the product selectivity, thus a higher yield of sorbitol is obtained (Wang et al. 2020).

In summary, conventionally two reaction steps are required for the production of sorbitol from cellulose: cellulose to sugar conversion by acid hydrolysis, and the consequent conversion of intermediate sugars to polyols by hydrogenation or hydrogenolysis over transition metal catalyst under H_2 at 4.0–12.0 MPa pressure. Along with sorbitol, mannitol is also produced through the epimerization of glucose to mannose.

12.3.3 Conversion of Polyols and Carbohydrates to C_5/C_6 Alkanes

12.3.3.1 Monosaccharides

A novel APR process was reported by Davda and Dumesic (2003) in the formation of paraffins over Pt/SiO_2-Al_2O_3. In the process C_5/C_6 alkanes were successfully produced from sorbitol at temperature and selectivity of 225–265 °C and 58–89% respectively. During the process, sorbitol first underwent dehydration to produce isosorbide (intermediate), which was further converted into 1, 2-hexanediol, after that hexane was produced over the Pt/SiO_2-Al_2O_3 catalyst through sequential reactions of ring opening and repeated dehydration and hydrogenation reactions. Through the cleavage of C–C bonds in the ARP reaction, mixed polyols can be produced from sorbitol. By means of decarbonylation, hydrogenation and dehydration reactions, these polyols can be converted into 1–5 carbon alkanes. Different Pt-based catalysts such as Pt/SiO_2-Al_2O_3, Pt/Al_2O_3 and Pt/Al_2O_3 mechanically mixed with SiO_2-Al_2O_3 were investigated for the APR reaction of sorbitol. Among these catalysts Pt/SiO_2-Al_2O_3 exhibits excellent catalytic activity (Huber et al. 2004). A hybrid catalyst system of platinum and zirconium oxide and mixed oxides, for example, ZrO_2-WO_x, Al_2O_3-WO_x and TiO_2-WO_x, was also adopted for production of lighter alkanes from sorbitol. Pt/ZrO2-TiO_2-WO_x was recommended as the best possible oxide as it showed steady performance in hydrothermal environments (Vilcocq et al. 2014). Researchers also used support materials of HZSM-5, ZSM-5 and/or MCM-41 in Ni-based catalysts, for conversion of xylitol and sorbitol to pentane and hexane (Zhang et al. 2011, 2012; Sun et al. 2017).

12.3.3.2 Cellulose and Biomass

We have discussed the production of C_5/C_6 alkanes from cellulose by hydrodeoxygenation. However, converting the cellulose to C6 alkane through this process needs an added step i.e. hydrolysis of cellulose. Microcrystalline cellulose can directly be converted into *n*-hexane through one-pot method using a heterogeneous catalyst under mild reaction conditions (Liu et al. 2015a). Heterogeneous catalytic mixtures such as $Ru/C + LiNbMoO_6 + H_3PO_4$ were reported as proficient catalytic system designed for the APR of cellulose. With the phosphoric acid catalysis, cellulose was hydrolysed to produce glucose, which can be hydrogenated over Ru/C catalyst to sorbitol. The released glucose and/or sorbitol can intercalate between layered $LiNbMoO_6$. Hexanes were formed in the active interlayer sites from glucose and/or sorbitol through consecutive C–O cleavage reactions. Though, isosorbide formation from sorbitol essentially involves tetrahydrofuranic moieties formation, a hardly occurred reaction within the inner layers of the catalyst. The total hexanes yield was up to 72% with inhibiting effect of the isosorbide formation, an unwanted intermediate. Many studies are available on one-pot conversion of cellulose and biomass to alkanes using different feedstock, catalysts and reaction conditions with different percentage of alkanes yield (Chen et al. 2013; de Beeck et al. 2015; Xia et al. 2016).

12.4 Jet Fuels from Lignocellulosic Biomass

Jet fuel is an intricate mixture of hydrocarbons (HCs) with carbon chains ranging from C_9 to C_{16}. Usually 80% of jet fuel contains a mixture of linear alkanes, slightly branched alkanes and cycloalkanes, whereas the remaining 20% is aromatics. The relative content of these four types of hydrocarbons in the fuel varies depending on place and season. Jet fuels have very stringent specifications and physicochemical properties such as high flash points, low freeze points, high energy density and superior sealing properties, among others. These physicochemical properties depend upon the relative content of the different types of hydrocarbons (Díaz-Pérez and Serrano-Ruiz 2020). Nowadays jet fuel is derived from petroleum resources (mixture of fully deoxygenated hydrocarbons of different weight, branching and cycling), after fractional distillation and simple catalytic process. LCB is another resource for the production of jet fuel, requiring complex chemical transformations such as selective deoxygenation followed by molecular weight and structural engineering. As shown in Figure 12.2, many routes are available for converting biomass to jet fuel, all united at a common step i.e. to lessen the structural complication of biomass and make the oxygenated components available for further transformation. Biomass can be converted to jet fuel through four different courses. These are sugars/platform molecules to jet fuels (STJ), oil to jet fuels (OTJ), gas to jet fuels (GTJ) and alcohols to jet fuels (ATJ) (Díaz-Pérez and Serrano-Ruiz 2020).

12.4.1 Production of Jet Fuels from Sugars and Platform Molecules

Sugars and sugar derivatives obtained after pretreatment, hydrolysis and dehydration of LCB can also be used for jet fuels production. The conversion of these sugars and derivatives to liquid hydrocarbon fuels is a complex catalytic process and involves intense

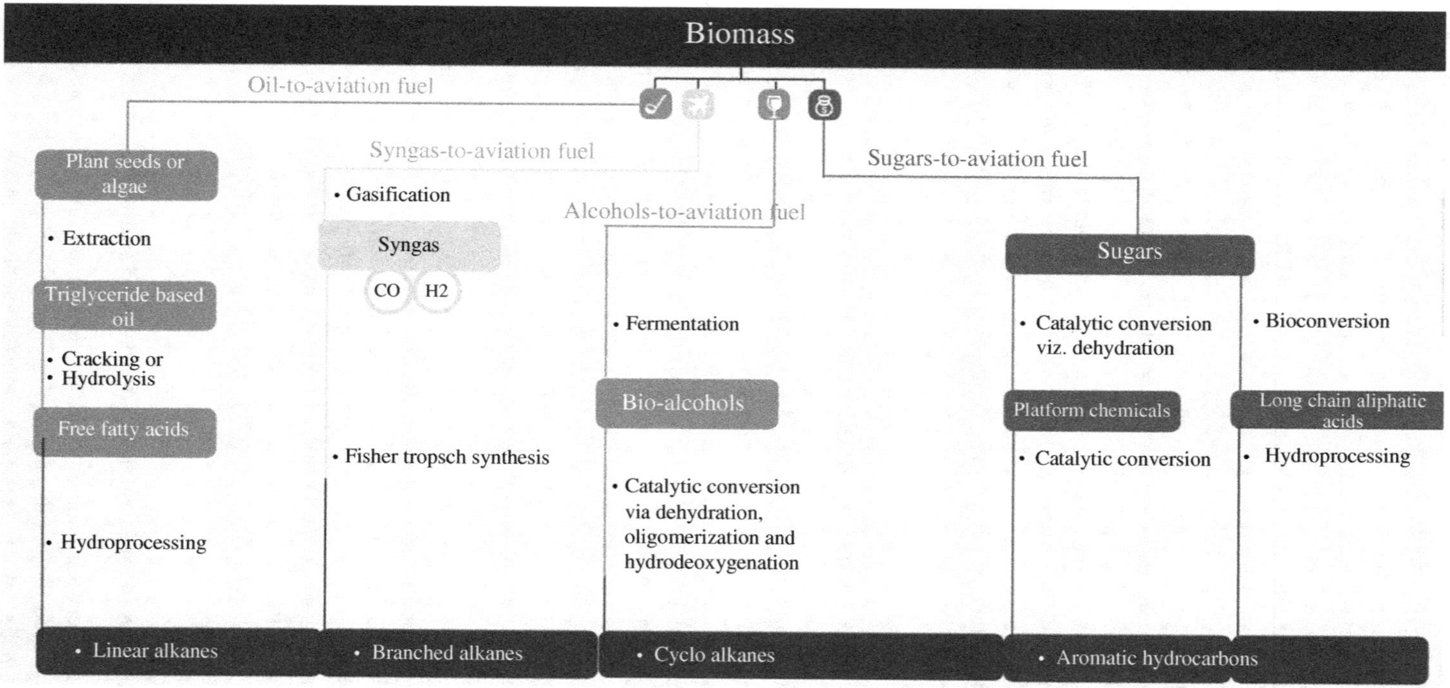

Figure 12.2 Route for the synthesis of jet (aviation) fuel from biomass.

chemical manipulations. Chemically, the derived sugars and other platform molecules contain many oxygen-containing functional groups e.g. alcohols, carbonyl group and carboxylic acids, and their maximum carbon chain length is up to 6, as these are derived from LCB monomers. Contrary to this, the length of the carbon chain in aviation fuel is C_9–C_{16} with no oxygen content. Therefore, the synthesis of aviation fuel from these sugars is a two-step process, first is the oxygen removal step (e.g. dehydration, hydrogenation and hydrogenolysis) and second involves C–C coupling reactions (e.g. aldol condensation, ketonization and oligomerization). Both reactions are performed in aqueous phase because of high water solubility of these molecules (Serrano-Ruiz et al. 2011). At moderate temperature about 200 °C, Pt-Re/C catalyst was used for partial deoxygenation of sugars (up to 80%) to produce a mixture of C_4–C_6 hydrocarbons such as acids, alcohols, ketones and heterocycles. Each family of these monofunctional compounds undergoes C–C coupling reactions to produce higher alkanes. For example, by performing aldol condensation with basic catalysts, ketones can be converted to C_8–C_{12} jet fuel components having low levels of branching (Kunkes et al. 2008; Pereira et al. 2020).

Mineral-acid-mediated dehydration of aqueous sugars produces furanics such as furfural and HMF (Chheda et al. 2007). These compounds can be converted to jet fuel by using the process of dehydration, hydrogenation and aldol-condensation reactions (Huber et al. 2005). These furanics condensed together or with other carbonyl compounds to produce larger carbon chain length compounds, which require further hydrogenation to obtain liquid hydrocarbon fuel with carbon chain lengths of C_9–C_{15} for HMF and C_8–C_{13} for furfural (West et al. 2008). Furanics can also be used for producing aromatic components of jet fuel. Chang et al. (2014) demonstrated the formation of *p*-xylene by cycloaddition of dimethylfuran with ethylene.

Levulinic acid (LA) is obtained through acid hydrolysis of cellulose from LCB. Jet fuel can be prepared from LA by catalytic methods including dehydration/ hydrogenation and ketonization processes. Formation of jet fuel components from LA is a multi-step reaction initiated with the formation of γ-valerolactone (GVL), an important intermediary, obtained by hydrogenation of LA (Díaz-Pérez and Serrano-Ruiz 2020). Then 5-nonanone ($C_9H_{18}O$) is produced from aqueous GVL through a pentanoic acid intermediate. This C_9 compound serves as a base molecule for the production of C_9–C_{18} branched alkanes. GVL can also be transformed into C_4 alkenes through a two-step process initiated with its decarboxylation over acidified silica/alumina to generate butene isomers and CO_2 (Bond et al. 2010). The produced butene is further oligomerized over an acid catalyst to generate a range of higher alkenes.

12.4.2 Production of Oil to Jet Fuels

Plant-derived oils can also be used as raw material for the production of jet fuel range liquid hydrocarbon through the processes of catalytic hydrotreating (oxygen removal at temperatures 250–400 °C and under high hydrogen pressures of 20–100 bar) (Serrano-Ruiz et al. 2012). The obtained fuel through this process is termed as hydroprocessed renewable jet (HRJ) fuel. Vegetable oils are composed of triglycerides, produced by esterification of glycerol and three fatty acids. Catalytic aqueous treatment of triglycerides releases the constituent fatty acids and a propane molecule. The released fatty acids can undergo the oxygen removal (in the form of water) step through hydrogenation/dehydration reactions (hydrodeoxygenation, HDO) to produce *n*-alkanes (Donnis et al. 2009). A part of fatty acids undergoes decarbonylation/decarboxylation reactions (hydrodecarbonlylation/ decarboxylation, HDC) which are step-down processes and result in *n*-alkanes with one carbon less than the parent molecules, where oxygen is removed as CO or CO_2, respectively (Donnis et al. 2009). The most applicable vegetable oils for biofuel contain C_{16} or C_{18} fatty acid which produces C_{15}–C_{17} alkanes after hydrotreating process. Therefore, further isomerization and hydrocracking processes are required to alter the carbon chain length to C_9–C_{16} in the jet fuel range. Hydrotreating process of vegetable oils is carried out using alumina-supported metal (e.g. Co, Mo, Ni, W) sulphides as catalyst (Khan et al. 2019). This route of jet fuel production can be applied for a large variety of feedstock such as triglycerides, fatty acids, vegetable oils, animal fats, waste cooking oils and trapped greases. Vegetable oils (e.g. camelina, jatropha etc.) containing high degree of unsaturations are particularly suitable feedstock for jet fuel production since these unsaturated fatty acids produce branched HCs (apt for cold-flow properties) during hydrotreating. Camelina and Jatropha are the important non-edible oil-producing plants which are commonly used for jet fuel production (Shonnard et al. 2010; Kasim and Harvey 2011). Oil produced from algae can also be used as feedstock for jet fuels production with yields (up to 77%) with suitable branched alkane content (Verma et al. 2011). Hydrotreating of oils is a very useful method for producing linear and slightly branched alkanes, whereas this method fails in producing other important components of jet fuels, namely cycloalkanes and aromatics. Therefore, an important advancement was carried out in the process where HRJ produced from Jatropha oil using NiMo and NiW catalysts supported on a silicoaluminophosphate (SAPO) by Verma et al. (2015). Interestingly, in addition to jet fuel alkanes, a significant quantity of aromatics (8%) was also obtained over the acidic SAPO support. Rabaev et al. (2015) also exploited SAPOs to convert soybean oil to jet fuels (48% yields) with 12% aromatic content. The HRJ was the first approved biojet fuel conversion process used for testing flights, and many companies have already adopted the process for its commercial production.

12.4.3 Production of Gas to Jet Fuels

Production of GTJ requires two unlike processes, namely gasification of biomass to syngas (H_2/CO), followed by Fischer–Tropsch (FT) conversion of syngas into jet fuel HCs (Gruber et al. 2019). In the gasification process, any carbon source (natural gas, coal and biomass) can be converted into a mixture of CO, H_2, CO_2, CH_4 under high temperature and controlled oxidizing conditions. Composition of the gas stream produced after the gasification process is varied according to biomass source, particle size, gasifier design and gasification conditions (Serrano-Ruiz and Dumesic 2012). The utilization of LCB resources has two important consequences for the downstream FT process. First is the presence of a larger number of impurities (N, Cl, S, ashes, tars, alkali) associated with the produced gas stream (CO and H_2) and ultimately causing deactivation of FT catalysts. Second is high oxygen content of LCB results in H_2/CO ratios (c. 0.5) which are lower than required value, for the FT synthesis of hydrocarbons (c. 2). Thus, before the FT process the produced syngas was cleared via additional cleaning and compositionally adjusted using water gas shift (WGS) processes (Dry 2004). This additional step significantly increases the additional cost and complexity of the process. The FT synthesis is an exothermic industrial technique that converts the syngas into a mixture of hydrocarbons ranging from C_1 to C_{50}, at moderate temperatures and pressures, over metal-based catalysts. The catalyst composition and reaction condition (temperature and pressure) during the process determine the nature of hydrocarbon produced. For example, use of cobalt-based catalysts produces linear alkanes (Saib et al. 2010; Ellis et al. 2019), whereas iron-based catalysts can operate at lower H_2/CO ratios and are more selective towards olefins. With the purpose of producing hydrocarbons in jet fuel range (C_9–C_{16}), first the process is operated at temperature around 230 °C to produce heavy hydrocarbons (waxes), which was followed by controlled cracking and isomerization steps to obtain jet fuel components (Dry 2004). Therefore, after FT reactor conventional petrochemical processes such as hydrocracking, isomerization and fractionation are usually required to adjust the hydrocarbons to the jet fuel range. The complexity and cost associated with GTJ route are an important limiting factor for commercialization of this technology.

12.4.4 Production of Alcohol to Jet Fuels

Production of jet fuel from alcohol requires three main steps: (i) Bioalcohol is dehydrated to their corresponding olefin, (ii) oligomerization of produced olefin to higher olefin, (iii) hydrogenation of olefins to alkanes. Ethanol (C_2) and butanol (C_4) are the two bioalcohols commonly used for producing jet fuels. These alcohols can be produced through microbial fermentation of sugars derived from LCB. The detailed process for the production of ethanol from LCB has been discussed above. Direct use of ethanol obtained from LCB as a fuel in blended form has many disadvantages such as high water absorption, high volatility, low flash point, low energy density and corrosiveness (Wang and Tao 2016). Through the ATJ route, these oxygenated fuels are converted into a mixture of hydrocarbons in jet fuel range. Recently Eagan et al. (2019) published a detailed review on catalytic routes available for converting ethanol to jet fuels and biodiesel. Butanol is better than ethanol in terms of water miscibility and energy content but still it is not suitable for jet fuel applications directly. Acidic catalysts such as silica-alumina, zeolites, silicoaluminophosphates and heteropolyacids can be used for dehydrating ethanol into ethylene (Zhan et al. 2010; Silvester et al. 2015). Dehydration of butanol can also be achieved by using acid catalysts at moderate temperatures, but after dehydration a mixture of olefins is obtained such as isobutylene, *n*-butene and 2-butene. The isobutylene and linear olefins formation can be controlled by the catalyst composition and porosity (Armstrong et al. 1993). Silica-alumina catalyst with 95% selectivity at 380 °C temperature was used to produce 1-butene from *n*-butanol. For oligomerization of produced olefins, well-known commercial homogeneous and heterogeneous catalysts were used. Like a commercial Ziegler–Natta catalysts was used for ethylene oligomerization to linear α-olefins. Many reports are available on oligomerization of different alkenes using different catalysts (Díaz-Pérez and Serrano-Ruiz 2020). Oligomerized olefins of suitable chain length were further hydrogenated to produce a mixture of alkanes suitable for jet fuel applications.

12.5 Conversion of Lignin to Hydrocarbons

Lignin is a phenol-based biopolymer contributing 10–25 wt% in LCB and has higher energy density compared with cellulose and hemicellulose. For converting lignin into fuel, its depolymerization and upgrading are required (Han et al. 2019). Pyrolysis is the simplest method for breakdown of lignin into smaller fragments and obtaining liquid products.

However, this method has some disadvantages such as low yield of liquid products, phenolic monomers and generation of high amounts of solid residues.

To avoid these troubles, various solvents (for example, water, ethanol, butanol, phenol, water–phenol mixtures and water–alcohol mixtures) can be used in mild reaction conditions for lignin depolymerization (Okuda et al. 2004). Use of alcohol is more appropriate than water because the phenolic monomers are miscible with solvent and thus avoid further recombination. After depolymerization of lignin, monomers mixture of phenolic compounds, including syringol, phenol, guaiacol and derivatives was obtained. Because of complex structure and high oxygen content of lignin, HDO was essential for the conversion of phenolic compounds to conventional liquid alkanes, Figure 12.3. Earlier reports on

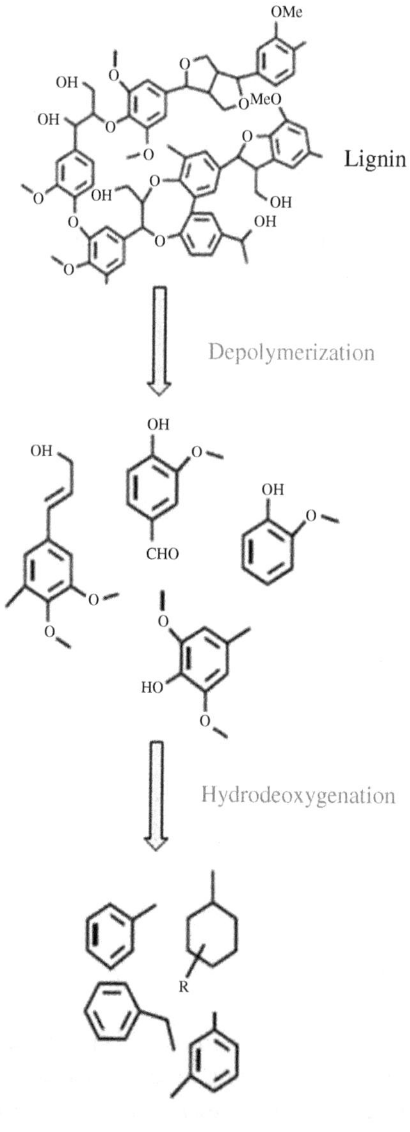

Figure 12.3 Conversion of lignin to hydrocarbon fuel.

HDO reaction were mainly engrossed on the conversion of phenol, guaiacol and syringol (lignin derivatives) to cycloalkanes or aromatic HCs (Han et al. 2019).

For efficient conversion of lignin into desired hydrocarbon, model compounds should be investigated by establishing proper reaction conditions. Selective breakdown of aryl ethers C–O bonds into phenolic dimers has attracted extensive attention, because it could provide better understanding for lignin depolymerization (Yan et al. 2008; Sergeev and Hartwig 2011). There are a number of phenolic dimer model compounds, having C–C and C–O–C linkages, such as α-*O*-4, β-*O*-4, 5-5, 4-*O*-5, β-β, β-1 and β-5; in which, β-*O*-4 linkage (alkyl-aryl ether) is the richest bond in lignin and phenolic dimers, accounting up to 55% (Zhao and Lercher 2012). There are few recent studies on HDO of the lignin and phenolic dimers to produce cycloalkanes and aromatic hydrocarbons using different functional catalysts.

Zhao and Lercher (2012) studied the combination of catalysts for producing cycloalkanes from phenolic dimers obtained from lignin. The use of dual catalysts Pd/C with HZSM-5 was demonstrated for removing oxygen-containing functional groups. The results show that 46% cyclohexane (C_6) and 54% ethyl cyclohexane (C_8) were obtained quantitatively from β-*O*-4 linkage model compound at 200 °C and 5 MPa H_2 for 2 hours. This catalyst combination on same reaction conditions was also applied for conversion of various phenolic dimers to produce cyclohexane (C_6) and alkyl cyclohexane (C_8). In addition, C_7–C_9 alkanes can also be obtained from α-*O*-4 dimer with yield of 93.6% over Pd/C catalyst in aqueous solution of phosphoric acid @ 4 MPa H_2 and 250 °C.

In the same way, Zhang et al. 2014 also projected a dual-functioning catalyst Ru/HZSM-5 (Si/Al) with metal active sites and acid to remove functional groups containing oxygen from phenolic dimers in water at 200 °C. Many scientists have reported new routes for producing jet or diesel fuel range alkanes from lignin. Such as (Bi et al. 2015; Liu et al. 2015b) discovered a new pathway for the one-pot conversion, and Shao et al. 2017 worked on a new way to develop aromatics efficiently from organosolv lignin.

In conclusion, conversion of phenolic compounds to hydrocarbons through HDO has been studied extensively in recent years. Some advancement was also made in the HDO of lignin-derived phenolic compounds.

12.6 Conclusion

The conversion of LCB to various fuel alternatives such as bioethanol and hydrocarbons was reviewed. Because of current environmental issues and oil reserve shortage, an alternative solution must be explored for producing renewable fuels. Because of huge reserve and economic sustainability, LCB can be exploited as a potential substrate to produce bioethanol and liquid hydrocarbons as sustainable fuels in recent years.

Extensive research has been performed on technological aspects for producing lignocellulosic ethanol. However, LCB to ethanol production is still a complicated task that restricts its utilization into commercial production. Industrialization of bioethanol production from LCB was limited because of many reasons such as complexity of biomass, expensive pretreatment methods, high cost of enzymes and limited microbial fermentation of C_5 sugar etc.

Research on the conversion of LCB to hydrocarbon fuels through integrated processes is developing nowadays. All types of monosaccharide produced after hydrolysis of cellulose and hemicellulose can act as intermediate for production of sugar alcohols through hydrogenation, furans by means of hydration and C_5/C_6 alkanes over bi-functional catalyst by HDO. Aqueous-phase HDO, over a combined catalyst system containing hydrogenation catalyst and acids, produces C_5/C_6 alkanes directly from cellulose, hemicellulose and biomass.

Production of jet fuel range hydrocarbons was also reviewed in present study. The main catalytic routes for producing jet fuels include OTJ, GTJ, ATJ and STJ. OTJ was performed using vegetable oils by hydrotreating and fractionation reaction steps over alumina-supported metal sulphide catalysts. In GTJ route, lignocellulosic feedstock was converted through gasification, FT and fractionation steps over Co- and Fe-based catalysts. Alcohol derived from biomass was converted to jet fuel through the ATJ route by dehydration, oligomerization, hydrogenation and fractionation steps on heterogeneous and homogeneous acid catalysts. Sugars, furanics and platform molecules through STJ route produce jet fuels by deoxygenation, C–C coupling, hydrogenation and fractionation reactions using a wide range of heterogeneous catalysts.

The chemical composition of jet fuel also includes cycloalkanes and aromatics hydrocarbons. These components were principally derived from lignin through depolymerization linked with the HDO reactions. Thus, all the components of lignocellulosic resources can be efficiently utilized for the production of liquid transportation fuels. Still there are many scopes to upgrade the existing conversion technology and make the process more viable on technological and economical angles.

References

Amezcua-Allieri, M. A., Sánchez Durán, T., & Aburto, J. (2017). Study of chemical and enzymatic hydrolysis of cellulosic material to obtain fermentable sugars. *Journal of Chemistry*, 2017. 9pp. Article ID 5680105, https://doi.org/10.1155/2017/5680105

Armstrong, P. A., Bhatt, B., Heydorn, E. C., & Toseland, B. A. (1993). *Isobutanol Dehydration: A Key Step in Producing MTBE from Syngas*. Air Products and Chemicals, Inc.

Balat, M. (2011). Production of bioethanol from lignocellulosic materials via the biochemical pathway: a review. *Energy Conversion and Management*, 52(2), 858–875.

de Beeck, B. O., Dusselier, M., Geboers, J., Holsbeek, J., Morré, E., Oswald, S., . . . & Sels, B. F. (2015). Direct catalytic conversion of cellulose to liquid straight-chain alkanes. *Energy & Environmental Science*, 8(1), 230–240.

Betancourt, P., Rives, A., Hubaut, R., Scott, C. E., & Goldwasser, J. (1998). A study of the ruthenium–alumina system. *Applied Catalysis A: General*, 170(2), 307–314.

Bi, P., Wang, J., Zhang, Y., Jiang, P., Wu, X., Liu, J., . . . & Li, Q. (2015). From lignin to cycloparaffins and aromatics: directional synthesis of jet and diesel fuel range biofuels using biomass. *Bioresource Technology*, 183, 10–17.

Binder, J. B., & Raines, R. T. (2010). Fermentable sugars by chemical hydrolysis of biomass. *Proceedings of the National Academy of Sciences*, 107(10), 4516–4521.

Bond, J. Q., Alonso, D. M., Wang, D., West, R. M., & Dumesic, J. A. (2010). Integrated catalytic conversion of γ-valerolactone to liquid alkenes for transportation fuels. *Science*, 327(5969), 1110–1114.

Brodeur, G., Yau, E., Badal, K., Collier, J., Ramachandran, K. B., & Ramakrishnan, S. (2011). Chemical and physicochemical pretreatment of lignocellulosic biomass: a review. *Enzyme Research*, 2011. Article ID 787532, 17pp. https://doi.org/10.4061/2011/787532

Cardona, C.A., Sánchez, O.J., Montoya, M.I., and Quintero, J.A. (2005). Analysis of fuel ethanol production processes using lignocellulosic biomass and starch as feedstocks. *Seventh World Congress of Chemical Engineering*, Glasgow, Scotland, UK.

Cavka, A., & Jönsson, L. J. (2013). Detoxification of lignocellulosic hydrolysates using sodium borohydride. *Bioresource Technology*, 136, 368–376.

Chang, C. C., Green, S. K., Williams, C. L., Dauenhauer, P. J., & Fan, W. (2014). Ultra-selective cycloaddition of dimethylfuran for renewable p-xylene with H-BEA. *Green Chemistry*, 16(2), 585–588.

Chen, K., Tamura, M., Yuan, Z., Nakagawa, Y., & Tomishige, K. (2013). One-pot conversion of sugar and sugar polyols to *n*-alkanes without C–C dissociation over the Ir-ReO_x/SiO_2 catalyst combined with H-ZSM-5. *ChemSusChem*, 6(4), 548–548.

Chheda, J. N., Román-Leshkov, Y., & Dumesic, J. A. (2007). Production of 5-hydroxymethylfurfural and furfural by dehydration of biomass-derived mono-and poly-saccharides. *Green Chemistry*, 9(4), 342–350.

Chiaramonti, D., Prussi, M., Ferrero, S., Oriani, L., Ottonello, P., Torre, P., & Cherchi, F. (2012). Review of pretreatment processes for lignocellulosic ethanol production, and development of an innovative method. *Biomass and Bioenergy*, 46, 25–35.

Davda, R. R., & Dumesic, J. A. (2003). Catalytic reforming of oxygenated hydrocarbons for hydrogen with low levels of carbon monoxide. *Angewandte Chemie International Edition*, 42(34), 4068–4071.

Díaz-Pérez, M. A., & Serrano-Ruiz, J. C. (2020). Catalytic production of jet fuels from biomass. *Molecules*, 25(4), 802.

Donnis, B., Egeberg, R. G., Blom, P., & Knudsen, K. G. (2009). Hydroprocessing of bio-oils and oxygenates to hydrocarbons. Understanding the reaction routes. *Topics in Catalysis*, 52(3), 229–240.

Dry, M. E. (2004). Present and future applications of the Fischer–Tropsch process. *Applied Catalysis A: General*, 1(276), 1–3.

Eagan, N. M., Kumbhalkar, M. D., Buchanan, J. S., Dumesic, J. A., & Huber, G. W. (2019). Chemistries and processes for the conversion of ethanol into middle-distillate fuels. *Nature Reviews Chemistry*, 3(4), 223–249.

Elgharbawy, A. A., Alam, M. Z., Moniruzzaman, M., & Goto, M. (2016). Ionic liquid pretreatment as emerging approaches for enhanced enzymatic hydrolysis of lignocellulosic biomass. *Biochemical Engineering Journal*, 109, 252–267.

Ellis, P. R., Enache, D. I., James, D. W., Jones, D. S., & Kelly, G. J. (2019). A robust and precious metal-free high performance cobalt Fischer–Tropsch catalyst. *Nature Catalysis*, 2(7), 623–631.

Fatma, S., Hameed, A., Noman, M., Ahmed, T., Shahid, M., Tariq, M., . . . & Tabassum, R. (2018). Lignocellulosic biomass: a sustainable bioenergy source for the future. *Protein and Peptide Letters*, 25(2), 148–163.

Gírio, F. M., Fonseca, C., Carvalheiro, F., Duarte, L. C., Marques, S., & Bogel-Łukasik, R. (2010). Hemicelluloses for fuel ethanol: a review. *Bioresource Technology*, 101(13), 4775–4800.

Gruber, H., Groß, P., Rauch, R., Reichhold, A., Zweiler, R., Aichernig, C., . . . & Hofbauer, H. (2019). Fischer-Tropsch products from biomass-derived syngas and renewable hydrogen. *Biomass Conversion and Biorefinery*, 1–12. https://doi.org/10.1007/s13399-019-00459-5.

Han, X., Guo, Y., Liu, X., Xia, Q., & Wang, Y. (2019). Catalytic conversion of lignocellulosic biomass into hydrocarbons: a mini review. *Catalysis Today*, 319, 2–13.

Hoffer, B. W., Crezee, E., Devred, F., Mooijman, P. R. M., Sloof, W. G., Kooyman, P. J., . . . & Moulijn, J. A. (2003). The role of the active phase of Raney-type Ni catalysts in the selective hydrogenation of d-glucose to d-sorbitol. *Applied Catalysis A: General*, 253(2), 437–452.

Huber, G. W., & Dumesic, J. A. (2006). An overview of aqueous-phase catalytic processes for production of hydrogen and alkanes in a biorefinery. *Catalysis Today*, 111(1-2), 119–132.

Huber, G. W., Shabaker, J. W., & Dumesic, J. A. (2003). Raney Ni-Sn catalyst for H_2 production from biomass-derived hydrocarbons. *Science*, 300(5628), 2075–2077.

Huber, G. W., Cortright, R. D., & Dumesic, J. A. (2004). Renewable alkanes by aqueous-phase reforming of biomass-derived oxygenates. *Angewandte Chemie*, 116(12), 1575–1577.

Huber, G. W., Chheda, J. N., Barrett, C. J., & Dumesic, J. A. (2005). Production of liquid alkanes by aqueous phase processing of biomass-derived carbohydrates. *Science*, 308(5727), 1446–1450.

Jönsson, L. J., & Martín, C. (2016). Pretreatment of lignocellulose: formation of inhibitory by-products and strategies for minimizing their effects. *Bioresource Technology*, 199, 103–112.

Jönsson, L. J., Alriksson, B., & Nilvebrant, N. O. (2013). Bioconversion of lignocellulose: inhibitors and detoxification. *Biotechnology for Biofuels*, 6(1), 16.

Jurasz, J., Canales, F. A., Kies, A., Guezgouz, M., & Beluco, A. (2020). A review on the complementarity of renewable energy sources: concept, metrics, application and future research directions. *Solar Energy*, 195, 703–724.

Kamei, I., Hirota, Y., & Meguro, S. (2012). Integrated delignification and simultaneous saccharification and fermentation of hard wood by a white-rot fungus, *Phlebia* sp. MG-60. *Bioresource Technology*, 126, 137–141.

Karimi, K., Shafiei, M., & Kumar, R. (2013). Progress in physical and chemical pretreatment of lignocellulosic biomass. In *Biofuel Technologies* (pp. 53–96). Springer, Berlin, Heidelberg.

Kasim, F. H., & Harvey, A. P. (2011). Influence of various parameters on reactive extraction of *Jatropha curcas* L. for biodiesel production. *Chemical Engineering Journal*, 171(3), 1373–1378.

Khan, S., Lup, A. N. K., Qureshi, K. M., Abnisa, F., Daud, W. M. A. W., & Patah, M. F. A. (2019). A review on deoxygenation of triglycerides for jet fuel range hydrocarbons. *Journal of Analytical and Applied Pyrolysis*, 140, 1–24.

Kumar, A. K., & Sharma, S. (2017). Recent updates on different methods of pretreatment of lignocellulosic feedstocks: a review. *Bioresources and Bioprocessing*, 4(1), 7.

Kumar, R., Tabatabaei, M., Karimi, K., & Sárvári Horváth, I. (2016). Recent updates on lignocellulosic biomass derived ethanol-a review. *Biofuel Research Journal*, 3(1), 347–356.

Kumari, D., & Singh, R. (2018). Pretreatment of lignocellulosic wastes for biofuel production: a critical review. *Renewable and Sustainable Energy Reviews*, 90, 877–891.

Kunkes, E. L., Simonetti, D. A., West, R. M., Serrano-Ruiz, J. C., Gärtner, C. A., & Dumesic, J. A. (2008). Catalytic conversion of biomass to monofunctional hydrocarbons and targeted liquid-fuel classes. *Science*, 322(5900), 417–421.

Liu, X., Wang, X., Yao, S., Jiang, Y., Guan, J., & Mu, X. (2014). Recent advances in the production of polyols from lignocellulosic biomass and biomass-derived compounds. *RSC Advances*, 4(90), 49501–49520.

Liu, Y., Chen, L., Wang, T., Zhang, Q., Wang, C., Yan, J., & Ma, L. (2015a). One-pot catalytic conversion of raw lignocellulosic biomass into gasoline alkanes and chemicals over $LiTaMoO_6$ and Ru/C in aqueous phosphoric acid. *ACS Sustainable Chemistry & Engineering*, 3(8), 1745–1755.

Liu, Y., Chen, L., Wang, T., Zhang, X., Long, J., Zhang, Q., & Ma, L. (2015b). High yield of renewable hexanes by direct hydrolysis–hydrodeoxygenation of cellulose in an aqueous phase catalytic system. *RSC Advances*, 5(15), 11649–11657.

Lu, Y., Wang, Y., Xu, G., Chu, J., Zhuang, Y., & Zhang, S. (2010). Influence of high solid concentration on enzymatic hydrolysis and fermentation of steam-exploded corn stover biomass. *Applied Biochemistry and Biotechnology*, 160(2), 360–369.

Machuca, A., & Ferraz, A. (2001). Hydrolytic and oxidative enzymes produced by white-and brown-rot fungi during Eucalyptus grandis decay in solid medium. *Enzyme and Microbial Technology*, 29(6-7), 386–391.

Martínez, Á.T., Speranza, M., Ruiz-Dueñas, F.J., et al. (2005). Biodegradation of lignocellulosics: microbial, chemical, and enzymatic aspects of the fungal attack of lignin. *International Microbiology* 8:195–204.

Mussatto, S.I., and Teixeira, J.A. (2010). Lignocellulose as raw material in fermentation processes. *Current Research, Technology and Education Topics in Applied Microbiology and Microbial Biotechnology* 2: 897–907

Nanda, S., A Kozinski, J., & K Dalai, A. (2016). Lignocellulosic biomass: a review of conversion technologies and fuel products. *Current Biochemical Engineering*, 3(1), 24–36.

Okuda, K., Umetsu, M., Takami, S., & Adschiri, T. (2004). Disassembly of lignin and chemical recovery—rapid depolymerization of lignin without char formation in water–phenol mixtures. *Fuel Processing Technology*, 85(8-10), 803–813.

Panwar, N. L., Kaushik, S. C., & Kothari, S. (2011). Role of renewable energy sources in environmental protection: a review. *Renewable and Sustainable Energy Reviews*, 15(3), 1513–1524.

Pereira, S. C., Souza, M., Pinto, J., Esteves, L. M., Lam, Y. L., e Miranda, L. S. D. M., & Pereira, M. M. (2020). Sugar ketals as a platform molecule to overcome the limitation of converting biomass into green-hydrocarbons in a typical refinery. *Sustainable Energy & Fuels*, 4(3), 1312–1319.

Rabaev, M., Landau, M. V., Vidruk-Nehemya, R., Koukouliev, V., Zarchin, R., & Herskowitz, M. (2015). Conversion of vegetable oils on Pt/Al_2O_3/SAPO-11 to diesel and jet fuels containing aromatics. *Fuel*, 161, 287–294.

Rahman, M. M., Liu, R., & Cai, J. (2018). Catalytic fast pyrolysis of biomass over zeolites for high quality bio-oil–a review. *Fuel Processing Technology*, 180, 32–46.

Saib, A. M., Moodley, D. J., Ciobîcă, I. M., Hauman, M. M., Sigwebela, B. H., Weststrate, C. J., . . . & Van de Loosdrecht, J. (2010). Fundamental understanding of deactivation and regeneration of cobalt Fischer–Tropsch synthesis catalysts. *Catalysis Today*, 154(3-4), 271–282.

Saidur, R., Abdelaziz, E. A., Demirbas, A., Hossain, M. S., & Mekhilef, S. (2011). A review on biomass as a fuel for boilers. *Renewable and Sustainable Energy Reviews*, 15(5), 2262–2289.

Sampathkumar, K., Kumar, V., Sivamani, S., & Sivakumar, N. (2019). An insight into fungal cellulases and their industrial applications. In *Approaches to Enhance Industrial Production of Fungal Cellulases* (pp. 19–35). Springer, Cham.

Sánchez, C. (2009). Lignocellulosic residues: biodegradation and bioconversion by fungi. *Biotechnology Advances*, 27(2), 185–194.

Schutyser, W., Van den Bosch, S., Dijkmans, J., Turner, S., Meledina, M., Van Tendeloo, G., . . . & Sels, B. F. (2015). Selective nickel-catalyzed conversion of model and lignin-derived phenolic compounds to cyclohexanone-based polymer building blocks. *ChemSusChem*, 8(10), 1805–1818.

Sergeev, A. G., & Hartwig, J. F. (2011). Selective, nickel-catalyzed hydrogenolysis of aryl ethers. *Science*, 332(6028), 439–443.

Serrano-Ruiz, J. C., & Dumesic, J. A. (2012). Catalytic production of liquid hydrocarbon transportation fuels. In *Catalysis for Alternative Energy Generation* (pp. 29–56). Springer, New York, NY.

Serrano-Ruiz, J. C., Luque, R., & Sepúlveda-Escribano, A. (2011). Transformations of biomass-derived platform molecules: from high added-value chemicals to fuels via aqueous-phase processing. *Chemical Society Reviews*, 40(11), 5266–5281.

Serrano-Ruiz, J. C., Ramos-Fernández, E. V., & Sepúlveda-Escribano, A. (2012). From biodiesel and bioethanol to liquid hydrocarbon fuels: new hydrotreating and advanced microbial technologies. *Energy & Environmental Science*, 5(2), 5638–5652.

Shao, Y., Xia, Q., Dong, L., Liu, X., Han, X., Parker, S. F., . . . & Wang, Y. (2017). Selective production of arenes via direct lignin upgrading over a niobium-based catalyst. *Nature Communications*, 8(1), 1–9.

Shonnard, D. R., Williams, L., & Kalnes, T. N. (2010). Camelina-derived jet fuel and diesel: sustainable advanced biofuels. *Environmental Progress & Sustainable Energy*, 29(3), 382–392.

Shuai, L., Yang, Q., Zhu, J. Y., Lu, F. C., Weimer, P. J., Ralph, J., & Pan, X. J. (2010). Comparative study of SPORL and dilute-acid pretreatments of spruce for cellulosic ethanol production. *Bioresource Technology*, 101(9), 3106–3114.

Silvester, L., Lamonier, J. F., Faye, J., Capron, M., Vannier, R. N., Lamonier, C., . . . & Dumeignil, F. (2015). Reactivity of ethanol over hydroxyapatite-based Ca-enriched catalysts with various carbonate contents. *Catalysis Science & Technology*, 5(5), 2994–3006.

Sindhu, R., Binod, P., & Pandey, A. (2016). Biological pretreatment of lignocellulosic biomass–an overview. *Bioresource Technology*, 199, 76–82.

Singh, M. K., Kumar, M., & Thakur, I. S. (2017). Proteomic characterization and schizophyllan production by *Schizophyllum commune* ISTL04 cultured on *Leucaena leucocephala* wood under submerged fermentation. *Bioresource Technology*, 236, 29–36.

Sun, Y., & Cheng, J. (2002). Hydrolysis of lignocellulosic materials for ethanol production: a review. *Bioresource Technology*, 83(1), 1–11.

Sun, F., Chen, L., Weng, Y., Wang, T., Qiu, S., Li, Q., . . . & Ma, L. (2017). Transformation of biomass polyol into hydrocarbon fuels in aqueous medium over Ru-Mo/CNT catalyst. *Catalysis Communications*, 99, 30–33.

Thiribhuvanamala, G., Kalaiselvi, G., Parthasarathy, S., Madhavan, S., & Prakasam, V. (2017). Extracellular secretion of lignocellulolytic enzymes by diverse white rot basidiomycetes fungi. *Annals of Phytomedicine*, 6(1), 20–29.

Tian, J., Wang, J., Zhao, S., Jiang, C., Zhang, X., & Wang, X. (2010). Hydrolysis of cellulose by the heteropoly acid $H_3PW_{12}O_{40}$. *Cellulose*, 17(3), 587–594.

US Energy Information Administration. International Energy Outlook. (2019). https://www.eia.gov/outlooks/ieo/pdf/ieo2019.pdf (accessed 15 January 2020).

Vane, L. M. (2008). Separation technologies for the recovery and dehydration of alcohols from fermentation broths. *Biofuels, Bioproducts and Biorefining*, 2(6), 553–588.

Verma, D., Kumar, R., Rana, B. S., & Sinha, A. K. (2011). Aviation fuel production from lipids by a single-step route using hierarchical mesoporous zeolites. *Energy & Environmental Science*, 4(5), 1667–1671.

Verma, D., Rana, B. S., Kumar, R., Sibi, M. G., & Sinha, A. K. (2015). Diesel and aviation kerosene with desired aromatics from hydroprocessing of jatropha oil over hydrogenation catalysts supported on hierarchical mesoporous SAPO-11. *Applied Catalysis A: General*, 490, 108–116.

Vilcocq, L., Koerin, R., Cabiac, A., Especel, C., Lacombe, S., & Duprez, D. (2014). New bifunctional catalytic systems for sorbitol transformation into biofuels. *Applied Catalysis B: Environmental*, 148, 499–508.

Wang, W. C., & Tao, L. (2016). Bio-jet fuel conversion technologies. *Renewable and Sustainable Energy Reviews*, 53, 801–822.

Wang, C., Pan, J., Li, J., & Yang, Z. (2008). Comparative studies of products produced from four different biomass samples via deoxy-liquefaction. *Bioresource Technology*, 99(8), 2778–2786.

Wang, S., Dai, G., Yang, H., & Luo, Z. (2017). Lignocellulosic biomass pyrolysis mechanism: a state-of-the-art review. *Progress in Energy and Combustion Science*, 62, 33–86.

Wang, M., Dewil, R., Maniatis, K., Wheeldon, J., Tan, T., Baeyens, J., & Fang, Y. (2019). Biomass-derived aviation fuels: challenges and perspective. *Progress in Energy and Combustion Science*, 74, 31–49.

Wang, C., Zhang, X., Liu, Q., Zhang, Q., Chen, L., & Ma, L. (2020). A review of conversion of lignocellulose biomass to liquid transport fuels by integrated refining strategies. *Fuel Processing Technology*, 208, 106485.

West, R. M., Liu, Z. Y., Peter, M., & Dumesic, J. A. (2008). Liquid alkanes with targeted molecular weights from biomass-derived carbohydrates. *ChemSusChem: Chemistry & Sustainability Energy & Materials*, 1(5), 417–424.

Xia, Q., Chen, Z., Shao, Y., Gong, X., Wang, H., Liu, X., . . . & Wang, Y. (2016). Direct hydrodeoxygenation of raw woody biomass into liquid alkanes. *Nature Communications*, 7, 11162.

Yan, N., Zhao, C., Dyson, P. J., Wang, C., Liu, L. T., & Kou, Y. (2008). Selective degradation of wood lignin over noble-metal catalysts in a two-step process. *ChemSusChem: Chemistry & Sustainability Energy & Materials*, 1(7), 626–629.

Ying, Q., Na, T., Chengkai, Z., Haining, N., & Jin, Z. (2018). High efficient hydrolysis of cellulose into sugar by chemical catalytic method. *Progress in Chemistry*, 30(9), 1415.

Yu, Y., Lou, X., & Wu, H. (2008). Some recent advances in hydrolysis of biomass in hot-compressed water and its comparisons with other hydrolysis methods. *Energy & Fuels*, 22(1), 46–60.

Zabed, H., Sahu, J. N., Boyce, A. N., & Faruq, G. (2016). Fuel ethanol production from lignocellulosic biomass: an overview on feedstocks and technological approaches. *Renewable and Sustainable Energy Reviews*, 66, 751–774.

Zhan, N., Hu, Y., Li, H., Yu, D., Han, Y., & Huang, H. (2010). Lanthanum–phosphorous modified HZSM-5 catalysts in dehydration of ethanol to ethylene: a comparative analysis. *Catalysis Communications*, 11(7), 633–637.

Zhang, Q., Qiu, K., Li, B., Jiang, T., Zhang, X., Ma, L., & Wang, T. (2011). Isoparaffin production by aqueous phase processing of sorbitol over the Ni/HZSM-5 catalysts: effect of the calcination temperature of the catalyst. *Fuel*, 90(11), 3468–3472.

Zhang, Q., Jiang, T., Li, B., Wang, T., Zhang, X., Zhang, Q., & Ma, L. (2012). Highly selective sorbitol hydrogenolysis to liquid alkanes over Ni/HZSM-5 catalysts modified with pure silica MCM-41. *ChemCatChem*, 4(8), 1084–1087.

Zhang, W., Chen, J., Liu, R., Wang, S., Chen, L., & Li, K. (2014). Hydrodeoxygenation of lignin-derived phenolic monomers and dimers to alkane fuels over bifunctional zeolite-supported metal catalysts. *ACS Sustainable Chemistry & Engineering*, 2(4), 683–691.

Zhao, C., & Lercher, J. A. (2012). Selective hydrodeoxygenation of lignin-derived phenolic monomers and dimers to cycloalkanes on Pd/C and HZSM-5 catalysts. *ChemCatChem*, 4(1), 64–68.

Zheng, Y., Pan, Z., & Zhang, R. (2009). Overview of biomass pretreatment for cellulosic ethanol production. *International Journal of Agricultural and Biological Engineering*, 2(3), 51–68.

13

Sustainable Solution for Future Energy Challenges Through Microbes

Sumit Sahni, Manoj Kumar Singh, and Anita Narang

Department of Botany, Acharya Narendra Dev College, University of Delhi, New Delhi, India

13.1 Introduction

Unprecedented growth in population and increase in the socio-economic stature of the middle class globally have created an incessant demand for energy. Today, most of the global energy demand is catered by fossil fuels which have limited reserves on earth and are rapidly depleting. Hence, these cannot sustain the burden of energy demands for more than two or three decades. This led the United Nations to add energy generation and distribution in its 17-point sustainable development goals (SDGs) in which it emphasized to increase the share of renewable energy progressively. Biofuels are an important one among the many available renewable energy sources which have the potential to satiate the ever-increasing energy demand and prove as a sustainable source of energy. Anything which can be used as fuel and has its origin from living organisms can be considered as biofuel. Biofuels in the form of wood, wood chips, charcoal etc. have been in use since time immemorial but cannot take centre stage due to their own limitations. Alternative biofuels include bioethanol, biodiesel, biogas, biohydrogen and bioelectricity which are more usage-ready and can be generated from the biomass available. Based on the resources used to produce them, biofuels have been classified into four generations i.e. first, second, third and fourth. Each generation has its own merits and demerits. Some are well studied, and technologies have been developed to produce them efficiently but are competing with food crops; others have no competition with food crops; however, the technologies involved in their generation are in infancy and need lots of research for their commercialization. There are some roadblocks which deter to prove them as sustainable energy sources which will be overcome in coming years. The major organisms involved in biofuel generation are not the higher organisms but microorganisms such as bacteria, fungi and algae which contribute at each step of biofuel production ranging from presenting themselves as biomass to treatment of biomass or as producers of catalysing enzymes in myriads of biochemical reactions involved.

Energy: Crises, Challenges and Solutions, First Edition. Edited by Pardeep Singh, Suruchi Singh, Gaurav Kumar, and Pooja Baweja.

This chapter presents the glimpse of various generations of biofuels, technologies involved in their production, utilization of various microbes at different stages of biofuels generation and future prospects of biofuels.

13.2 Importance of Energy and Energy Statistics

Energy has emerged as the most essential commodity post industrial revolution. It has become imperative to address all the major global challenges such as economic and social upliftment, sustainable development, human well-being and global security. Energy consumption is explained in two different terms i.e. primary and final energy consumption. Primary energy consumption (PEC) includes the total energy consumed including all the losses or energy consumption during transformation, whereas final energy consumption (FEC) or delivered energy includes the energy reaches to the point of retail for delivery to final users and excludes inputs and losses involved in the energy transformation sectors. Population, economic activity and technology are the three gears which drive energy demand, and in future an exponential growth is predicted in all the three. The world population which was 2 billion in 1930 will reach 9.7 billion by 2050 which means addition of 7.7 billion people, a surge of 385% and proportionate increase in energy demand (UNDESA 2019). Figure 13.1 illustrates the trend in global energy demand across the world. At the same time, due to rapid transition in the world economic scenario, the most populous countries become the drivers of the world economy which results in billions of people ready to cross poverty line corroborating the fact of future energy challenges. These concepts are further getting consolidated by the data. The total primary energy consumption in the world is increasing by more than 4% per annum. At such pace, world's energy consumption will reach from 14017 Mtoe per annum in 2020 to 16706 Mtoe per annum in 2040. Major contributor to this increase is

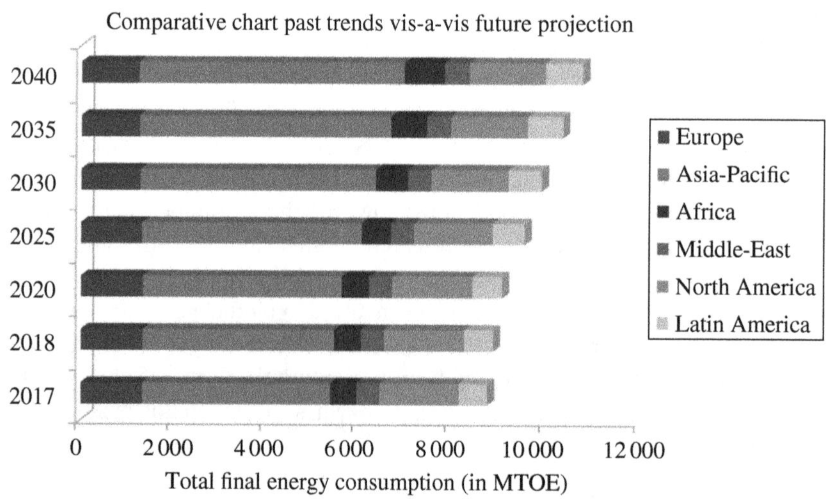

Figure 13.1 Global consumption of total final energy (EnerOutlook 2020).

the Asia-Pacific region with slightly higher than 7% per annum. A similar trend is observed in total final energy consumption as well (EnerOutlook 2020).

Of the total PEC globally, now approximately 80% is contributed by fossil fuels (coal, oil and gas); however, there is only limited supply of fossil fuels reserves remaining for e.g. 50–100 years of coal and 40 years of oil and gas. Moreover, 80% of the proven oil reserves are located in middle-east and central Asia which are politically and socially unstable. Above all, the price volatility of oil due to geo-political conflicts in 1973 and 1979 propelled the world economies to increase the stake of renewable energy substantially in energy production by the concentrated efforts of all countries. Of the 17 SDGs identified for the member nations by the United Nations for 2030, affordable and clean energy for all is one. To achieve this goal, multi-pronged strategy has been devised i.e. to increase the production of energy vis-à-vis greater contribution from renewables, improvement in the technology for better energy efficiency, increasing the financial flow to developing countries for R&D in renewable energy production (https://sustainabledevelopment.un.org/sdg7, 2020).

Although the pie of renewable energy in the total energy consumption has increased by 18% from the 2000 to 2015 but even at this pace the share of renewable energy will reach 21% by 2030, a substantial shortfall in UN SDG goals. Another parameter, global primary energy intensity (ratio of energy used per unit of GDP), is also showing similar patterns. It has improved from 5.9 to 5.1 in a span of 7 years (2010–2016), but is still short of the 2.7 per cent annual rate goal (https://sustainabledevelopment.un.org/sdg7, 2020). Moreover, till today, only 55% of the renewable energy share is contributed by modern forms, rest is derived from traditional methods of biomass use such as charcoal and fuel wood.

Of the many available renewable energy sources, only few can be considered as a sustainable fuel source for future needs,namely hydroelectric power, wind energy, solar heat energy, photovoltaic cells, passive energy systems, biofuels and geothermals. Despite their renewable nature, sustainability and environmental benefits, most of these sources have associated limitations ranging from huge land requirements, mankind displacements and ecological imbalances due to hydroelectric projects, non-availability of sites with sufficient wind speeds and death of birds through wind turbines, release of toxic chemicals in production of solar cells and production of hydrogen sulphide gas through geothermal plants (Pimental 2008) with exception in the case of biofuels.

13.3 Brief History of Biofuels

Biofuels are the energy source from biological materials which are either used directly and classified as primary biofuels such as fire wood, wood chips, wood pellets, charcoal or derived through technological modifications and known as secondary biofuels, namely bio-alcohols, biodiesel, biogas and bio-hydrogen gas (Alia et al. 2019). Primary biofuels have been in use since discovery of fire, but the large-scale use of secondary biofuels initiated with Nikolaus August Otto, who in 1860 promoted the use of ethanol as a fuel in early internal combustion engine. Later, Rudolph Diesel in 1897 designed a prototype engine to run directly on peanut oil. The famous model-T by Henry Ford, which paved the way to modern automobile industry, was designed to run on hemp-based ethanol (Guo et al. 2015). Surging need of fuels in the era of world war, huge production cost of bioethanol and availability of cheaper alternatives, namely petroleum, coal and gas, the use of biofuel became obsolete (Packer et al. 2016). It was in 1973 when Organization of the Petroleum Exporting Countries (OPEC) imparted oil embargo, and later due to Iranian revolution in 1979, there was huge volatility in oil prices which unleashed the interest of the world in biofuels (Difiglio 2014).

13.4 Classification of Biofuels

Biofuels, in addition to being classified as primary and secondary on the basis of their physical nature, have also been classified as first (1G), second (2G), third (3G) and fourth (4G) generation, based on the type of feedstock used and technologies involved in its conversion. The higher generation does not mean improvised fuels because in most of the generations, the fuels in use are the same i.e. alcohols, methyl or ethyl esters, methane gas, hydrogen gas etc. Rather, it is innovation and improvisation in technology to extract fuels from new kind of feedstock. New methods and technologies are being devised to produce fuels from non-conventional sources. Each generation has its pros and cons which are discussed below.

13.4.1 First Generation (1G)

The 1G biofuels are produced from feedstocks that belong to edible category such as bioethanol produced from potato, corn, wheat, rice, barley sugarcane, sugar beet, sweet sorghum, palm juice, fruits etc. or biodiesel produced from soybean, rapeseed and palm oil. These crops have been classified on the basis of nature of biomass they provide, namely starch and sugar-rich or oil-rich crops. Conversion of sugar to ethanol is a simple process which only involves fermentation; however, conversion of starch to ethanol directly is not possible using conventional technology due to the long-chain polymer structure (Zabed et al. 2016). These crops are subject to either physical processes such as grinding, mixing, cooking or biological processes involving treatment with microbes such as *Saccharomyces cerevisiae*, *Saccharomyces uvarum*, *Kluyveromyces*, and *Schizosaccharomyces pombe* or enzymatic treatment with amylase, pullulanase and glucoamylase to convert it into glucose, maltose or iso-maltose, and the process is known as saccharification (Naik et al. 2010). After this process, it follows same fermentation pathway as simple sugars. The pullulanase is obtained commercially from *Bacillus sp.* AN-7, *B. acidopullulyticus*, *B. deramificans*, *B. cereus FDTA-13*, *Klebsiella planticola* and *Geobacillus stearothermophilus* (Hii et al. 2012) and glucoamylase from *Aspergillus niger*, *Aspergillus awamori*, *Aspergillus terreus*, *Clostridium* sp. (James and Lee 1997). Of the diverse edible biomass available, sugarcane is the most used crop in bioethanol production (Rulli et al. 2016), and United States is the leader of the bioethanol market with 50% share in production of bioethanol globally followed by Brazil and China (FAO-OECD 2020). Regardless of the material, after digestion both sugar and starch undergo fermentation process. Of the multiple fermentation schemes known, the scheme which yields higher ethanol per mole of hexose was chosen for industrial production. *Saccharomyces cerevisiae*, and various species of the genus *Zymomonas*, convert the pyruvate obtained via Embden–Meyerhof pathway or Entner–Doudoroff pathway respectively to alcohol via pyruvate decarboxylase (EC 4.1.1.1)/alcohol dehydrogenase enzymes (ADH) (EC 1.1. 1.1) (Elshahed 2010). Ethanol production using *S. cerevisiae* has been the most thoroughly studied industrial fermentation process. Every aspect of the process has been thoroughly researched whether it is identification of strains capable of producing ethanol from multiple sugars or strains that can withstand relatively higher ethanol levels, to sustain production at high substrate levels, optimization of achieving high turnover ratio or near stoichiometric conversion of glucose to ethanol (Walker 1998; Liu et al. 2008).

Biodiesel, a yellowish liquid developed by trans-esterification of vegetable oil or animal fat, came into existence by George Chavanne in 1937, who was pioneer in presenting the procedure of alcoholysis of vegetable oil. It took 40 years from there to develop first industrial process for production of Biodiesel by Expedito Parente and another 12 years to operate the first industrial production of biodiesel from rapeseed oil in Austria (Guo et al. 2015). Since then, the biodiesel production has seen an upward trend as more countries adopted the technologies and initiated the research and development in the field. Since the year 2000, the world biodiesel production has increased consistently from 0.2 billion gallons to 6.3 billion gallons in 2013 and projected to rise up to 11.62 billion gallons by the year 2028 (Guo et al. 2015; OECD-FAO 2020). European Union has the maximum share in production of biodiesel with 36% followed by United States (19%) and Brazil (12%), and the most favoured feedstock is rapeseed and soybean oil (FAO-OECD 2020).

Trans-esterification of oils or fats to biodiesel can be achieved through three routes, namely non-catalysed reaction, chemical-catalysed reactions and enzyme-catalysed reaction. Several inherent limitations associated with non-catalysed and chemical-catalysed reactions (side reactions of saponification, expensive acid pretreatment, recovery and purification of catalyst, energy-intensive nature and requirement of waste water treatment) led the researchers to explore enzyme-catalysed reactions. Adoption of enzyme-based technology can either eliminate or minimize these problems (Norjannah et al. 2016). Lipases (EC3.1.1.3) belong to class 'hydrolases' and sub-classified as triacylglycerol acylhydrolases. It can perform hydrolysis of triacyl glycerols (TGs) as well as re-synthesis of long-chain acyl glycerols (or biodiesel), hence also termed as carboxylesterases. Lipases are ubiquitously found in plants, insects, animals and microbes but are commercially obtained from microbes due to their stability and lower cost of production (Christopher et al. 2014). The active site of lipase is constituted by serine, aspartic/glutamic acid and histidine. They get highly activated in the presence of substrate, a unique feature termed as 'interfacial activation' (Brzozowski et al. 1991; Ondul et al. 2015). Christopher et al. (2014) provided an exhaustive list of microorganisms which ranges from bacteria, fungi and yeasts from where lipases have been produced. The important members from each class have been listed in the Table 13.1. Of the many species, *Candida* sp., *Pseudomonas* sp. and *Rhizopus* sp. are still the preferred source of lipases. Lipases catalyse the trans-esterification reaction via formation of acyl-enzyme intermediates followed by donation of the acyl group to produce fatty acid alkyl esters (Christopher et al. 2014). It is unlikely for lipases from different sources to show the same activity on different substrates due to structural dissimilarities, hence process optimization is needed. Lipases have been classified into three groups based on their substrate specificity i.e. 1, 3-specific lipases, fatty-acid-specific lipases, and non-specific lipases. The 1, 3-specific lipases catalyse the ester bonds via hydrolysis at first and third position of TGs. The fatty-acid-specific lipases hydrolyse only those esters which have double bonds at C9 in

Table 13.1 Selected sources of microbial Lipases.

Bacteria	Fungi	Yeasts
Achromobacter lipolyticum	*Alternaria brassicicola*	*Candida rugosa*
Aeromonas hydrophila	*Aspergillus niger*	*Candida deformans*
Aeromonas caviae	*Mucor miehei*	*Rhodotorula mucilaginosa*
Acinetobacter sp.	*Penicillium cyclopium*	*Saccharomyces lipolytica*
Bacillus subtilis	*Humicola lanuginose*	*Trichosporon asahii*
Burkholderia glumae	*Rhizopus microsporus*	*Yarrowia lipolytica*
Staphylococcus sp.	*Rhizopus oryzae*	*Yamadazyma terventina*
Moraxella sp.	*Thermomyces lanuginosus*	*Geotrichum candidum*
Thalassospira permensis	*Rhizomucor miehei*	*Pichia silvicola*
Pseudomonas alcaligenes	*Streptomyces exfoliates*	*Pichia xylose*
Yersinia enterocolitica	*Metarhizium anisopliae*	*Cryptococcus sp*

cis-position. The non-specific lipases according to their name randomly cleave the acyl-glycerols in free fatty acids (FFA) (Christopher et al. 2014).

For ideal biodiesel production, lipases should be able to convert all three forms of glycerides (mono-, di- and tri-glycerides) to biodiesel. Hence, non-specific lipases are ideal in efficiently catalysing the esterification of FFA as it can cleave all types of glycerols.

Although the biodiesel yield from the trans-esterification reaction is usually similar in volume as the feedstock oil, however, reaction conditions and purification process greatly affect the yield and quality of the final product. There are many permutations and combinations of different feedstock oil, enzyme source and acyl acceptors used to produce biodiesel which is exquisitely compiled by Norjannah et al. (2016).

Some researchers also created an intermediate generation 1.5G between 1G and 2G keeping the sustainability of raw material in mind where 1G biofuels are obtained from food crops (starch and sugar-based), and 2G are non-food-based (cellulose based), 1.5G of crops are non-food grade but still rely on sugar or starch-based material i.e. cassava or sweet sorghum (Lackner 2015). To make the understanding lucid, we are not elaborating this generation in the current chapter.

13.4.2 Second Generation (2G)

By 2008, world biofuels production reached 83 billion litres with contribution of bioethanol being approximately 82% and rest by biodiesel. Apparently, all of these biofuels were of first generation i.e. sources from food crops such as sugarcane, maize, canola, sunflower etc. (Eisentraut 2010). Increasing demand of fuels compelled the farmers to grow food crops more for energy production. This led to exorbitant increase in prices of food commodities which is followed by huge public backlash against using food crops for energy production. As a result, scientists started finding opportunities in non-food crops as alternative sources to produce biofuels. This resulted in evolution of 2G biofuels which are based on non-edible biomass such as lignocellulosic material and non-edible oil crops. The lignocellulosic materials were sourced from forestry feedstock, residues from agricultural and food processing industries to produce bioethanol, biobutanol or biogasoline and various non-edible oil sources such as jatropha, karanja waste cooking oil, tallow, lard, restaurant grease, salmon oil, etc. for biodiesel production. Ahorsu et al. (2018) and Baruah et al. (2018) compiled general biomass types according to their origin and major components.

The agricultural and forestry feedstocks which include straw, hulls, stems etc. are also known as lignocellulosic material as it is composed of cellulose, hemicellulose, lignin and some amount of starch as well in decreasing order of their abundance. Although the composition of constituents in lignocellulosic biomass (LCB) varies as per the source but as a general consideration cellulose ranges from 35 to 50%, hemicellulose by 20–35% and lignin ranges from 10 to 25%. Baruah et al. (2018) compiled an exhaustive list of 19 plant materials with their cellulose, hemicellulose and lignin percentage. Out of three celluloses and hemicelluloses are polysaccharides, but lignin is an alkyl-aromatic polymer. Cellulose is a polymer of D-glucose joined by β-1,4-glycosidic bonds (Alalwan et al. 2019, hemicellulose is a heteropolymer composed of pentose (D-xylose, L-arabinose), hexoses (D-glucose, D-galactose and D-mannose) and uronic acid (Dahman et al. 2019) and lignin, also an aromatic heteropolymer composed of p-coumaryl alcohol, coniferyl alcohol and sinapyl alcohol (Linger et al. 2014; Dahman et al. 2019). Converting LCB into biofuels is challenging as the three fibril compounds i.e. cellulose, hemicellulose and lignin are firmly braided making cellulose apparently unexposed to the enzymes for their hydrolysis and fermentation to produce ethanol. Pretreatment of feedstock increases the exposed surface area of cellulose and hemicellulose for microbial degradation and hence increases the yield. Apart from this, pretreatment also reduces the volume of the raw material as well as the 3-D lattice nature of cellulose, a deterrent in fermentation process. A wide range of pretreatment methods are available, namely ammonia fibre explosion (AFEX), ammonia recycled percolation (ARP), alkali treatment, concentrated and dilute acid treatment, ionic liquid, mechanical, organosolv, ozonolysis, steam explosion, supercritical fluid technology and wet oxidation all with their specific advantages and disadvantages. The composition and structure of every biomass are different so a common strategy cannot be applied for the pretreatment of any type of biomass. Baruah et al. (2018) and Aditiya et al. (2016) provide in-depth insight on various pretreatment methods and other downstream processes and can be consulted for further details. Pretreatment increases the hydrolytic efficiency of LCB where more than 90% yield can be achieved (Sindhu et al. 2016). Different stages in treatment of LCBs to produce biofuels are elucidated in Figure 13.2.

As sustainable solutions to energy needs are the core of this chapter, we are hereby emphasizing only on the most sustainable pretreatment process i.e. 'biological pretreatment' as only it requires mild condition means no energy consumption to produce one. Moreover, the process is selective, no additional chemical requirements, low severity and most importantly it removes antimicrobial substances which inhibit downstream processes i.e. hydrolysis and fermentation (Aditiya et al. 2016; Ummalyma et al. 2019). Biological pretreatment involves microorganisms as a whole cell or utilizing their enzymes to pretreat LCB. Many classes of bacteria (*Bacillus*, *Actinomycetes* and *Streptomyces*) and fungi

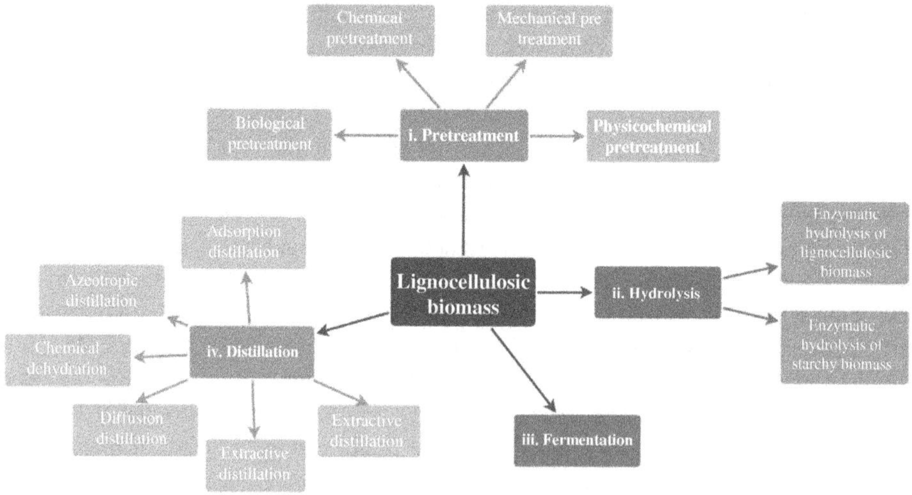

Figure 13.2 Different stages of treatment of LCBs (Aditiya et al. 2016).

(*Candida*, *Aspergillus*) exhibiting broad spectrum biodegradation have been reported which can be used in pretreatment of LCB, but fungi have proved to be best suited for the process as they have the capability to simultaneously degrade cellulose, hemicellulose and lignin. White rot, brown rot and soft rot fungi are the most commonly used fungi of which white rot is the most preferred due to increased sugar yield (Baruah et al. 2018). LCB degradation by fungi occurs extracellularly because of its insoluble nature. Two types of extracellular enzymatic systems participate in fungal degradation of LCB i.e. hydrolytic system and ligninolytic system which degrade polysaccharides and lignin respectively. Interestingly, bacteria have limited ligninolytic activity but the ability to decompose cellulose is widespread. Nevertheless, recent reports confirmed a sizable role of bacteria in lignin valorization (Lee et al. 2019). These bacteria belong to mainly three classes: Actinomycetes, α-proteobacteria and γ-proteobacteria (Janusz et al. 2017). Lee et al. (2019) enlisted the bacterial strains used in lignin degradation. The white rot fungi can be classified into two groups on the basis of their preferred substrate: one that can simultaneously degrade lignin and polysaccharides resulting in sugar loss and other which can specifically degrade lignin. The group with lignin-specific activity is generally used for better yields (Baruah et al. 2018). Some commonly available white rot fungi involved in LCB pretreatment are *Pycnoporus cinnabarinus*, *Phanerochaete chrysosporium*, *Irpex lacteus*, *Punctularia sp.*, *Ceriporia lacerata*, *Ceriporiopsis subvermispora*, *Cyathus stercoreus*, *Pleurotus ostreatus*, *Phlebia subserialis*, *Postia placenta*, *Gloeophyllum trabeum*, *Echinodontium taxodii*, *Pholiota adiposa*, *Armillaria gemina* and *Trametes hirsuta* (Sindhu et al. 2016; Baruah et al. 2018; Ummalyma et al. 2019). Apart from these, even protists have also been identified in lignin degradation (Li et al. 2020). Table 13.2 summarizes different biological agents used in pretreatment of LCB as whole-cell biocatalysts.

Naturally, angiospermic hardwood is preferred by white rot fungi, whereas for brown rot fungi softwood of coniferous ecosystems is the favoured food (Janusz et al. 2017). Majority of these microorganisms produce extracellular enzymes which help in degradation of

Table 13.2 Different classes of microorganisms, representative members and their effects in pretreatment of lignocellulosic biomass.

Microorganisms	Feedstock	Major effects
Fungi		
White Rot		
Punctularia sp.	Bamboo culms	50% lignin removal
Irpex lacteus	Corn stalks	82% breakdown of lignin
Pleurotus ostreatus/ Pleurotus pulmonarius	*Eucalyptus grandis*	20-fold increase in hydrolysis
Phanerochaete chrysosporium	Rice husk	Yield of 458 mg of reducing sugars per gm of rice husk
Ceriporiopsis subvermispora	Wheat straw/corn stover	Minimal cellulose loss/threefold increase in reducing sugar yield
Phanerochaete chrysosporium	Corn stover silage	39% lignin removal
Fusarium sp.	Rice straw	17.1% decrease in lignin content, 10.8% decrease in silica content
Trametes versicolor	Corn silage	70% upsurge in lignin degradation
Pleurotus ostreatus	Rice straw	33% lignin removal
Ceriporiopsis subvermispora	Sugarcane bagasse	47% recovery of sugar
Pleurotus pulmonarius	Saw dust	Increased sugar concentration
Armillaria gemina	Aspen biomass	62% saccharification efficiency
Pycnoporus sanguineus	Cereal straw	More than 96% lignin removal
Pycnoporus cinnabarinus	Wheat straw	Lignin degradation 50%
Fusarium concolor	Wheat straw	13.07% loss in lignin
Echinodontium taxodii	Hardwood	45.6% lignin loss
Soft Rot		
Tricoderma reesei	Rice straw	23.6% lignin removal
Tricoderma reesei	Sisal leaf decortication residue	21% increase in cellulose content
Trichoderma viride	Sugarcane trash	56% Lignin digestion
Brown Rot		
Gloeophyllum trabeum	Wheat straw	37.6% hemicellulose removal
Gloeophyllum trabeum	Moso bamboo	Preferential degradation of hemicellulose
Laetiporus sulphureus	*Pinus radiata*	7.5% hemicellulose degradation
	Eucalyptus globulus	6% hemicellulose degradation with partial hydrolysis of lignin

(*Continued*)

Table 13.2 (Continued)

Microorganisms	Feedstock	Major effects
Serpula lacrymans	Corn leaves	32% weight loss; 32.75% increased phenol release
	Corn cob	16% weight loss;86.79% increased phenol release
	Sugarcane bagasse	16% weight loss;74.29% increased phenol release
	Cacao pod	7% weight loss;22.22% increased phenol release
	Rice straw	4% weight loss;118.5% increased phenol release
Bacteria		
Streptomyces viridosporus	Wheat straw biomass	Increased carbohydrate/lignin ratio
Streptomyces coelicolor	Grass lignocellulose	Lignin-derived acid precipitated polymeric lignin (APPL) formation
Amycolatopsis sp.	Soft and hard wood	Significant loss in lignin
Thermobifida fusca	Switch grass and corn stover	Survive only on lignin as the sole carbon source
Clostridium thermocellum	Populus lignin	Decreased β-O-4linkage and increased S/G index
Caldicellulosiruptor bescii	Switch grass	Release of lignin derived aromatic compounds
Azotobacter vinelandii	Kraft lignin	Lignin mineralization and solubilization
Protists		
Thraustochytrium striatum	Kraft pine lignin	Max of 25% lignin degradation

phenolic and non-phenolic lignin via oxidation (Sindhu et al. 2016). The common enzymes involved in lignin degradation are lignin peroxidases and laccases. These enzymes degrade lignin through lipid peroxidation of phenolic as well as non-phenolic compounds and oxidation of phenolic compounds and aromatic amines respectively. Whole-cell pretreatment has an inherent disadvantage of being an extremely slow process which may take upto months to attain desired results. Enzymatic pretreatment is an alternative to the whole-cell pretreatment which drastically increases the rate of reaction.

13.4.2.1 Enzymatic Pretreatment Process

Enzymes involved in the degradation of lignin have been classified into two broad categories i.e. lignin-modifying enzymes (LME) and lignin-degrading axillary enzymes (LDAs) (Janusz et al. 2017). LDAs are not directly involved in degradation of lignin but assist LME in the

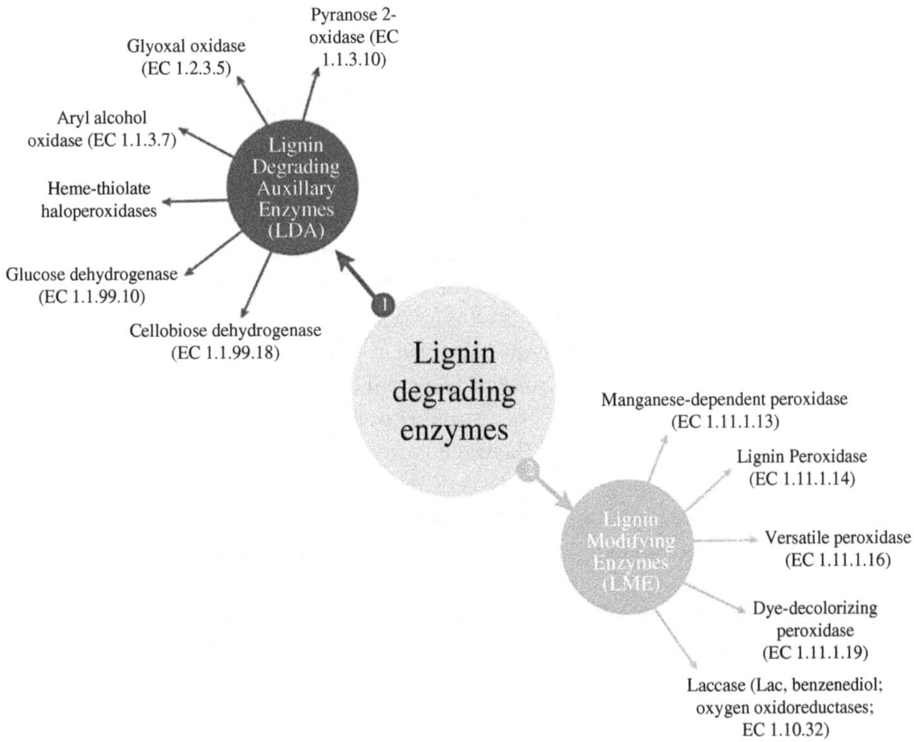

Figure 13.3 Classification of enzymes based on their mode of action in lignin degradation process (Janusz et al. 2017).

process. LMEs' can be further distinguished into lignin-modifying peroxidases (LMPs) and Laccase. LMPs belong to super family of catalase-peroxidase and each member has protoporphyrin IX as prosthetic group. The constituent members of this class include: lignin peroxidase, manganese-dependent peroxidase, versatile or multifunctional peroxidase and a new heme peroxidases family i.e. dye-decolourizing peroxidase. Laccase or Lac, benzenediol: Oxygen oxidoreductases are phylogenetically unrelated to LMPs but are important LMEs which oxidize a range of aromatic moieties using molecular oxygen. Figure 13.3 illustrates different enzymes in each category. LDAs involve in lignin degradation through serial action of several enzymes in the generation of hydrogen peroxide which facilitate the peroxidases, in detoxification by reducing quinones. This group includes glyoxal oxidase, aryl alcohol oxidase, cellobiose dehydrogenase, pyranose-2-oxidase, glucose dehydrogenase and heme thiolate haloperoxidases (HTP), a recently identified class of enzymes. Although there is confusion in classifying this newly discovered class in LME or LDA due to its shared catalytic properties with members from both groups. Due to unavailability of experimental evidence to prove that these alone can degrade lignin, Janusz et al. (2017) classified them in LDAs. For further details of different enzymes, their structure, further classification and functioning, Janusz et al. (2017) can be referred.

To improve the efficacy of biological pretreatment methods, combination strategies have been tried and found to be effective. These includes biological pretreatment–hydroliquifaction mild physical/chemical–biological pretreatment, stream explosion–biological pretreatment, biological–liquid hot water pretreatment, mild acid–biological pretreatment, ultrasonic–biological pretreatment, biological–organosolv, hydrogen peroxide–biological pretreatment, biological–AFEX pretreatment and biological–alkaline/oxidative pretreatment (Ummalyma et al. 2019).

After lignin, crystalline nature of cellulose and hemicellulose is the other deterrent in the production of bioethanol from 2G feedstocks. The next step in processing of LCB is the hydrolysis of hemicellulose and cellulose. It is also known as saccharification. Although enzymes for hydrolysis of hemicellulose are complex, but hemicellulose, due to its amorphous nature, is easier to hydrolyse. Hemicellulose is a heteropolymer of pentoses, hexoses and sugar acids and the entirety of enzymes involved in its hydrolysis are collectively called as hemicellulases. Presence of entire suite of hemicellulases in a single organism to completely hydrolyse the hemicellulose is less common but few organisms, namely *Penicillium capsulatum, Talaromyces emersonii, Thermomonospora fusca* and *Caldicellulosiruptor saccharolyticus* can show complete hydrolysis of hemicellulose into xylose. Xylose finally gets metabolized to pyruvate through pentose phosphate pathway. There are variety of microorganisms found which can naturally metabolize xylose, namely yeast, *Pichia stapis,* anaerobic fungi and several members of the order Thermoanaerobacteriales (Elshahed 2010). However, since hemicellulose and cellulose are always present in plant material, a single organism or a standardized consortium of microbes capable of efficiently and simultaneously metabolizing both sugars would increase the yield.

After hemicellulose, hydrolysis of cellulose is the next step in the production of bioethanol from LCB. Hydrolysis of cellulose needs simultaneous action of three different enzymes i.e. endo1,4-β-gluconase, exo-1,4-β-gluconase and β-galactosidase which attacks internal β-glucosidic linkages, removes cellobiose units and converts cellobiose to glucose respectively (Elshahed 2010).

The next step is fermentation of the hydrolysate to ethanol using variety of microorganisms. Fermentation is a critical process whose success depends on conditions (temperature and pH) affecting metabolic activity of microorganisms selected. For example, bacteria prefer pH 6.5-7.5, whereas fungi proliferate in acidic environment ranging from 3.5 to 5.0. The most common bacteria employed for the process is *Zymomonas mobilis*. Other species from genera *Klebsiella* and *Erwinia* have also been investigated for alcohol production from five and six carbon sugars and enzymatically hydrolysed lignocellulosic materials (Elshahed 2010). *S. cerevisiae*, akin *Z. mobilis*, is the most commonly used fungi in fermentation process of hexose but does not have affinity towards xylose. There are species which have been identified for their ability to convert pentose sugars (xylose) to alcohol, namely *Pichia stipites, Candida shehatae* and *Pachysolen tannophilus* (Aditiya et al. 2016). In the production of ethanol from LCB, two different approaches are followed i.e. separate hydrolysis and fermentation (SHF) or simultaneous saccharification and fermentation (SSF). SSF can improve ethanol production as there is no time lag and is cost-efficient (Cheng et al. 2017). In addition to these two, there are other hybrid systems being developed which include simultaneous saccharification and co-fermentation (SSCF) and consolidated bioprocessing (CBP). CBP combines all three steps of LCB production (enzyme production, enzyme saccharification and sugar fermentation) within a single recombinant microbe. This process results in decreasing cost and helps avoid the enzyme production costs. Lynd group is the pioneer one to initiate research on CBP. The group utilized thermophilic *Firmicutes* in bioprocessing (Lynd et al. 2005). The rate of cellulose metabolism is known to increase with temperature and thermophilic bacteria e.g. *Clostridium thermocellum* is known to have the highest cellulose degradation rates. Moreover, many clostridia can convert sugar produced from cellulose degradation to ethanol. This dual ability of the bacteria of orders Clostridiales and Thermoanaerobiale makes them ideal candidates for CBP (Elshahed 2010). Apart from *Clostridia* sp., *Moorella* strains also showed high ethanol production and tolerance that makes it commercially viable option in CBP. The main disadvantage of CBP method is difficulty in creation of optimum conditions for each enzyme simultaneously. Parisutham et al. (2014) in their review explained the feasibility of CBP thoroughly.

The final step in the production of 2G bioethanol is distillation of ethanol from the mixture. Industrial requirement of ethanol emphasized on anhydrous ethanol but ethanol forms azeotropic mixture with water and is difficult to remove. Aditiya et al. (2016) compiled the common methods for production of anhydrous ethanol such as (i) membrane process, (ii) diffusion distillation, (iii) chemical dehydration, (iv) azeotropic distillation, (v) vacuum distillation, (vi) adsorption process and (vii) extractive distillation.

13.4.2.2 2G Biodiesel

The biodiesel production from 2G feedstock is not much different from the 1G except the source of feed. The feedstock is composed of jatropha, karanja, rubber tree, tobacco seed, waste cooking oil and lipids of animal origin. Animal fats are preferable feedstocks for their properties such as higher-octane numbers, non-corrosiveness, wasteless and sustainability. However, the main drawback of this generation is the lack of technologies for the management of waste generated by biodiesel production. Furthermore saturated fatty acids in animal fats make the transesterification complex (Alalwan et al. 2019).

There are many other technologies to utilize LCB in producing variety of biofuels ranging from butanol, iso-butanol, propanol, pentanol, higher alkanes which proves the sustainability of the technologies involved in producing such fuels (Bhatia et al. 2017). The detailed explanation of the technological interventions in conversion of ligno-cellulosic biomass to various biofuels has been thoroughly explained in the previous chapter of the book by Singh et al. (2020).

13.4.3 Third Generation (3G)

Although the 2G biofuels have overcome the limitations of the 1G in terms of food–energy conflict and escalated food cost, they have their own challenges. The most important ones are competition with food crops for land and water, high production costs and requirement of sophisticated downstream processing techniques. This led to the conception of 3G biofuels in late 1950s by Oswald and Golueke i.e. production of fuels from microorganisms (Alalwan et al. 2019). The feedstock for the 3G fuels include algae, cyanobacteria, bacteria, yeast and fungi, but the most commonly used microorganism is microalgae. The production of fuels from algae can solve the challenges which were surfaced with the 2G biofuels as algae need minimal raw materials, namely CO_2, some nutrients, sunlight and water. Moreover algae can grow on waste water, hence, eliminate the requirement of land or even freshwater (Dahman et al. 2019). Above all, algae can be 200 times more productive than any land-based feedstock source (Lackner 2015) which means more fuel per acre of area. Microalgae show different modes of nutrition, namely autotrophic, heterotrophic, mixotrophic or photoheterotrophic and are even capable of switching between as a response to environmental changes.

Carbohydrates present in biomass such as algal starch, cellulose in cell wall or other accumulating carbohydrates can be converted to ethanol via hydrolysis and the oil present can be converted into biodiesel. The mean oil content in dry biomass of algae ranges between 1 and 70% but under certain conditions can reach up to 90% way higher than many oil-producing plants (Demirbas 2011). Even the per hectare oil yield from algae can be 10 times more than common oil-yielding plants. The US Department of Energy (USDE) initiated the project to illustrate the potential of algae as biofuels in 1978–1996, where over 3000 species of microalgae from diverse environmental habitats were isolated and screened. Alalwan et al. (2019) compiled an exhaustive list of different microalgal strains with their oil content.

There are different industrial production systems for different types of algae, namely open ponds, photobioreactors (PBR) and closed systems. In early days, open systems were found to be more acceptable commercially as they are easier to set up. Lately, PBRs gained attention due to their inherent properties such as easy to regulate, less prone to contamination and high productivity. Apart from these, many hybrid systems have also been tried successfully which can be studied from Demirbas (2011). The nutrition to the open pond is introduced through waste/sewage water or runoff water and to PBR through continuous stream of sterilized water containing required nutrients. After growth, harvesting is the next step which costs about 20–30% of the total process. Most common harvesting methods involve flocculation followed by sedimentation, centrifugation, filtration, ultrafiltration (Mata et al. 2010; Dutta et al. 2014). Harvesting leads the algae through any or all of the three processes i.e. oil extraction, biochemical treatment or thermochemical treatment to produce biodiesel, ethanol/biomethane/biobutanol or syngas, aviation fuel/bioenergy respectively. Algae fuel is also known as Oilgae and is composed of triacyl glycerides (TAG) which can be converted to biodiesel through transesterification as discussed in 1G biofuels above. Algae, if undergo fermentation, can produce bioethanol or biogas depending on the conditions, thereby eliminating the need of biomass drying. The common algal species used are *Chlorella* sp., *Dunaliella* sp., *Pavlova*, *Skeletonema* sp., *Spirulina platensis*, to name a few. Mata et al. (2010) explained a detailed comparative of lipid content and productivity of different microalgae species. The research in this area is still in its toddler stage.

Apart from algae, various cyanobacteria and fungal members have also been used in production of biofuels such as *Mucor indicus*, *Aspergillus oryzae*, *Cunninghamella echinulata*, *Mortierella isabellina*, *Humicola lanuginosa*, *Mucor circinelloides* and *Mortierella vinacea*. The microbial species with more than 20% of lipid content in their dry biomass weight are known as oleaginous species (Patel et al. 2020). Oleaginous yeasts and fungi have the benefits of converting non- or underutilized biomasses such as lignocellulosic materials from agricultural and industrial activities as their by-products to lipids, thereby not affecting the food supply.

13.4.4 Fourth Generation (4G)

There are two entirely different fuels compiled in fourth-generation (4G) biofuels by different researchers, namely genetically engineered microorganisms such as microalgae, bacteria, fungi as feedstock for the production of bioethanol, biodiesel, biohydrogen etc. (Alalwan et al. 2019; Kumar and Chandra 2020; Ziolkowska 2020) and use of microorganisms for the production of electricity via microbial fuel cells and solar biofuels (Aro 2016). Process and technology to produce biofuels from the 4G are exactly the same as used in the previous generations. The major difference is in the carbon neutrality of the process as they consume more of the CO_2 than they emit during the production and combustion. Microbial fuel cells have been thoroughly explained by Chaturvedi and Verma (2016) where they explained the technology with examples how to use microorganisms to transform chemical energy of organic compounds into electricity. This is still in its nascent stage and needs a lot of research in next few years, but this generation holds a promise for sustainable solution to the next-generation energy needs.

In addition to producing biofuels from microorganisms, there is another potential of microorganisms i.e. to enhance the production of fossil fuels from existing resources, hence creating a niche of biological approach for extraction of fossil fuels. These processes can be used in oil well where the production has either declined or ceased recently. This approach has been further subdivided into two approaches i.e. microbially enhanced oil recovery (MEOR) and microbially enhanced energy recovery (MEER). MEOR involves microbes in oil wells which have reached their tertiary stage of production where oil production has reached the level from where it becomes unviable. MEER involves microbes which can convert one form of hydrocarbons to other more efficient form, namely consortium of methanogenic bacteria has been used to convert unrecoverable hydrocarbons into methane gas (Elshahed 2010). Youssef et al. (2007), (2009) and Gieg et al. (2008) have very well described these processes.

13.5 Conclusions

As elucidated above, the role of microorganisms starts from being a catalyst or producer of one, a machinery to convert simple sugars into bioethanol in 1G and 2G biofuels to being itself used as a feedstock in 3G and 4G biofuels. Now, even these microorganisms are being used as generators to produce electricity. The microorganisms have touched each and every realm of energy production and their roles are still evolving. There are few takeaway points.

- Undoubtedly, microorganisms will outlive humans whatsoever be the conditions may prevail on earth. This implies that as far as humans live, there will be continuous supply of microorganisms for sure. Therefore if microorganisms can be utilized as energy resource, there will be ample energy generation which is sustainable too.
- Microorganisms, since life existed on earth, are already playing their part in harnessing solar energy either as cyanobacteria and algae via photosynthesis, as energy mobilizer by decomposition or producing energy through chemical reactions by chemosynthesis.

- Many efforts have been put to streamline the process of energy generation through microorganisms. Production of microbes through fermenters needs constant supply of biomass which is abundant on earth. The entire biomass production on earth is approximately 150 billion tons of organic dry matter including both land and aquatic biomass. Maximum 1.25% is used as food and traces in feed, energy and industrial raw materials. The rest of the biomass is non-utilized, under-utilized or recycled into the earth system, which can be used as substrate for microbial production (Naik et al. 2010).
- The waste generated by humans, may it be solid or waste water, is a rich source of nutrients, poses problems in their management, creating environmental threats, and has no commercial value. This can be used as a nutrient source for microbe production and hence can be efficiently managed as well without any cost addition.
- In addition to making energy generation sustainable, the newer generations of biofuels from microbes are also making the world net carbon neutral, solving another major problem of global warming.

Globally, there are 4D's: decarbonization, digitalization, decentralization and democratization which are defining energy shift. With the constantly increasing global demand for clean energy and even more during the COVID-19 pandemic due to work-from-home (WFM) culture, these tiny organisms are incessantly producing energy. We just need to develop better technologies to get energy from them sustainably to meet future demands.

References

Aditiya, H. B., Mahlia, T. M. I., Chong, W. T., Nur, H., & Sebayang, A. H. (2016). Second generation bioethanol production: a critical review. *Renewable and Sustainable Energy Reviews*, 66, 631–653.

Ahorsu, R., Medina, F., & Constantí, M. (2018). Significance and challenges of biomass as a suitable feedstock for bioenergy and biochemical production: a review. *Energies*, 11(12), 3366.

Alalwan, H. A., Alminshid, A. H., & Aljaafari, H. A. (2019). Promising evolution of biofuel generations. Subject review. *Renewable Energy Focus*, 28, 127–139.

Alia, K. B., Rasul, I., Azeem, F., Hussain, S., Siddique, M. H., Muzammil, S., Riaz, M., Bari, A., Liaqat, S. & Nadeem, H. (2019). Microbial production of ethanol. In Inamuddin, Mohammad Faraz Ahmer, Abdullah M. Asiri *Microbial Fuel Cells: Materials and Applications*. Materials Research Forum LLC., Pennsylvania (pp. 307–334).

Aro, E. M. (2016). From first generation biofuels to advanced solar biofuels. *Ambio*, 45(1), 24–31.

Baruah, J., Nath, B. K., Sharma, R., Kumar, S., Deka, R. C., Baruah, D. C., & Kalita, E. (2018). Recent trends in the pretreatment of lignocellulosic biomass for value-added products. *Frontiers in Energy Research*, 6, 141.

Bhatia, S. K., Kim, S. H., Yoon, J. J., & Yang, Y. H. (2017). Current status and strategies for second generation biofuel production using microbial systems. *Energy Conversion and Management*, 148, 1142–1156.

Brzozowski, A. M., Derewenda, U., Derewenda, Z. S., Dodson, G. G., Lawson, D. M., Turkenburg, J. P., Bjorkling, F., Huge-Jensen, B., Patkar, S.A. & Thim, L. (1991). A model for interfacial activation in lipases from the structure of a fungal lipase-inhibitor complex. *Nature*, 351(6326), 491–494.

Chaturvedi, V., & Verma, P. (2016). Microbial fuel cell: a green approach for the utilization of waste for the generation of bioelectricity. *Bioresources and Bioprocessing*, 3(1), 38.

Cheng, N., Koda, K., Tamai, Y., Yamamoto, Y., Takasuka, T. E., & Uraki, Y. (2017). Optimization of simultaneous saccharification and fermentation conditions with amphipathic lignin derivatives for concentrated bioethanol production. *Bioresource Technology*, 232, 126–132.

Christopher, L. P., Kumar, H., & Zam Dahman e, V. P. (2014). Enzymatic biodiesel: challenges and opportunities. *Applied Energy*, 119, 497–520.

Dahman, Y., Syed, K., Begum, S., Roy, P., & Mohtasebi, B. (2019). Biofuels: their characteristics and analysis. In Deepak Verma, Elena Fortunati, Siddharth Jain, Xiaolei Zhang *Biomass, Biopolymer-Based Materials, and Bioenergy* (pp. 277–325). Woodhead Publishing.

Demirbas, M. F. (2011). Biofuels from algae for sustainable development. *Applied Energy*, 88(10), 3473–3480.

Difiglio, C. (2014). Oil, economic growth and strategic petroleum stocks. *Energy Strategy Reviews*, 5, 48–58.

Dutta, K., Daverey, A., & Lin, J. G. (2014). Evolution retrospective for alternative fuels: first to fourth generation. *Renewable Energy*, 69, 114–122.

Eisentraut, A. (2010). *Sustainable Production of Second-generation Biofuels: Potential and Perspectives in Major Economies and Developing Countries*. IEA Energy Papers, No. 2010/01, OECD Publishing, Paris. https://doi.org/10.1787/5kmh3njpt6r0-en.

Elshahed, M. S. (2010). Microbiological aspects of biofuel production: current status and future directions. *Journal of Advanced Research*, 1(2), 103–111.

EnerOutlook 2050: In-Depth Look at Future Energy Scenarios (2020). https://eneroutlook.enerdata.net/forecast-world-final-energy-consumption.html.

Gieg, L. M., Duncan, K. E., & Suflita, J. M. (2008). Bioenergy production via microbial conversion of residual oil to natural gas. *Applied and Environmental Microbiology*, 74(10), 3022–3029.

Guo, M., Song, W., & Buhain, J. (2015). Bioenergy and biofuels: history, status, and perspective. *Renewable and Sustainable Energy Reviews*, 42, 712–725.

Hii, S. L., Tan, J. S., Ling, T. C., & Ariff, A. B. (2012). Pullulanase: role in starch hydrolysis and potential industrial applications. *Enzyme Research*, 2012.

James, J. A., & Lee, B. H. (1997). Glucoamylases: microbial sources, industrial applications and molecular biology – a review. *Journal of Food Biochemistry*, 21(6), 1–52.

Janusz, G., Pawlik, A., Sulej, J., Świderska-Burek, U., Jarosz-Wilkołazka, A., & Paszczyński, A. (2017). Lignin degradation: microorganisms, enzymes involved, genomes analysis and evolution. *FEMS Microbiology Reviews*, 41(6), 941–962.

Kumar, A., & Chandra, R. (2020). Ligninolytic enzymes and its mechanisms for degradation of lignocellulosic waste in environment. *Heliyon*, 6(2), e03170.

Lackner, M. (2015). 3rd-generation biofuels: bacteria and algae as sustainable producers and converters. In: Chen WY., Suzuki T., Lackner M. (eds). *Handbook of Climate Change Mitigation and Adaptation*, Springer, New York, NY. https://doi.org/10.1007/978-1-4614-6431-0_90-1.

Lee, S., Kang, M., Bae, J. H., Sohn, J. H., & Sung, B. H. (2019). Bacterial valorization of lignin in biorefinery processes. *Frontiers in Bioengineering and Biotechnology*, 7, 209.

Li, X., Li, M., Pu, Y., Ragauskas, A. J., & Zheng, Y. (2020). Simultaneous depolymerization and fermentation of lignin into value-added products by the marine protist, Thraustochytrium striatum. *Algal Research*, 46, 101773.

Linger, J. G., Vardon, D. R., Guarnieri, M. T., Karp, E. M., Hunsinger, G. B., Franden, M. A., Johnson, C.W., Chupka, G., Strathmann, T.J., Pienkos, P.T. & Beckham, G. T. (2014). Lignin valorization through integrated biological funneling and chemical catalysis. *Proceedings of the National Academy of Sciences*, 111(33), 12013–12018.

Liu, Z. L., Saha, B. C., & Slininger, P. J. (2008). Lignocellulosic biomass conversion to ethanol by Saccharomyces. In Judy D. Wall, Caroline S. Harwood, Arnold Demain *Bioenergy* (pp. 17–36). American Society of Microbiology.

Lynd, L. R., Van Zyl, W. H., McBride, J. E., & Laser, M. (2005). Consolidated bioprocessing of cellulosic biomass: an update. *Current Opinion in Biotechnology*, 16(5), 577–583.

Mata, T. M., Martins, A. A., & Caetano, N. S. (2010). Microalgae for biodiesel production and other applications: a review. *Renewable and Sustainable Energy Reviews*, 14(1), 217–232.

Naik, S. N., Goud, V. V., Rout, P. K., & Dalai, A. K. (2010). Production of first and second generation biofuels: a comprehensive review. *Renewable and Sustainable Energy Reviews*, 14(2), 578–597.

Norjannah, B., Ong, H. C., Masjuki, H. H., Juan, J. C., & Chong, W. T. (2016). Enzymatic transesterification for biodiesel production: a comprehensive review. *RSC Advances*, 6(65), 60034–60055.

OECD Publishing. (2020). OECD-FAO agricultural outlook 2019-2028. Organisation for Economic Co-operation and Development OECD.

Ondul, E., Dizge, N., Keskinler, B., & Albayrak, N. (2015). Biocatalytic production of biodiesel from vegetable oils. *Biofuels – Status and Perspective*; Biernat, K., Ed.; IntechOpen: London, UK, 21–37.

Packer, M. A., Harris, G. C., & Adams, S. L. (2016). Food and feed applications of algae. In Faizal Bux, Yusuf Chisti *Algae Biotechnology* (pp. 217–247). Springer, Cham.

Parisutham, V., Kim, T. H., & Lee, S. K. (2014). Feasibilities of consolidated bioprocessing microbes: from pretreatment to biofuel production. *Bioresource Technology*, 161, 431–440.

Patel, A., Karageorgou, D., Rova, E., Katapodis, P., Rova, U., Christakopoulos, P., & Matsakas, L. (2020). An overview of potential oleaginous microorganisms and their role in biodiesel and omega-3 fatty acid-based industries. *Microorganisms*, 8(3), 434.

Pimental, D. (2008). Renewable and solar energy technologies: energy and environmental issues. In *Biofuels, Solar and Wind as Renewable Energy Systems* (ed. D. Pimental). Springer Science, Ithaca, NY, pp. 1– 18.

Rulli, M. C., Bellomi, D., Cazzoli, A., De Carolis, G., & D'Odorico, P. (2016). The water-land-food nexus of first-generation biofuels. *Scientific Reports*, 6(1), 1–10.

Sindhu, R., Binod, P., & Pandey, A. (2016). Biological pretreatment of lignocellulosic biomass–an overview. *Bioresource Technology*, 199, 76–82.

Singh, M. K., Sahni S. & Narang, A. (2020). *Production of Liquid Biofuels from Lignocellulosic Biomass*, Chapter 12.

Ummalyma, S. B., Supriya, R. D., Sindhu, R., Binod, P., Nair, R. B., Pandey, A., & Gnansounou, E. (2019). Biological pretreatment of lignocellulosic biomass – current trends and future perspectives. In Angelo Basile and Francesco Dalena *Second and Third Generation of Feedstocks* (pp. 197–212). Elsevier.

United Nations (2019). World population prospects 2019: highlights. *Department of Economic and Social Affairs, Population Division*. https://sustainabledevelopment.un.org/content/documents/2019_Tracking_SDG7_Report.pdf'

Walker, G. M. (1998). *Yeast Physiology and Biotechnology*. Wiley.
Youssef, N., Simpson, D. R., Duncan, K. E., McInerney, M. J., Folmsbee, M., Fincher, T., & Knapp, R. M. (2007). In situ biosurfactant production by Bacillus strains injected into a limestone petroleum reservoir. *Applied and Environmental Microbiology*, 73(4), 1239–1247.
Youssef, N., Elshahed, M. S., & McInerney, M. J. (2009). Microbial processes in oil fields: culprits, problems, and opportunities. *Advances in Applied Microbiology*, 66, 141–251.
Zabed, H., Sahu, J. N., Boyce, A. N., & Faruq, G. (2016). Fuel ethanol production from lignocellulosic biomass: an overview on feedstocks and technological approaches. *Renewable and Sustainable Energy Reviews*, 66, 751–774.
Ziolkowska, J. R. (2020). Biofuels technologies: An overview of feedstocks, processes, and technologies. In Jingzheng Ren, Antonio Scipioni, Alessandro Manzardo, Hanwei Liang *Biofuels for a More Sustainable Future* (pp. 1–19). Elsevier.

14

Fungal Microbial Fuel Cells, an Opportunity for Energy Sources

Current Perspective and Future Challenges

Sudakshina Tiwari[1], *Deepali Koreti*[1], *Anjali Kosre*[1], *Pramod Kumar Mahish*[2], *S.K. Jadhav*[1], *and Nagendra Kumar Chandrawanshi*[1]

[1] *School of Studies in Biotechnology, Pt. Ravishankar Shukla University, Raipur (C.G.), Raipur, Chhattisgarh, India*
[2] *Department of Biotechnology, Govt. Digvijay Autonomous P.G. College Rajnandgaon (C.G.), Rajnandgaon, Chhattisgarh, India*

14.1 Introduction

Recently, fossil fuels are the main energy sources; however, this source is limited and also extensive use of this causes many types of the environmental issues including global warming, pollution and instability of economy and biodiversity, which stimulate the research for alternative resources of energy that overcome this drawback (Khan and Hussain 2017). The production of energy by the current method is not sustainable, causes global warming that directly affects humankind, plants and animals etc. To overcome this global problem, there is a need to search for new, renewable, sustainable and green energy. In the present scenario, various alternative energy sources are available. Among these alternatives, microbial fuel cells (MFCs) are gaining much interest in recent years because they produce renewable energy forms such as bioelectricity, without combustion and wastewater generation (Kaushik and Jadhav 2017; Meshram and Jadhav 2019). MFCs are a hybrid bioelectrochemical system that converts chemical energy into electrical energy via microbial catalysed redox reaction under an ambient temperature (Beyene et al. 2018). In this redox reaction, electrons and protons produce anodes that are transferred to the cathode through the outer circuit and proton exchange membrane (PEM). This electrical energy is also called bioelectricity; it can be used as sustainable green energy (Cheng et al. 2006). Mostly bacterial-based MFCs were used because their electron transfer mechanism is known. Low power production and less knowledge about electron transfer mechanism in fungi-based MFCs led to lower interest. However, fungi can act as both anode catalyst and cathode catalyst and also is not pathogenic and is easy to handle; therefore many researchers are working on these fungi-based MFCs. The recent output suggests that the electron transfer mechanism in fungi is similar to bacteria when lactate dehydrogenase or ferricyanide reductase redox protein is present in the cell membrane of fungi (Sekrecka-Belniak and Toczylowska-Mamińska 2018). Furthermore, lithium-based batteries are a more attractive

Energy: Crises, Challenges and Solutions, First Edition. Edited by Pardeep Singh, Suruchi Singh, Gaurav Kumar, and Pooja Baweja.

candidate for the research due to their storage capacity as compared with other batteries, and also they do not cause environmental pollution. This book chapter reviews and discusses the current progression in the field of fungi-based MFCs and the role of fungi in the formation of battery electrodes.

14.2 General Introduction of Microbial Fuel Cells (MFCs)

MFCs are a bio-electrochemical system that is prepared by creating electrical flow from the oxidation of natural substances. Electrical flow is generated by the arrival of electrons outside the cells from the metabolic movement of microorganisms from wastewater or specific substrate (Ramadan 2017). Only limited microorganisms (exoelectrogens) can release electrons during direct contact with anode or cathode, by nanowires and mediator. Anode electrodes are utilized as an acceptor, and cathode acts as a donor of electrons. MFCs can autonomously treat wastewater as well as other waste material (Logan 2008). Green energy can be generated by MFCs as more sustainable for the environment. Chemical compounds, water, biohydrogen and electricity can also be generated by using MFCs (Oh et al. 2010). Low-level electricity is produced from MFCs, this issue is confronted by researchers to enhance the performance of MFCs and need to focus on continuing development (Table 14.1) towards higher efficiency (Ieropoulos et al. 2012). MFCs work on the basic principle that the chemical energy of the substrate is converted into electrical energy by the

Table 14.1 Research contribution related to development of MFCs.

Research contribution	References
Generated electrical energy by inserting a platinum electrode in living culture of *E. coli* and Yeast	Potter (1911)
Generated more than 35 V from a batch of biological fuel cells	Cohen (1931)
Coined the term bioelectricity by Cole	Cole (1972)
Addition of synthetic mediators accelerated the rate of e^- transfer from microorganism to the anode	Bennetto (1990)
First observed the bioelectric phenomenon by Galvani	Bresadola (1998)
Mediatorless MFCs	Kim et al. (2002)
Endogenous mediators	Bond and Lovley (2003)
Bioelectricity generation by utilizing MFCs in wastewater treatment technologies	Gude (2016)
The major focus was on microbiological and non-microbiological aspects	Li et al. (2017)
Use of porous nickel oxide/graphene nanocomposite anode in MFCs	Wu et al. (2018)
Ionic liquid-based membrane separator used in MFCs	Koók et al. (2019)
Use of numerical modeling based on lattice Boltzmann method (LBM) for the analysis of electrochemical performance of MFCs	Di Ilio and Falcucci (2020)

use of electrogenic microorganisms. These microorganisms consume organic substrates and produce CO_2 and H_2O, in the presence of oxygen (Ebadinezhad et al. 2019).

$$\rightarrow C_6H_{12}O_6 + 6H_2O + O_2 \quad 6CO_2 + 12H_2O$$

However, in the absence of O_2, they produce CO_2 in the anode chamber and H_2O in the open air in the cathode chamber.

$$\rightarrow C_6H_{12}O_6 + 6H_2O \quad 6CO_2 + 24H^+ + 24e^- (\text{anode})$$

$$\rightarrow 24H^+ + 24e^- + 6O_2 \quad 12H_2O \text{ (Mohan et al. 2014)}$$

14.2.1 FCs

The MFC chamber can be made by non-reactive plastic containers or glass, polycarbonate, or plexi glass material (Rhoads et al. 2005; Kaushik and Jadhav 2017). It consists of two anodic and cathodic chambers that are separated by a PEM. A single-compartment MFC can also be obtained, in which cathodic chamber is eliminated by exposing the cathode directly to the air (Rhoads et al. 2005; Du et al. 2007).

14.2.2 Electrode of MFCs

The electrode materials are one of the important factors determining MFCs' efficiency and performance (Tursun et al. 2016). An electrode material should contain qualities such as high conductivity, high specific surface area, low oxidization, porosity, low cost and suitable for microbial growth etc. Carbon, carbon cloth, zinc, platinum, platinum black, graphite felt and graphite are some of the anode electrodes used (Tharali et al. 2016; Meshram and Jadhav 2019). Mostly air cathode can be used in the cathodic chamber for the electron accepter; however, some modification can also be done by chemical catholyte such as potassium ferricyanide ($K_3Fe(CN)_6$), potassium permanganate ($KMnO_4$) and potassium dichromate ($K_2Cr_2O_7$); however, these chemicals are not sustainable on large scale for the production of higher electricity. Similarly, platinum and platinum-black cathode electrode can also be used; however, most of these cathodes that can be used in MFCs making the process unsustainable, and because they also have operation costs thus, to overcome this problem, biocathode might be used in MFCs; the use of microbes as catalysts holds great attention for bioremediation and waste treatment etc. Furthermore, the use of biocathode is inexpensive, self-regenerating as well as sustainable (Clauwaert et al. 2007).

14.2.3 Proton Exchange Membrane

If the anode and cathode are not separated by the general ion exchange membrane, a problem may occur in MFCs such as crossover can occur between oxygen and substrate that will inhibit the anaerobic microorganism on the anode surface. The separator should have some qualities such as low membrane resistance, low oxygen permeability, chemical stability and cost-effective. Hence, the separators that have this type of quality should be used (Koók et al. 2019). Thus a huge number of the ion exchange membranes have been already used in MFCs such as anion exchange membrane, PEM, cation exchange membrane (CEM), J-cloth, ultrafiltration (UF), microfiltration membranes (MFM), glass fibres and non-woven fabric filter (NWF) salt bridge etc. (Choi et al. 2013; Kaushik and Jadhav 2017).

14.2.4 Microorganisms and Their Electron Transfer Mechanism

There are three different ways followed by microorganisms for electron transfer to the anode electrode: direct electron transfer, use of mediators and use of nanowires produced by microorganisms. Two types of mediators can be used by microorganisms: firstly, exogenous mediators such as humic acid, thionine, methylene blue or neutral red; and secondly, endogenous mediators have been produced by microorganisms for the direct transfer of an electron to the electrode from respiratory enzymes (Bond and Lovley 2003).

Microorganisms that have been reported previously, such as *Clostridium butyricum* (Niessen et al. 2004), *Saccharomyces cerevisiae* (Reed and Nagodawithana 1991, pp. 261–314) and *Proteus vulgaris* (Bennetto 1990), are used as a mediator for transfer of an electron to anode surface in MFCs, while *Rhodoferax ferrireducens*, *Shewanella putrefaciens* and *Geobacter sulfurreducens* have been exposed to generate electricity in mediatorless MFCs (Bond and Lovley 2003). An electron can be transferred to the anode surface by the use of redox protein which is present on the outer membrane of the microorganism. Thus these microorganisms worked as a biocatalyst in mediatorless MFCs; this is an example of direct electron transfer in MFCs (Mathuriya and Sharma 2009).

14.3 Factor Affecting the MFCs' Performance

MFCs' performance is affected by some factors such as the configuration of the reactor, buffer, substrate, electrolyte resistance and so on (Kim et al. 2008).

14.3.1 Configuration of Reactor

Designs of MFCs are the main crucial factor in the production of electricity. To enhance the power output from these systems, many modifications, as well as replacement of the basic materials, have been reported by researchers. The following sections deal with some commonly used designs:

14.3.1.1 Single-Chamber MFCs

The designs of single-chamber MFCs are very simple; it has only a single anodic chamber including anode and cathode electrode (could be an air cathode or without air cathode) (An et al. 2020). The advantages of this single-chamber MFCs are to overcome the internal ohmic resistance of fuel cell and easy to handle (Du et al. 2007).

14.3.1.2 Dual-Chamber MFCs

The dual-chamber MFCs (Figure 14.1) have two different anode and cathode chambers separated by polymeric PEM or salt bridge. This separation allows movement of the protons across the cathode where oxygen is combined with protons and forms water (Jatoi et al. 2020).

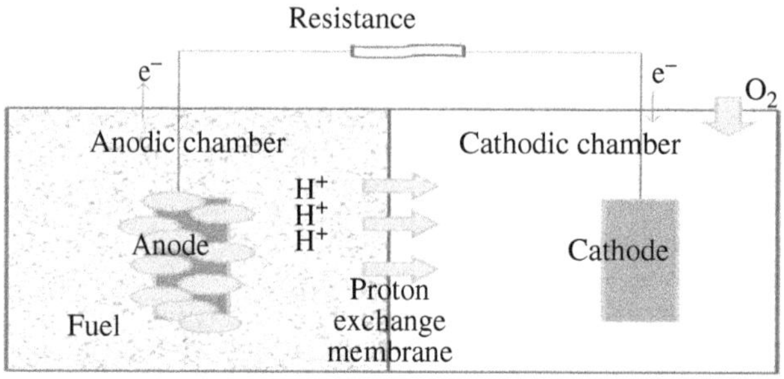

Figure 14.1 Dual-chamber microbial fuel cell (Du et al. 2007).

Many types of problems occurred in dual-chamber MFCs such as overpotentials at the cathode (Rabaey et al. 2004), organic loading rate, (Mohan et al. 2010) and the competence of the PEM for transporting protons (Rahimnejad et al. 2014). So, to overcome these problems, many types of modifications have been done in MFCs such as ceramics-based MFCs, tubular MFCs, sediment-type MFCs, stacked air cathode microbial fuel etc. (Choi and Ahn 2013; Gajda et al. 2015; Rahimnejad et al. 2015; Bhargavi et al. 2018).

14.3.2 Buffer

The pH variation has also affected the performance of MFCs. If constant pH is not obtained, then proton will not be transferred to the cathode and oxygen reaction rate ceases (Nimje et al. 2012), and it is also observed that the internal resistance of MFCs has been affected by the differences of pH between the anodic and cathodic solutions. If the pH difference is increased between anodic and cathodic solution, then internal resistance will be decreased (Jadhav and Ghangrekar 2009). Electrochemically active bacteriaalso maintained an anolyte pH by the use of inorganic carbon that is also a final metabolite of these bacteria. Rossi et al. (2020) observed that the phosphate buffer solution (PBS) concentration was increased ranging from 50 to 200 Mm; then anode resistance decreased by 64% and cathode resistance was not changed. Kaushik and Jadhav (2017) found that neutral pH is very good for bioelectricity production.

14.3.3 Substrate

The substrates in MFCs are a very crucial factor because it provides a nutrient and energy source for microorganisms. Various substrates in MFCs yield different power outputs, because the organic substrates present in the wastewater serve as recalcitrant contaminants. These could harshly destroy the capability of electron transfer and respiratory process of microorganisms (Ahn and logan 2010). Therefore Yu et al. (2019) and coworkers used the substrate such as chlorophenols as well as co-substrate such as acetate for improving the bioelectricity production; thus co-substrates could endorsement the survival of biofilm and can be used as primary nutrients. They improve microbial activity, induce 4-CP sensitive extracellular electron transfer (EET) enzymes' activity under 4-CP stress, such as succinate dehydrogenase (complex II) and cytochromes c OmcA and OmcB, and create a more vigorous biofilm community (Yu et al. 2019). In another work, four external carbon resources such as glucose, sucrose, sodium acetate and ethanol were used as co-substrates for the removal of phenol and generation of power. Among these, acetate was reported as a good co-substrate with initial phenol degradation (78.8%) and the voltage output (389.0 mV) (Shen et al. 2020).

The concentration of the substrate also affects the MFCs' performance. Catal (2019) and his colleague found that when concentrations of benzoylecgonine were increased from 1000 ng/ml in MFCs, then the electricity generation was decreased. However, if the concentration of substrate was increased in the anodic chamber, more electricity was generated and also columbic efficiency was increased and maximum power and current densities 50.7 and 570 mA/m^2 were reported respectively at the 10 000 mg/l concentration of COD. This study showed that the MFCs were affected by the substances that are present in the substrates (Rahmani et al. 2020).

14.3.4 Electrolyte Resistance

Electrolyte solutions affect the current generation and power output of MFCs due to oxidation–reduction potentials (ORP) of anode and cathode solutions are directly associated with the metabolic actions of microorganisms. Therefore the capability of electron acceptance and generation are dependent on electrolyte solutions. If there are large differences of ORP between anode and cathode solutions, then power output increases because electrons move from lower ORP to higher ORP (Li et al. 2010), and type of ions, ionic concentration, temperature and the geometry of the projected area across which is the current passed to the main component, that is, the resistance of the ionic solution (electrolyte) aredefined in the following formula

$$R = \rho\ l/A$$

Where 'ρ'= resistivity of the solution and its reciprocal term of conductivity (κ)
'l'= length of anodic/cathodic chamber contains the electrolyte solution
'A'= area of the anodic/cathodic chamber containing the solution

The main important parameters that affect the conductivity of the electrolyte are the current flow path, the geometry of the electrolyte solution and microorganisms in wastewater that might contribute to the electrolyte conductivity (Sekar and Ramasamy 2013).

14.4 Fungal Microbial Fuel Cells

Fungi are the more promising candidate for the production of electricity in MFCs, because fungi are easy to handle (Schaetzle et al. 2008). Fungi can be used as cathodic catalyst and anodic catalyst in different configurations of MFCs with the use of a mediator or without the use of mediators. Although the electron transfer mechanism in fungi-based MFCs (Figure 14.2) is little understood, nowadays they are used as good biocatalysts due to the production of the degradative enzyme for particular degrading waste material. Fungi can use two mechanisms for electron transfer to the anode: first is the direct electron transfer in which redox protein is used; the second one is indirect electron transfer which can use an exogenous mediator for the electron transfer (Sekrecka-Belniak and Toczylowska-Mamińska 2018).

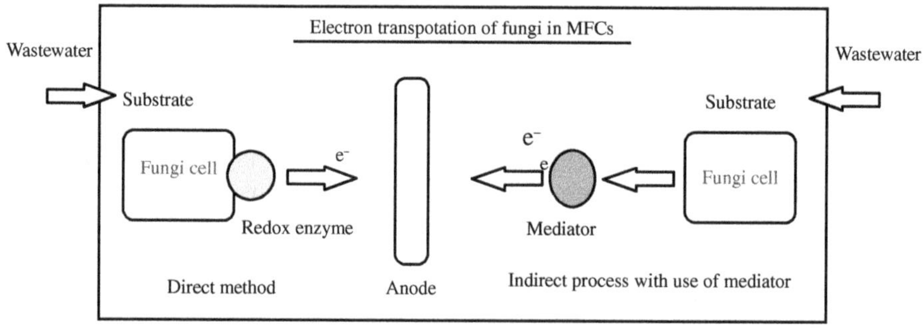

Figure 14.2 Electron transfer mechanism of fungi (direct and indirect) when fungi work as anode catalysts in MFCs (Sekrecka-Belniak and Toczylowska-Mamińska 2018).

14.4.1 Saccharomyces cerevisiae

Few eukaryotes have been used as the catalyst for MFCs because many researches have been conducted on only prokaryotes that can be used as electricigens. Due to the clear genetic background and fast-growing rate of yeast they were first used as elecricigen microbial sources in MFCs (Cao et al. 2018). *Saccharomyces cerevisiae* was first reported as electricigen by Potter (1911); however, it has been given yielded low power output compared with bacteria. The maximum power output was obtained when an engineered strain of *S. cerevisiae* was used in MFCs (Fishilevich et al. 2009). When the yeast extract was used as an electron mediator in *S. cerevisiae* MFCs, the current density and power density obtained were 300 mA/cm^2 and 70 mW/cm^2, respectively (Sayed et al. 2015). Indirect *S. cerevisiae* fuel cell was constructed by Halme and Zhang (1995), who cultivated yeast in glucose fermented and supernatant detached to the MFC anode. Thus the yeast cell secreted a soluble redox molecule which was used for the electron transfer to the anode and 120 μWcm^{-3} power density was obtained, and *S. cerevisiae* MFCs also yielded 1.5 Wm^{-2} power density, when the methylene blue mediator was used in anode chamber which was given similar result as bacteria (Ganguli and Dunn (2009). Whereas *S. cerevisiae* was grown in mediatorless MFC; 3 mWm^2 power densities were obtained due to their work as anode catalyst and also generated power by the adhesion of *S. cerevisiae* on the anode surface, thus surface-confined species was responsible for electron transfer (Sayed et al. 2012). Some types of methodologies can useful for the optimization of fungal MFCs' performance. Christwardana et al. (2018) have used the response surface methodology for the optimization of glucose/yeast ratio and concentration of glucose in *S. cerevisiae*–based MFCs. They also identified the effect of a mediator in electricity production in MFCs. Furthermore, they found 340.9 mWm^{-2} power density without using mediator and 374.4 mWm^{-2} with the utilization of mediator. High glucose concentration and low glucose/yeast ratio had yielded the maximum output without the use of mediators, but low glucose concentration and high glucose/yeast ratio yielded the highest output with the use of mediators, also analysed by the use of response surface methodology. Immobilization of microorganisms is the best technique for generating green energy in MFCs, which can improve the lifespan and stability of microorganisms. Furthermore, *S. cerevisiae* was immobilized within the alginate matrix for enhancing the performance of MFCs. Thus the glucose oxidation and electron transfer were improved in MFCs and maximum current was generated (Mardiana et al. 2019).

14.4.2 Candida melibiosica

C. melibiosica is facultative anaerobic yeast. *C. melibiosica* 2491 is a strain of *C. melibiosica* that has a phytase activity for the degradation of plant waste material. It may be used as biocatalyst in MFCs because it can transfer an electron without the use of artificial mediators (Babanova et al. 2011). However, the exogenous mediator also was used for investigating the performance of *C. melibiosica* 2491–based MFCs found by Hubenova and Mitov (2010) and their colleagues, and they obtained 0.18 Wm^{-3} power densities when they used methylene blue mediator in anode and ferricyanide in the cathode chamber. Other mediators are neutral red methyl red (MR), cresol red, Eriochrome Black T, methyl orange (MO), bromocresol green, tropaeolin, bromothymol blue, bromocresol purple bromophenol blue, murexide, congo red, eosin and methyl yellow etc. When these artificial mediators were used in *C. melibiosica* yeast-based fuel cell, the current and the power density were 6–7 times increased compared with the mediatorless yeast-based fuel cell. Among these mediators, methylene blueyielded the best result (Babanova et al. 2011).

C. melibiosica 2491–based MFCs can be also considered with the use of modified anode material. Mostly anode made by carbon felt was used in MFCs; however, for improving the efficiency, nano-modified nickel–iron (N-I) and nickel–iron–phosphorous (N-I-P) carbon felt anode was used. The anode modifications have been done by the galvanostatic or potentiostatic pulse plating technique. It was observed that N-I-P modified electrode yielded the highest power densities of 155 ± 6 and 260 ± 8 mW/m^2 by the modification of electrode with galvanostatically and potentiostatically pulse plating technique respectively and N-I modified electrode yielded 52 ± 9 mW/m^2 which was lower power output compared with N-I-P modified electrode in MFCs (Hubenova et al. 2011).

14.4.3 Hansenula anomala

Hansenula anomala immobilized cells were also investigated as a biocatalyst in MFCs with using glucose substrate. The immobilization of cells on the anode surface was done by the use of covalent linkage as well as physical adsorption. *H. anomala* has the ability for direct electron transfer to anode surface. So electrons could be transferred by the use of ferricyanide reductase and lactate dehydrogenase redox protein which is present on the outer membrane of *H. anomala* in MFC, and the contrasting anode was used in the mediator for fewer MFCs such as graphite felt, graphite and polyaniline–Pt composite-coated graphite. It was reported that (Prasad et al. 2007) polyaniline–Pt composite-coated graphite yieled maximum power density which is 2.9 W/m^3 as compared with other electrodes.

14.5 Other Fungi Used as a Biocatalyst in Microbial Fuel Cells

Mbokou (2017) and his colleague for first time used ***Scedosporium dehoogii*** which is a filamentous fungus as a biocatalyst or microbial organism in a anode chamber for the degradation of acetaminophen (APAP) and 6.5 mWm^{-2} power density was obtained. Pontié et al. (2019) with coworkers also found 50 mWm^{-2} power density during the degradation of APAP pollutant with the use of *S. dehoogii* fungi in MFCs. Through this process

S. dehoogii fungi were used as the bioanode in MFCs and carbon felt was used as the cathode. This result showed that *S. dehoogii* fungi have catalysed dual activity such as bioremediation of pollutants as well as biochemical reaction electricity generation. Nowadays fungus–bacterium combination has been used for wastewater treatment because the bacterium was used in the fungus mycelium network to carry the electron to the anode. The sterile condition is needed for pure culture in MFCs to degrade organic waste; however, this time mixed culture can be used for the treatment of complex substrate existing in wastewater because their efficiency is higher than pure culture for the production of electricity (Kim et al. 2007a). One example is shown here; *Trametes versicolor* was cultivated with *Shewanella oneidensis* and energy (1000 mV) was generated in MFCs with the use of electro-Fenton technology. In the electro-Fenton technology, H_2O_2 is produced from the degradation of organic pollutants in MFCs (De dios et al. 2013). The most effective yeast *Arxula adeninivorans* was another choice as biocatalyst for MFCs because it can be cultivated at temperatures up to 48 °C, with extensive pH tolerance and high salt concentration and also has the ability to utilize a very extensive substrate range. Thus *A. adeninivorans* acts as electyricigen in MFCs to produce bioelectricity. First-time dual-chamber MFCs were used for the investigation of the bioelectrochemical activity of *A. adeninivorans* in continuous mode and 28 mWm^{-2} power density was obtained. A soluble electro active redox molecule was secreted by *A. adeninivorans,* which is supposed to induce electron generation in MFCs (Haslett et al. 2011). Additionally, Kaneshiro (2014) and other researchers have used seven fungi such as *Pichia pastoris, Kluyveromyces lactis, Schizosaccharomyces pombe, Candida glabrata, S. cerevisiae, K. marxianus* and *Hansenula polymorpha* in millilitre-scale MFCs, and among these fungi, *K. marxianus* fungus yielded the best power output. The 850 W/m^2 maximum power densities were obtained in millilitre-scale MFCs. Fungi can be also used as biocathode in MFCs, so they improve the oxygen reduction activity in the cathode chamber for exceptional electricity generation. ***Coriolus versicolor*** (white-rot fungus) can produce laccase that can be used in the reduction of the oxygen into the water. Due to its laccase production property, it was being used as biocathode in MFCs for improving the production of electricity (Wu et al. 2012). ***Ganoderma lucidum*** BCRC 36123 is also a white-rot fungus that can be used as biocathode in MFCs. It decolorized the azo dye acid orange as well as 699 mV electricity was generated in MFCs simultaneously (Lai et al. 2017). ***Galactomyces reessii*** is another laccase-secreting fungus that was also used as a biocathode in MFCs faded with rubber wastewater in the anode chamber. 59 mW/m^2 power density was obtained by biocathode-based FMFCs (Chaijak et al. 2018). This type of result showed that fungi are the best biocatalyst sources in MFCs for the degradation of organic compounds along with electricity generation.

14.6 Batteries Design with the Use of Fungal Electrode

14.6.1 Batteries Design

Increasing human population day by day increases the consumption of energy and needs more energy. The fossil fuels are a highly ranked energy source, which is second mostconsumed globally. Fossil fuels are profusely exploited worldwide, thus released large amounts of greenhouse gases and CO_2, etc., which contribute to the rising of global warming problems. The energy crisis was resolved with the use of non-renewable fossil fuel approaches. The fossil fuels are limited in nature; therefore renewable sources of energy such as biomass, the solar, wind, MFCs and photovoltaic have been used. The energy generated from these sources is always changeable due to the environmental condition; and these have limited storage capacity and are not available for an unlimited time. Therefore a new energy storage system is needed which should have much qualities such as high power, high energy density and almost 100% efficiency. One of the most applicable systems for energy storage is lithium-based batteries having high storage efficiency (Wagner et al. 2013). Carbon/lithium cobalt oxide ($LiCoO_2$) cell is the first commercialized lithium-ion battery which was firstly used in 1991. Many other types of batteries also have been identified such as Ca, Zn, Al, Mg and Na-based batteries; however, their efficiency is comparatively lower than Li-based batteries (Li et al. 2016). In recent years MFCs-based bio-battery is used for energy storage due to the electron storage capability of activated carbon bioanode. MFCs-based bio-battery works on bi-directional electron transfer; hence generated energy can be stored for different time domains. Bio-battery is composed of an anode, cathode, separator and electrolyte and on the oxidation /reduction reaction of organic compound present in the anodic chamber. Electron and proton are produced by the metabolic activity of microorganisms that transfer to the cathode through the outer circuit and separator respectively, and they produce water by reduction of oxygen (Sudirjo et al. 2019).

14.6.2 Structure and Composition of Lithium-Based Batteries

All types of lithium-based batteries have a similar structure that consists of a negative and a positive electrode that are separated by a permeable insulating separator which is permeable for ions and electrolytes. Functionally the negative electrode acts as an anode and the positive electrode works as a cathode. So, different types of lithium batteries are available in the market such as Li-S battery, Li-ion battery and Li-Air battery etc. (Mahmood and Hou 2014).

14.6.3 Lithium–Sulphur (Li-S) Batteries

Li-S batteries are a new efficient storage system due to their high storage capacity as well as being cheaper than other lithium-based batteries, which have an extensive application in renewable energy storage, portable electronic equipment and electric vehicles (Wu et al. 2017; Chen et al. 2020). Li-S batteries consist of sulphur-containing cathode, lithium-based anode and a lithium salt-dissolved electrolyte. Multi-step lithium-ion (Li^+)-coupled electro reduction/oxidation electrochemical reactions have been induced. During the discharge reaction, polysulphides are produced as the intermediate between S_8 and lithium sulphide (Li_2S). The discharge reaction is given below:

$$S_8 + Li_2 + e^- \rightarrow - Li_2S_n (2.4 - 2.1\ V)$$

$$Li_2S_n + Li_2 + e^- \rightarrow Li_2S_2 \text{ or } Li_2S\ (< 2.1\ V)$$

The performances of batteries are based on electrochemical cycling and one cycle is different from another cycle and alters the steps of charge and discharge (Zhang et al. 2017).

14.6.4 Lithium-Ion Batteries

The Li-ion battery contains an anode and a cathode electrode, where the anode reacts as the negative electrode and the cathode reacts as positive as well as electrolytes act as an insulator for electron which has the capability of ionic conduction. A permeable separator has the ability of transportation of the ion from the anode and cathode chamber but hinders the electron conduction to avert the short-circuiting of the batteries. Consequently, when an electron moves from a negative to a positive electrode via an external circuit, electricity is generated and stored in batteries. In extensive discharging action, Li-ion moves from the negative electrode to the positive electrode via the use of electrolyte but in the charging process, both Li-ion and electron move from the positive to negative electrode (He et al. 2012).

14.6.5 Lithium-Air Batteries

The lithium–air batteries normally contain a metallic Li anode (act as a negative electrode) and O_2 electrode (positive electrode) with various electrolytes (Girishkumar et al. 2010). Oxygen (O_2) is donated from atmospheric air set out as the active matter for lithium–air batteries in that way giving an enormous benefit in terms of energy density over predictable batteries. Diverse types of lithium–air batteries have been reported by researchers such as hybrid, aqueous, non-aqueous and solid-state batteries (Mahmood and Hou 2014). During the process of discharge, Li^+ is released from the negative electrode and reacts with dissolved O_2 in the electrolyte on the surface of the positive electrode catalyst. The reverse reaction can occur during the charging process, Li^+ returns onto the surface of the negative electrode and O_2 gas is developed from a catalyst electrode (Jung et al. 2016).

14.6.6 Role of Fungi in Batteries Design

Lithium-based batteries are the better candidate for the recent work due to their efficiency for energy storage, but the main problem with the Li-S batteries is the shuttle effect which lowers the coulombic efficiency and cycling performance. The shuttle effect is a phenomenon occurring in batteries due to the dissolving of polysulphide in electrolytes, which transport to the cathode through the separator and by the insulating nature of sulphur which lowers the rate capability and cycling performance of batteries (Manthiram et al. 2014). A lot of methods have been used to evade the aforesaid challenges which mostly improve the cathode activity. Some examples include infiltration of sulphur into reduced graphene oxide, porous carbon and carbon nanotubes material using a metal framework for the storage of sulphur. Apart from this method, another method was implied such as covalently bonding or coating the sulphur with conducting polymers, which is effective in suppressing the dissolution and stopping the transportation of polysulphides to the cathode and enhancing cyclability. However, this type of fabrication cannot be used for a long time due to its complexity and expensive nature. Therefore microbial-based fabrication is used for conductive porous carbon film to overcome the problem that occurred in the physical fabrication of conductive polymers. Mostly filamentous fungi work as carbonizable binder for attaching the graphene nanosheets within the carbon film, and fungal-based fabrication is safe and sustainable (Chai et al. 2015). Another way, fungi mycelium is also used for the synthesis of an electrode material such as graphite and other activated carbon, which acts as an anode in Li-based batteries. Graphite is the most used anode material in Li-ion batteries, because of its low cost and long cycle life. However, graphite carbon had some deficiencies such as high tenderness to the electrolyte, charge capacity is low, can confine the electro-chemical performance of Li-based batteries. Thus, carbon nanotubes and manganese oxides can show excellent electrochemical properties and be developed as other anode material for improving the capacity and power density of lithium-based batteries (Xia et al. 2013).

At present, the fungal bio-mineralization process is also used for the synthesis of the electrode material. The earliest carbonized fungal biomass–mineral composite (MycMnOx/C) electrode material was synthesized by *Neurospora crassa* mycelium (Li and Gadd 2017). *Auricularia* species was used for the synthesis of activated carbon (AC) electrode by using a hydrothermal-assisted pyrolysis procedure. The activated carbon electrodes have many qualities such as energy/power density, stability, surface capacitance and volumetric capacitance, which were used in Li-based batteries (Zhu et al. 2011). Among these *Aspergillus campestris* was also used for the synthesis of various carbon electrodes due to the fibrous nature. The fungi-based synthesis of electrode material is more green, consistent and inexpensive and can be easily adorned during the synthesis process. However, at this time, further research on fungus-based electrode synthesis is needed.

14.7 Application of MFCs

MFCs are a versatile technique for further research that is used in many areas. It has the ability for the degradation of pollution, bioelectricity generation and the production of value-added products for further use. Here are some important applications of MFCs described (Figure 14.3).

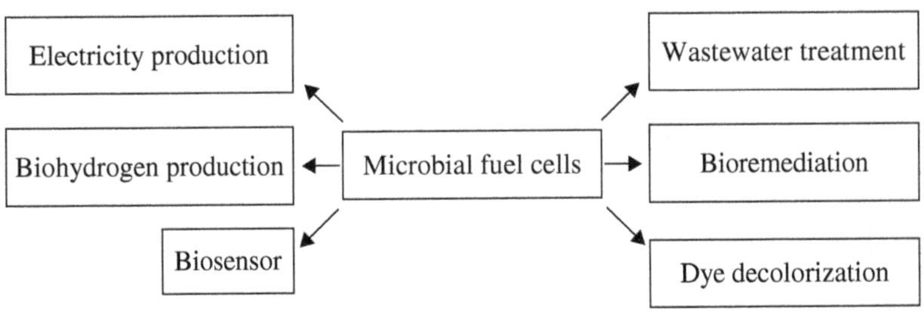

Figure 14.3 Application of microbial fuel cells.

14.7.1 Bioelectricity Production

Anaerobic condition is very suitable for the generation of electric current. For this reason, MFCs are very suitable system for the production of energy by the use of electrogenic microorganisms (Table 14.2), but the efficiency of production of current is very low, that's why some workers used the capacitors for the storage of energy. EcoBot is a biologically inspired robot that has features such as motion, sensing, computing and communication powered by the use of MFCs. In the present scenario, many scientists have worked with

Table 14.2 Bioelectricity production in MFCs.

Waste material	Microorganisms	Types of MFCs	Power out put (mW/m^2)	References
Marine sediment	*Deltaproteobacterium*	Two-chamber	14.00	Bond et al. (2002)
Sewage sludge	*Escherichia coli (E. coli)*	Single-chamber	91.00	Park and Zeikus (2003)
Acetate	*G. sulfurreducens*	Two-chamber	48.40 ± 0.3	Jung and Regan (2007)
Lactate	*Geobacter* species	Two-chamber	52.00 ± 4.7	Jung and Regan (2007)
Glucose	*E. coli*	Single-chamber	760.00	Zhang et al. (2007)
Cellulose	Cellulose-degrading bacteria	Two-chamber	188.00	Hassan et al. (2012)
Ethanol	*Betaproteo*bacterium	Two-chamber	40.00 ± 2	Kim et al. (2007b)
Glucose	*S. cerevisiae*	Two-chamber	16.00	Jung and Regan (2007)
Glucose	*Geobacter* species *(Firmicutes)*	Two-chamber	40.30 ± 3.9	Jung and Regan (2007)
Cyctenin	*Gammaproteo* and *Shewanella affinis* (KMM3586)	Two-chamber	36.00	Zhou et al. (2013)
Food waste	*Chlorella vulgaris*	Two-chamber	191.51	Hou et al. (2016)
Tannery effluent waste and sweet lime waste	*Geobacter metallireducens*	Two-chamber	396.70	Sindhuja et al. (2018)
Chocolate wastewater	*Achromobacter insuavis* strain BT1 (MF346036.1), *A. insuavis* strain B3 (MF346037.1), *Bacillus encimensis* strain B4 (MF346038.1) and *Kocuria flava* strain B5	Two-chamber	104.90	Subha et al. (2019)
Acetate	*Geobacter* species	Single-chamber	195.07	Ai et al. (2020)

miniature MFCs that will be implanted in the human body to power an implantable medical device (Du et al. 2007). In the present time, MFCs have been used for powering the microcomputer, and the minimum voltage required for the running of a microcomputer is 1.850 V and MFCs produced 2.550 V that is higher than that for running a microcomputer (Walter et al. 2020).

Some other application: The major use of MFCs is bioelectricity production. Moreover, it also has other applications such as biohydrogen production, biosensor, wastewater treatment, bioremediation and dye decolorization, etc.

14.7.2 Biohydrogen Production

Hydrogen gas is a new alternative source of energy, and it is largely used in vehicle fuel. Mostly hydrogen gas is produced by fossil fuels; however, these sources are limited and cause many types of environmental problems, so the production of hydrogen gas by other biodegradable organic material using the biological process (Cheng and logan 2007). Modified MFCs technologies have been used for the production of biohydrogen. Although in this process proton and electron are generated from microbial metabolic activity in anode transfer to the cathode for the production of hydrogen, it is thermodynamically unfavourable; so for this reason some researchers applied external potential to increase the potential of cathode that is favourable for the production of hydrogen (Du et al. 2007). Therefore, some workers have used a combined form of MFCs-MEC (microbial electrolysis cell) that overcomes the input of external electrical energy for hydrogen production and obtained 41% (10.1–14.3 mmol/g cellulose) hydrogen gas (Wang et al. 2011). At present, hydrogen is produced by using modified MFCs system from various waste materials such as carbohydrates, proteins, lignocellulosic containing waste, oil and grease, chlorides, sulphates chemical containing waste and sometimes heavy metals wastewater from food and sugar processing industries etc. According to Ramu et al. (2020), wastewater having higher organic matter is favoured for the production of biohydrogen.

14.7.3 Biosensor

The biosensor is an analytical device containing enzyme, antibody, phages, aptamers or single-stranded DNA as an immobilized biological material, which can exclusively interact with an analyte, which can be measured by producing physical, chemical or electrical signals with the use of the transducer. An analyte may be glucose, urea, drugs and other toxic material (Iqbal et al. 2000). MFCs can be used as a biosensor also, they are attached with the anode as a biofilm working as a biological recognition element and can recognize the sudden changes in analyte present in wastewater and give the response as electrical current by the use of transducer (Peixoto et al. 2011). Thus MFCs based on biosensors can useful for water quality analysis such as biological oxygen demand (BOD), chemical oxygen demand (COD), dissolved oxygen (DO), volatile fatty acids (VFAs), toxins, heavy metals in wastewater (Jin et al. 2017). MFCs-based biosensor has not been used as any additional transducer or external power sources. This technology has many benefits such as long-term stability, simple and cost-effective (Do et al. 2020). In recent times, paper-based MFCs sensors created by freeze-dried bacteria are also used for water quality monitoring (Cho et al. 2020).

14.7.4 Wastewater Treatment

Wastewater is highly variable depending on the industrial sources. Many types of toxic materials are present in wastewater, so treatment is needed for wastewater. Mainly chemical and physical methods are applied for wastewater treatment, but these methods are cost-effective and time-consuming. Due to these reasons, the biological method is very favourable and eco-friendly. MFCs are the most potent devices for the wastewater treatment along with bioelectricity production as well as are used for bioremediation etc. Pure substrates such as acetate, glucose and sucrose are present in wastewater producing more power output than complex substrates present in wastewater. Thus, a different type of wastewater is used for treatment in MFCs, which include domestic, textile industry, brewery, food processing, sugar industry (Ramu et al. 2020), dairy industry, biscuit industrial (Kumar et al. 2019), chocolate industry (Patil et al. 2009; Tharali et al. 2016) etc. along with the energy production. At present, many hybrid MFCs devices are being used for the treatment of wastewater such as microbial desalination cell, bioelectro-Fenton-MFCs, MFCs-electrosorption cell, microbial reverse-electrodialysis cell, microbial solar cell, wetland-MFCs and plant-MFCs, and these systems also produced bioelectricity as well as value-added product (Zhang et al. 2019).

14.7.5 Bioremediation

Bioremediation is the process in which microorganisms or enzymes are used to degrade the environmental pollutant, which is present in the water, soil and air etc. Many compounds present in the effluents from petroleum and chemical industries are very hazardous for health; therefore it is required to treat chemical compounds before they discharge off into the water bodies (Khan et al. 2019). MFCs are used for the bioremediation as well as electricity generation because of the microorganism present in the anodic chamber that degrades the waste material and produces electrons and protons that transfer to the cathodic chamber and generate electricity (Córdova-Bautista et al. 2020). Some researchers used the biocathodic treatment for the degradation of phenolic compounds in MFCs because it showed COD removal efficiency, power density and coulombic efficiency than bioanodic treatment (Khan et al. 2019). Morris and Jin (2007) applied the MFCs technology for the bioremediation of hydrocarbon in groundwater and obtained 120 mW/m^2 power densities. Microorganisms are the most important factor for bioelectricity production and petroleum hydrocarbon degradation. Thus diesel hydrocarbon is degraded by the new member of *Gamma proteobacteria* class *Vibrio* species strain E2 in high salt (3%–4%) and alkaline pH (8–9) mineral medium (Li et al. 2019). In recent years, sediment MFCs are used for the bioremediate toxic material present in sediment because sediment acts as a source or sink of contaminants that affect the whole aquatic ecosystem. In sediment MFCs, anode was fixed in sediment, which is an anoxic environment and cathode was placed in overlying water, which is an oxic environment and produced electricity as well as converted toxic material into less toxic form (Abbas et al. 2017).

14.7.6 Dye Decolorization

Wastewater from the textile industry contains aromatic amines, which is a very dangerous pollutant for the environment. Dyes contain azo bond (–N=N–) and associated chromophores when released in surface water that causes many health-related problems such as cancer and other diseases. Different methods are used for the azo dye-containing wastewater treatment, but it is an expensive and time-consuming process, so the MFCs are an effective technique to degrade dyes and produce the energy simultaneously. Microorganisms are used in MFCs for degrading dyes which can directly transfer an electron to an anode without the use of an electron shuttle. These microorganisms are *Proteus hauseri* ZMd44, *Pseudomonas aeruginosa* and *Geobacter sulfurreducens* etc. Some researchers used the mixed cultures of fungal and bacteria in MFCs for removing the dyes from wastewater because itproduced more energy than a single strain (Ilamathi and Jayapriya 2018). Simões et al. (2019) and his colleagues applied an immobilized fungal strain, *Pleurotus ostreatus* URM 4809, in the cathodic chamber that produced laccase and degraded Remazol brilliant blue dye, which is mostly used in the textile industry. The MFCs showed a significant role in the production of laccase which is assisted for the rapid decolorization of the effluent of textile dye.

14.8 Challenges and Future Prospective

Nowadays fungi-based MFCs have been used on a large scale, but some hurdles occurred during MFCs processing such as low power production. Single-chamber fungi-based MFCs should be used for further electricity production because higher internal resistance is found in dual-chamber FMFCs. Mostly ferricyanide mediators are used in FMFCs as electron acceptors, but it cannot be used on a large scale. Thus new fungi strain is being searched for electricity production and also to overcome the mediator used in FMFCs (Sekrecka-Belniak and Toczyłowska-Mamińska 2018). Further studies should be focused on a better characterization of the fungi involved in the degradation and also should be applied to nonmaterials and nanotechnology approaches, which will be used in bioremediation and energy production (Fogel and Limson 2016, pp. 551-575). Fungi can also be used as biocathode in MFCs for better electricity generation; therefore cathode-related research should be done. Substrates are also challenging factor for the performance of MFCs because MFCs are used for the degradation of waste material which is present in a different type of wastewater; some are complex and simple, but some are risk to human health; therefore good substrate should be used such as carbohydrate-rich substrate and containing high moisture for the power production in MFCs (Mbugua et al. 2020). Other challenging factors are electrode material and microorganisms themselves which play the most important role in MFCs. Typically the electrode material that is used should have elevated conductivity, low level of oxidization, more accurate surface area, porosity, suitable for microorganism growth and cheap etc. However, electrodes that have all these qualities are very expensive; for this reason, a good cheap electrode is required to attain this quality. Lastly, the microorganism also exhibited more challenges in MFCs because microorganisms release electrons through a metabolic activity that is transferred to the anode, then bioelectricity is generated. Although direct electron transfer can be used by some microbes, some use a mediator; therefore these mechanisms are not to be suitable for all microorganisms. Fungi are also used as a biocatalyst in MFCs, but their electron transfer mechanism is unknown. Due to this, further research should be done which improves the performance of the MFCs (Ivase et al. 2020).

14.9 Conclusion

As the sources of fossil energy are limited, energy limitation encouraged the scientists on the planet to think about different sources of energy. In addition, using fossil energy can cause pollution on earth. MFCs are a more significant, alternative green energy source and are a new technology to generate energy without any environmental pollution. The present chapter exposed that the fungal-based MFCs are a novel technology to generate bioelectricity was well as use of fungus for the fabrication of electrode material of lithium-based batteries. So it can be important for further research on fungal MFCs and lithium-based batteries. Fungal-based MFCs have dual activity; it can degrade waste material and produce bioelectricity simultaneously. Mostly *Saccharomyces cerevisiae* is used as anode catalyst or cathode catalyst in MFCs, but nowadays other fungi are also used for the production of bioelectricity. Fungi-based fabrication is eco-friendly, safe and cost-effective and also the shuttle effect does not arise in lithium-based batteries. Fungal MFCs are the best and valuable method for the production of bioelectricity as well as the synthesis of the new electrode material.

Acknowledgements

The authors would like to thank the School of Studies in Biotechnology, Pt. Ravishankar Shukla University Raipur (Chhattisgarh) for providing all the necessary facilities for this work.

References

Abbas, S. Z., Rafatullah, M., Ismail, N. et al. (2017). Enhanced bioremediation of toxic metals and harvesting electricity through sediment microbial fuel cell. *International Journal of Energy Research* 41(14): 2345–2355.

Ahn, Y., and Logan, B. E. (2010). Effectiveness of domestic wastewater treatment using microbial fuel cells at ambient and mesophilic temperatures. *Bioresource Technology* 101(2): 469–475.

Ai, C., Yan, Z., Hou, S., Zheng, X. et al. (2020). Effective treatment of acid mine drainage with microbial fuel cells: an emphasis on typical energy substrates. *Minerals* 10(5): 443.

An, B. M., Seo, S. J., Hidayat, S. et al. (2020). Treatment of ethanolamine and electricity generation using a scaled-up single-chamber microbial fuel cell. *Journal of Industrial and Engineering Chemistry* 81: 1–6.

Babanova, S., Hubenova, Y., and Mitov, M. (2011). Influence of artificial mediators on yeast-based fuel cell performance. *Journal of Bioscience and Bioengineering* 112(4): 379–387.

Bennetto, H. P. (1990). Electricity generation by microorganisms. *Biotechnology Education* 1(4): 163–168.

Beyene, H. D., Werkneh, A. A., and Ambaye, T. G. (2018). Current updates on waste to energy (WtE) technologies: a review. *Renewable Energy Focus* 24: 1–11.

Bhargavi, G., Venu, V., and Renganathan, S. (2018). Microbial fuel cells: recent developments in design and materials. *IOP Conference Series: Materials Science and Engineering* 330. https://doi.org/10.1088/1757-899X/330/1/012034.

Bond, D. R., and Lovley, D. R. (2003). Electricity production by *Geobacter sulfurreducens* attached to electrodes. *Applied and Environmental Microbiology* 69(3): 1548–1555.

Bond, D. R., Holmes, D. E., Tender, L. M. et al. (2002). Electrode-reducing microorganisms that harvest energy from marine sediments. *Science* 295(5554): 483–485

Bresadola, M. (1998). Medicine and science in the life of Luigi Galvani (1737–1798). *Brain Research Bulletin* 46(5): 367–380.

Cao, Y., Zhang, H., Liu, H. et al. (2018). Biosynthesis and production of sabinene: current state and perspectives. *Applied Microbiology and Biotechnology* 102(4): 1535–1544.

Catal, T., Kul, A., Atalay, V. E. et al. (2019). Efficacy of microbial fuel cells for sensing of cocaine metabolites in urine-based wastewater. *Journal of Power Sources* 414: 1–7.

Chai, L., Wang, J., Wang, H. et al. (2015). Porous carbonized graphene-embedded fungus film as an interlayer for superior Li–S batteries. *Nano Energy* 17: 224–232.

Chaijak, P., Sukkasem, C., Lertworapreecha, M. et al. (2018). Enhancing electricity generation using a laccase-based microbial fuel cell with yeast *Galactomyces reessii* on the cathode. *Journal of Microbiology and Biotechnology* 28(8): 1360–1366.

Chen, C., Jiang, J., He, W. et al. (2020). 3D Printed high-loading lithium-sulfur battery toward wearable energy storage. *Advanced Functional Materials* 30(10): 1909469.

Cheng, S., and Logan, B. E. (2007). Sustainable and efficient biohydrogen production via electrohydrogenesis. *Proceedings of the National Academy of Sciences* 104(47): 18871–18873.

Cheng, S., Liu, H., and Logan, B. E. (2006). Increased power generation in a continuous flow MFC with advective flow through the porous anode and reduced electrode spacing. *Environmental Science & Technology*, 40(7): 2426–2432.

Cho, J. H., Gao, Y., Ryu, J. et al. (2020). Portable, disposable, paper-based microbial fuel cell sensor utilizing freeze-dried bacteria for *in situ* water quality monitoring. *ACS Omega* 5(23): 13940–13947.

Choi, J. and Ahn, Y. (2013). Continuous electricity generation in stacked air cathode microbial fuel cell treating domestic wastewater. *Journal of Environmental Management* 130: 146–152.

Choi, S., Kim, J. R., Cha, J. et al. (2013). Enhanced power production of a membrane electrode assembly microbial fuel cell (MFC) using a cost effective poly [2, 5-benzimidazole] (ABPBI) impregnated non-woven fabric filter. *Bioresource Technology* 128: 14–21.

Christwardana, M., Frattini, D., Accardo, G. et al. (2018). Optimization of glucose concentration and glucose/yeast ratio in yeast microbial fuel cell using response surface methodology approach. *Journal of Power Sources* 402: 402–412.

Clauwaert, P., Van der Ha, D., Boon, N. et al. (2007). Open air biocathode enables effective electricity generation with microbial fuel cells. *Environmental Science & Technology*, 41(21):7564–7569.

Cohen, B. (1931). The bacterial culture as an electrical half-cell. *Journal of Bacteriology* 21(1): 18–19.

Cole, K. S. (1972). *Membranes, Ions and Impulses: A Chapter of Classical Biophysics*. University of California Press 1.

Córdova-Bautista, Y., Ramirez-Morales, E., Perez-Hernandez, B. et al. (2020). Electricity production and bioremediation from synthetic sugar industry wastewater by using microbial isolate in microbial fuel cell. *Sugar Technology*. https://doi.org/10.1007/s12355-020-00830-1.

De dios, M. Á. F., del Campo, A. G., Fernández, F. J. et al. (2013). Bacterial–fungal interactions enhance power generation in microbial fuel cells and drive dye decolourisation by an *ex situ* and in situ electro-Fenton process. *Bioresource Technology* 148: 39–46.

Di Ilio, G. and Falcucci, G. (2020). Multiscale methodology for microbial fuel cell performance analysis. *International Journal of Hydrogen Energy*. https://doi.org/10.1016/j.ijhydene.2020.04.057.

Do, M. H., Ngo, H. H., Guo, W. et al. (2020). Microbial fuel cell-based biosensor for online monitoring wastewater quality: a critical review. *Science of the Total Environment* 712: 135612.

Du, Z., Li, H. and Gu, T. (2007). A state of the art review on microbial fuel cells: a promising technology for wastewater treatment and bioenergy. *Biotechnology Advances* 25(5): 464–482.

Ebadinezhad, B., Ebrahimi, S., and Shokrkar, H. (2019). Evaluation of microbial fuel cell performance utilizing sequential batch feeding of different substrates. *Journal of Electroanalytical Chemistry* 836: 149–157.

Fishilevich, S., Amir, L., Fridman, Y. et al. (2009). Surface display of redox enzymes in microbial fuel cells. *Journal of the American Chemical Society* 131(34): 12052–12053.

Fogel, R., and Limson, J. L. (2016). Applications of nanomaterials in microbial fuel cells. *Nanomaterials for Fuel Cell Catalysis*, 551–575. https://doi.org/10.1007/978-3-319-29930-3_14.

Gajda, I., Greenman, J., Melhuish, C., and Ieropoulos, I. (2015). Simultaneous electricity generation and microbially-assisted electrosynthesis in ceramic MFCs. *Bioelectrochemistry* 104: 58–64.

Ganguli, R. and Dunn, B. S. (2009). Kinetics of anode reactions for a yeast-catalysed microbial fuel cell. *Fuel Cells* 9(1): 44–52.

Girishkumar, G., McCloskey, B., Luntz, A. C. et al. (2010). Lithium- air battery: promise and challenges. *The Journal of Physical Chemistry Letters* 1(14): 2193–2203.

Gude, V. G. (2016). Wastewater treatment in microbial fuel cells–an overview. *Journal of Cleaner Production* 122: 287–307.

Halme, A. and Zhang, X. C. (1995). Biofuel cell utilizing *Saccharomyces cerevisiae*-modelling of the process. *IFAC Proceedings* 28(3): 165–170.

Haslett, N. D., Rawson, F. J., Barriëre, F. et al. (2011). Characterisation of yeast microbial fuel cell with the yeast Arxula adeninivorans as the biocatalyst. *Biosensors and Bioelectronics* 26(9): 3742–3747.

Hassan, S. H., Kim, Y. S., and Oh, S. E. (2012). Power generation from cellulose using mixed and pure cultures of cellulose-degrading bacteria in a microbial fuel cell. *Enzyme and Microbial Technology* 51(5): 269–273.

He, P., Yu, H., and Zhou, H. (2012). Layered lithium transition metal oxide cathodes towards high energy lithium-ion batteries. *Journal of Materials Chemistry* 22(9): 3680–3695.

Hou, Q., Pei, H., Hu, W. et al. (2016). Mutual facilitations of food waste treatment, microbial fuel cell bioelectricity generation and *Chlorella vulgaris* lipid production. *Bioresource Technology* 203: 50–55.

Hubenova, Y. and Mitov, M. (2010). Potential application of *Candida melibiosica* in biofuel cells. *Bioelectrochemistry* 78(1): 57–61.

Hubenova, Y., Rashkov, R., Buchvarov, V. et al. (2011). Nanomodified NiFe-and NiFeP-carbon felt as anode electrocatalysts in yeast-biofuel cell. *Journal of Materials Science* 46(22): 7074–7081.

Ieropoulos, I., Greenman, J., and Melhuish, C. (2012). Urine utilization by microbial fuel cells; energy fuel for the future. *Physical Chemistry Chemical Physics* 14(1): 94–98.

Ilamathi, R. and Jayapriya, J. (2018). Microbial fuel cells for dye decolorization. *Environmental Chemistry Letters* 16(1): 239–250.

Iqbal, S. S., Mayo, M. W., Bruno, J. G. et al. (2000). A review of molecular recognition technologies for detection of biological threat agents. *Biosensors and Bioelectronics* 15(11-12): 549–578.

Ivase, T. J. P., Nyakuma, B. B., Oladokun, O. et al. (2020). Review of the principal mechanisms, prospects, and challenges of bioelectrochemical systems. *Environmental Progress & Sustainable Energy* 39(1): 13298.

Jadhav, G. S. and Ghangrekar, M. M. (2009). Performance of microbial fuel cell subjected to variation in pH, temperature, external load and substrate concentration. *Bioresource Technology* 100(2): 717–723.

Jatoi, A. S., Baloch, A. G., Jadhav, A. et al. (2020). Improving fermentation industry sludge treatment as well as energy production with constructed dual chamber microbial fuel cell. *SN Applied Sciences* 2(1): 9.

Jin, H., Gui, R., Yu, J., Lv, W. et al. (2017). Fabrication strategies, sensing modes and analytical applications of ratiometric electrochemical biosensors. *Biosensors and Bioelectronics* 91: 523–537.

Jung, S. and Regan, J. M. (2007). Comparison of anode bacterial communities and performance in microbial fuel cells with different electron donors. *Applied Microbiology and Biotechnology* 77(2): 393–402.

Jung, K. N., Kim, J., Yamauchi, Y. et al. (2016). Rechargeable lithium–air batteries: a perspective on the development of oxygen electrodes. *Journal of Materials Chemistry A* 4(37): 14050–14068.

Kaneshiro, H., Takano, K., Takada, Y. et al. (2014). A milliliter-scale Yeast-based fuel cell with high performance. *Biochemical Engineering Journal* 83: 90–96.

Kaushik, A. and Jadhav, S. K. (2017). Conversion of waste to electricity in a microbial fuel cell using newly identified bacteria: *Pseudomonas fluorescens. International journal of Environmental Science and Technology* 14(8): 1771–1780.

Khan, A. M. and Hussain, M. S. (2017). Conversion of wastes to bioelectricity, bioethanol, and fertilizer. *Water Environment Research* 89(8): 676–686.

Khan, N., Anwer, A. H., Ahmad, A. et al. (2019). Investigation of CNT/PPY-modified carbon paper electrodes under anaerobic and aerobic conditions for phenol bioremediation in microbial fuel cells. *ACS Omega* 5(1): 471–480.

Kim, H. J., Park, H. S., Hyun, M. S. et al. (2002). A mediator-less microbial fuel cell using a metal reducing bacterium, *Shewanella putrefaciens. Enzyme and Microbial Technology* 30(2): 145–152.

Kim, B. H., Chang, I. S., & Gadd, G. M. (2007a). Challenges in microbial fuel cell development and operation. *Applied Microbiology and Biotechnology* 76(3): 485.

Kim, J. R., Jung, S. H., Regan, J. M. et al. (2007b). Electricity generation and microbial community analysis of alcohol powered microbial fuel cells. *Bioresource Technology* 98(13): 2568–2577.

Kim, I. S., Chae, K. J., Choi, M. J. et al. (2008). Microbial fuel cells: recent advances, bacterial communities and application beyond electricity generation. *Environmental Engineering Research* 13(2): 51–65.

Koók, L., Kaufer, B., Bakonyi, P. et al. (2019). Supported ionic liquid membrane based on [bmim][PF6] can be a promising separator to replace Nafion in microbial fuel cells and improve energy recovery: a comparative process evaluation. *Journal of Membrane Science* 570: 215–225.

Kumar, S. A., Thamarai, P., and Pournan, L. (2019). Comparative study on the performance of double chambered microbial fuel cell for the treatment of confectionery and biscuit industrial effluents. *Caribbean Journal of Science* 53: 2478–2489.

Lai, C. Y., Liu, S. H., Wu, G. P. et al. (2017). Enhanced bio-decolorization of acid orange 7 and electricity generation in microbial fuel cells with superabsorbent-containing membrane and laccase-based bio-cathode. *Journal of Cleaner Production* 166: 381–386.

Li, Q. and Gadd, G. M. (2017). Fungal nanoscale metal carbonates and production of electrochemical materials. *Microbial Biotechnology* 10(5): 1131–1136.

Li, F., Sharma, Y., Lei, Y. et al. (2010). Microbial fuel cells: the effects of configurations, electrolyte solutions, and electrode materials on power generation. *Applied Biochemistry and Biotechnology* 160(1): 168.

Li, Q., Chen, J., Fan, L. et al. (2016). Progress in electrolytes for rechargeable Li-based batteries and beyond. *Green Energy & Environment* 1(1): 18–42.

Li, X., Lu, Y., Luo, H. et al. (2017). Microbial stratification structure within cathodic biofilm of the microbial fuel cell using the freezing microtome method. *Bioresource Technology* 241: 384–390.

Li, X., Zheng, R., Zhang, X. et al. (2019). A novel exoelectrogen from microbial fuel cell: bioremediation of marine petroleum hydrocarbon pollutants. *Journal of Environmental Management* 235: 70–76.

Logan, B. E. (2008). *Microbial Fuel Cells*. Hoboken, NJ: Wiley, ISBN: 978-0-470-23948-3.

Mahmood, N. and Hou, Y. (2014). Electrode nanostructures in lithium-based batteries. *Advanced Science* 1(1): 1400012.

Manthiram, A., Fu, Y., Chung, S. H. et al. (2014). Rechargeable lithium–sulfur batteries. *Chemical Reviews* 114(23): 11751–11787.

Mardiana, U., Innocent, C., Cretin, M. et al. (2019). Applicability of alginate film entrapped Yeast for microbial fuel cell. *Russian Journal of Electrochemistry* 55(2): 78–87.

Mathuriya, A. S. and Sharma, V. N. (2009). Bioelectricity production from paper industry waste using a microbial fuel cell by *Clostridium* species. *Journal of Biochemical Technology* 1(2): 49–52.

Mbokou, S. F., Tonle, I. K., and Pontie M. (2017). Development of a novel hybrid biofuel cell type APAP/O_2 based on a fungal bioanode with a *Scedosporium dehoogii* biofilm. *Journal of Applied Electrochemistry* 47(2): 273–280.

Mbugua, J. K., Mbui, D. N., Mwaniki, J. et al. (2020). Influence of substrate proximate properties on voltage production in microbial fuel cells. *Journal of Sustainable Bioenergy Systems* 10(2): 43.

Meshram, R. and Jadhav, S. K. (2019). Treatment of oil refinery wastewater simultaneously with bioelectricity production in mediator-less microbial fuel cell using native gram positive *Bacillus* sp. *Research Journal of Pharmacy and Technology* 12(4): 1953–1961.

Mohan, S. V., Mohanakrishna, G., Velvizhi, G. et al. (2010). Bio-catalyzed electrochemical treatment of real field dairy wastewater with simultaneous power generation. *Biochemical Engineering Journal* 51(1-2): 32–39.

Mohan, S. V., Velvizhi, G., Modestra, J. A. et al. (2014). Microbial fuel cell: critical factors regulating bio-catalyzed electrochemical process and recent advancements. *Renewable and Sustainable Energy Reviews* 40: 779–797.

Morris, J. M., and Jin, S. (2007). Feasibility of using microbial fuel cell technology for bioremediation of hydrocarbons in groundwater. *Journal of Environmental Science and Health, Part A* 43(1): 18–23.

Niessen, J., Schröder, U., and Scholz, F. (2004). Exploiting complex carbohydrates for microbial electricity generation–a bacterial fuel cell operating on starch. *Electrochemistry Communications* 6(9): 955–958.

Nimje, V. R., Chen, C. Y., Chen, H. R. et al. (2012). Comparative bioelectricity production from various wastewaters in microbial fuel cells using mixed cultures and a pure strain of *Shewanella oneidensis*. *Bioresource Technology* 104: 315–323.

Oh, S. T., Kim, J. R., Premier, G. C. et al. (2010). Sustainable wastewater treatment: how might microbial fuel cells contribute. *Biotechnology Advances* 28(6): 871–881.

Park, D. H. and Zeikus, J. G. (2003). Improved fuel cell and electrode designs for producing electricity from microbial degradation. *Biotechnology and Bioengineering* 81(3): 348–355.

Patil, S. A., Surakasi, V. P., Koul, S. et al. (2009). Electricity generation using chocolate industry wastewater and its treatment in activated sludge based microbial fuel cell and analysis of developed microbial community in the anode chamber. *Bioresource Technology* 100(21): 5132–5139.

Peixoto, L., Min, B., Martins, G. et al. (2011). In situ microbial fuel cell-based biosensor for organic carbon. *Bioelectrochemistry* 81(2): 99–103.

Pontié, M., Jaspard, E., Friant, C. et al. (2019). A sustainable fungal microbial fuel cell (FMFC) for the bioremediation of acetaminophen (APAP) and its main by-product (PAP) and energy production from biomass. *Biocatalysis and Agricultural Biotechnology* 22: 101376.

Potter, M. C. (1911). Electrical effects accompanying the decomposition of organic compounds. *Proceedings of the royal society of London*. Series B, Containing Papers of A Biological Character 84(571): 260–276.

Prasad, D., Arun, S., Murugesan, M. et al. (2007). Direct electron transfer with yeast cells and construction of a mediatorless microbial fuel cell. *Biosensors and Bioelectronics* 22(11): 2604–2610.

Rabaey, K., Boon, N., Siciliano, S. D. et al. (2004). Biofuel cells select for microbial consortia that self-mediate electron transfer. *Applied and Environmental Microbiology* 70(9): 5373–5382.

Rahimnejad, M., Bakeri, G., Najafpour, G. et al. (2014). A review on the effect of proton exchange membranes in microbial fuel cells. *Biofuel Research Journal* 1(1): 7–15.

Rahimnejad, M., Adhami, A., Darvari, S. et al. (2015). Microbial fuel cell as new technology for bioelectricity generation: a review. *Alexandria Engineering Journal* 54(3): 745–756.

Rahmani, A. R., Navidjouy, N., Rahimnejad, M. et al. (2020). Effect of different concentrations of substrate in microbial fuel cells toward bioenergy recovery and simultaneous wastewater treatment. *Environmental Technology* 1–30. https://doi.org/10.1080/09593330.2020.1772374.

Ramadan, B.S. (2017). Challenges and opportunities of microbial fuel cells (MFCs) technology development in Indonesia. *MATEC Web of Conferences*: *EDP Sciences 101*. doi: 10.1051/matecconf/201710102018.

Ramu, S. M., Thulasinathan, B., Hari, D. G. et al. (2020). Fermentative hydrogen production and bioelectricity generation from food based industrial waste: an integrative approach. *Bioresource Technology* 310: 123447.

Reed, G. and Nagodawithana, T. W. (1991). Baker's yeast production. In *Yeast Technology*, 261–314. https://doi.org/10.1007/978-94-011-9771-7_7.

Rhoads, A., Beyenal, H., and Lewandowski, Z. (2005). Microbial fuel cell using anaerobic respiration as an anodic reaction and biomineralized manganese as a cathodic reactant. *Environmental Science & Technology* 39(12): 4666–4671.

Rossi, R., Hall, D. M., Wang, X. et al. (2020). Quantifying the factors limiting performance and rates in microbial fuel cells using the electrode potential slope analysis combined with electrical impedance spectroscopy. *Electrochimica Acta*, 348. https://doi.org/10.1016/j.electacta.2020.136330.

Sayed, E. T., Tsujiguchi, T., and Nakagawa, N. (2012). Catalytic activity of baker's yeast in a mediatorless microbial fuel cell. *Bioelectrochemistry* 86: 97–101.

Sayed, E. T., Barakat, N. A., Abdelkareem, M. A. et al. (2015). Yeast extract as an effective and safe mediator for the baker's-yeast-based microbial fuel cell. *Industrial & Engineering Chemistry Research* 54(12): 3116–3122.

Schaetzle, O., Barrière, F., and Baronian, K. (2008). Bacteria and yeasts as catalysts in microbial fuel cells: electron transfer from micro-organisms to electrodes for green electricity. *Energy & Environmental Science* 1(6): 607–620.

Sekar, N.and Ramasamy, R. P. (2013). Electrochemical impedance spectroscopy for microbial fuel cell characterization. *Journal of Microbial and Biochemical Technology* 6(2) 1–14.

Sekrecka-Belniak, A. and Toczyłowska-Mamińska, R. (2018). Fungi-based microbial fuel cells. *Energies* 11(10): 2827.

Shen, J., Du, Z., Li, J. et al. (2020). Co-metabolism for enhanced phenol degradation and bioelectricity generation in microbial fuel cell. *Bioelectrochemistry*. https://doi.org/10.1016/j.bioelechem.2020.107527.

Simões, M. F., Maiorano, A. E., dos Santos, J. G. et al. (2019). Microbial fuel cell-induced production of fungal laccase to degrade the anthraquinone dye Remazol Brilliant Blue R. *Environmental Chemistry Letters* 17(3): 1413–1420.

Sindhuja, M., Harinipriya, S., Bala, A. C. et al. (2018). Environmentally available biowastes as substrate in microbial fuel cell for efficient chromium reduction. *Journal of Hazardous Materials* 355: 197–205.

Subha, C., Kavitha, S., Abisheka S. et al. (2019). Bioelectricity generation and effect studies from organic rich chocolaterie wastewater using continuous upflow anaerobic microbial fuel cell. *Fuel* 251: 224–232.

Sudirjo, E., Buisman, C. J., and Strik, D. P. (2019). Marine sediment mixed with activated carbon allows electricity production and storage from internal and external energy sources: a new rechargeable bio-battery with bi-directional electron transfer properties. *Frontiers in Microbiology* 10: 934.

Tharali, A. D., Sain, N., and Osborne, W. J. (2016). Microbial fuel cells in bioelectricity production. *Frontiers in Life Science* 9(4): 252–266.

Tursun, H., Liu, R., Li, J., Abro, R., Wang, X. et al. (2016). Carbon material optimized biocathode for improving microbial fuel cell performance. *Frontiers in Microbiology* 7: 6.

Wagner, R., Preschitschek, N., Passerini, S. et al. (2013). Current research trends and prospects among the various materials and designs used in lithium-based batteries. *Journal of Applied Electrochemistry* 43(5): 481–496.

Walter, X. A., Greenman, J., and Ieropoulos, I. A. (2020). Microbial fuel cells directly powering a microcomputer. *Journal of Power Sources* 446: 227328.

Wang, A., Sun, D., Cao, G. et al. (2011). Integrated hydrogen production process from cellulose by combining dark fermentation, microbial fuel cells, and a microbial electrolysis cell. *Bioresource Technology* 102(5): 4137–4143.

Wu, C., Liu, X. W., Li, W. W. et al. (2012). A white-rot fungus is used as a biocathode to improve electricity production of a microbial fuel cell. *Applied Energy* 98: 594–596.

Wu, H., Lin, J., Mou, J. et al. (2017). A sustainable hierarchical carbon derived from cultivated fibroid fungus for high performance lithium–sulfur batteries. *RSC Advances* 7(75): 47407–47415.

Wu, X., Shi, Z., Zou L. et al. (2018). Pectin assisted one-pot synthesis of three dimensional porous NiO/graphene composite for enhanced bioelectrocatalysis in microbial fuel cells. *Journal of Power Sources* 378: 119–124

Xia, Y., Xiao, Z., Dou, X. et al. (2013). Green and facile fabrication of hollow porous MnO/C microspheres from microalgaes for lithium-ion batteries. *ACS Nano* 7(8): 7083–7092.

Yu, Y., Ndayisenga, F., Yu, Z. et al. (2019). Co-substrate strategy for improved power production and chlorophenol degradation in a microbial fuel cell. *International Journal of Hydrogen Energy* 44(36): 20312–20322. https://doi.org/10.1016/j.ijhydene.2019.05.221

Zhang, T., Zeng, Y., Chen, S. et al. (2007). Improved performances of E. coli-catalyzed microbial fuel cells with composite graphite/PTFE anodes. *Electrochemistry Communications* 9(3): 349–353.

Zhang, G., Zhang, Z. W., Peng, H. J. et al. (2017). A toolbox for lithium–sulfur battery research: methods and protocols. *Small Methods* 1(7): 1700134(1–32).

Zhang, Y., Liu, M., Zhou, M. et al. (2019). Microbial fuel cell hybrid systems for wastewater treatment and bioenergy production: synergistic effects, mechanisms and challenges. *Renewable and Sustainable Energy Reviews* 103: 13–29.

Zhou, M., Yang, J., Wang, H., Jin, T. et al. (2013). Bioelectrochemistry of microbial fuel cells and their potential applications in bioenergy. In *Bioenergy Research: Advances and Applications* 131–152. https://doi.org/10.1016/B978-0-444-59561-4.00009-7.

Zhu, H., Wang, X., Yang, F., & Yang, X. (2011). Promising carbons for supercapacitors derived from fungi. *Advanced Materials* 23(24): 2745–2748.

15

Current Perspective of Sustainable Utilization of Agro Waste and Biotransformation of Energy in Mushroom

Anjali Kosre[1], *Deepali Koreti*[1], *Pramod Kumar Mahish*[2], *and Nagendra Kumar Chandrawanshi*[1]

[1] *School of Studies in Biotechnology, Pt. Ravishankar Shukla University, Raipur (C.G.), Raipur, Chhattisgarh, India*
[2] *Department of Biotechnology, Govt. Digvijay Autonomous P.G. College, Rajnandgaon (C.G.), Rajnandgaon, Chhattisgarh, India*

15.1 Introduction

Biomass is formed from residue contents derived from all organic materials including plants, fungi, algae and crops, etc. through life processes to produce carbohydrates. The conservation, management and utilization of biomass resources are most valuable sustainable development strategies such as produce energy, biomass, biochemicals and is safe for the environment (Gavrilescu 2020). Carbohydrates present in biomass forms the structural blocks of all organic components. Approximately less than 1% of the light energy is converted into chemical energy form via the photosynthesis process and hence stored. The chemical energy in the form of chemical bonds is stored in the structural components of biomass. The efficient processing of biomass by either chemical or biological method reacts with the chemical bonds that stores the energy. The process occurs cyclically. A particular variety of biomass has its specific components depending on their physical and chemical properties with different functions. As the renewable sources of energy have been overexploited by human beings through burning biomass as a fuel and consuming plants for their life process. Conversion of biomass into fossil fuels takes millions of years, and these fossilized fuels are not renewable. Overexploitation of fossil fuels and natural resources contributes to a greater extent of CO_2 emission which contributes to an increase in the greenhouse effect. Biomass-derived energy is the major source of energy and serves 10–14% of energy all over the world (McKendry 2002). Biomass-derived energy may compete directly with the food sector or indirectly if biomasses derived from agro harvests are grown on soil would be used to produce food, for instance mushroom. It may affect the price of food and food security if the demand increases. Typically, this problem has been a matter of concern for the biofuel sector, which utilizes food crops principally. Enhanced awareness in the field of biofuels production will furthermore increase the demand for biomass feedstock. Since agro wastes

Energy: Crises, Challenges and Solutions, First Edition. Edited by Pardeep Singh, Suruchi Singh, Gaurav Kumar, and Pooja Baweja.

and biomass products are thrown away, it can be sustainably used as raw material for the generation of biofuels (Chakravorty et al. 2009; Dar et al. 2021).

The emerging bioenergy sector is moderately a complex sector. Several types of biomass wastes such as a variety of liquid, solid and gaseous transporters of bioenergy are presently available. Various methods for its conversion to available energy services through technology are also present. Biomass-based marketplaces run often on unofficial organization, which causes complexity to follow the data properly. Data collection is performed by diverse institutions that are not always well-coordinated and organized or may contradict the survey report. As a result, national and global data on the utilization of biomass and bioenergy demand are comparatively difficult to compute. Upcoming future of renewable sources of energy contributes to 15–20% in conservative setup, 30–45% in moderate situations and 50–95% in highly renewable circumstances. The electricity sector has the highest shares among them. Scenarios of all energy sectors depict an assortment of energy supply technologies jointly with energy demand growth and efficient energy improvements (Hoekstra et al. 2010). Biomass-derived biofuels supplied 3% of the transport fuel globally in the year 2010 IEA, 2010. Biofuels such as bioethanol produced from biomass such as sugarcane and corn starch by yeast fermentation technology and biodiesel produced through from oil and lipids by transesterification process. On the other hand, second-generation biofuels face several challenges, such as costly saccharification, wastewater generation, low solvent and product tolerance to the catalyst and inefficient and expensive process of product separation (Klein-Marcuschamer et al. 2012).

The need of capital may be limited, as the manufacturing spot of raw materials for the generation of revolutionary bioenergy is likely to be developed mainly in the exterior of cultivated land. Around 100 million hectares (Ha) of land would be adequate to accomplish biofuel's targeted production in global transportation fuels up to 2050 (IEA 2018a). A vital stage in emergent biofuel production, its sustainability and the biofuel's generation on economic parameter from lignocellulosic biomass. Perennial grown woody crops for energy production typically cover high yields than vegetable, oil and grain crops utilized for modern biofuel transformation. The potential coverage of woodland and grassland lignocellulosic raw materials is approximately 1.75 billion ha globally. However, they both supply food and fuel for surviving to local population or used as livestock grazing ground extensively. Approximately 700–800 million ha are sufficient for cost-effective feasible production of lignocellulosic feedstock (Popp et al. 2014). Growth and expansion of markets for biofuels, its investment on building new sets have in low speed in different countries in reaction to several factors such as competition for feedstock, policy uncertainty and concern about contest for food production, impacts of drought conditions on productivity, for land and water resources, concerns about the sustainability of produced bioenergy, respectively (BNEF 2012).

Biologically derived energy or 'Bioenergy' is the significant resources of renewable energy source in the world are mostly based on combustion-based power plants for domestic/residential purposes and in the industrialized area. Presently, 50% of global power is utilized in the form of heat. Even though on theoretical basis, this demand of energy could be decreased at average stretch due to efficient energy implementations, such as energy-saving buildings which includes the near-zero energy buildings for which the realistic target potential is to reduce heat demand until 2035 (International Energy Agency 2010). A novel concept of an economy that counts on the environmental variables and recovery period of raw biomaterials residues and biowastes must be promoted. For this principle, assessments of the sustainability of bioenergy must cover the three columns: social, environmental and economic sustainability, respectively, so that the results can persuade decision makers in the direction of those chains which are most helpful and favourable for the environment, society and economical benefits. In this perspective, the sustainable use of bioenergy may boost the three-dimensional benefits achieved from the sustainable management of renewable resources in such emerging bioenergy. The most significant role that bioenergy is expected to play is the considerable reduction of carbon emissions and help to achieve the sustainable development goals (SDGs) (GBEP Sustainability Indicators for Bioenergy (GSI) 2011; Bioenergy for Sustainable Development 2017).

15.2 Sustainable utilization of Agro waste Through Mushroom Cultivation Technology

Agricultural waste or agro wastes are the wastes that remain after the harvesting of crops. They are derived from plants and contain cellulose and lignin as their structural unit, so named as lignocellulosic wastes. Mushroom cultivation technology is an emergent industry that has attracted interest all over the world. The biological transformation of lignocellulosic biomass into foodstuff has a favorable ecological impact and continues to enhance in upcoming years (Chang 2008; Chang and Wasser 2017). The overall production of medicinal and edible mushrooms was estimated around 34 million tons in 2013 and valued above 34 billion US dollars (Royse et al. 2017). Mushroom species cultivated all over the world are the *Agaricus bisporus*, *Pleurotus* spp. and *Lentinula edodes* commonly known as oyster mushroom, button mushroom and shitake, respectively (Suguimoto et al. 2001; Li 2012). Other species such as *P. sajor caju* (king oyster mushroom), *Pleurotus florida*, *Pleurotus flabellatus* (grey oyster mushroom), *P. eryngii* (white oyster mushroom), *Volvariella volvacea* (straw paddy mushroom) and other species of mushroom (Islam et al. 2009). The popularly known and cultivated species of *Pleurotus* spp. are *P. sajor caju*, *P. eryngii* and *P. florida*, etc. (Alam et al. 2008). The cultivation of this species is effortless and does not involve any skilled personal or specific arrangement and is a very well-known species in Asian countries because of its promising dietary compounds and its ability to grow quickly on a collection of farm wastes it has gained attention and popularity. *Agaricus bisporus* and *L. edodes* are also leading species of mushroom grown globally (Phan and Sabaratnam 2012).

Mushroom is also known as the edible fungus grows widely on lignocellulosic agro wastes such as wheat straw, rice straw and paddy straw which is a proficient biological way to recycle agro residues and transform into valuable bioenergy. Some of the widely accepted mushrooms are white bottom, shiitake, enoki, portobello, maitake and oyster mushroom based on their popularity. Being containing unique flavour and nutritional attributes and it is accepted as an essential food product worldwide (Rani et al. 2008). However, in the fruiting bodies harvesting, a considerable quantity of organic waste residue is generated. Macrofungi efficiently convert the lignocellulosic waste by digesting the substrate and releasing simple sugars and enzymes and transforming it into food product in the form of edible mushroom. The most commonly cultivated species of mushrooms is *Pleurotus ostreatus* which potentially degrades lignin compared to other fungi. Among the *Pleurotus* species, *Pleurotus sajor-caju* and *P. ostreatus* mushroom demonstrated to be the best species for the degradation of constitution of cell wall when cultivated on rice straw substrate (Jafari et al. 2007). *Pleurotus* species can successfully degrade various other lignocellulosic materials such as rice straw, wheat straw and sawdust compared to other white-rot fungi. Macro fungi cultivation using biomass or agro waste leads to the substrates's biodegradation by spreading mycelia colony all over the substrate and breaking chains of the monomeric unit of building polysaccharides. Mushroom also secretes enzymes that are oxidative and hydrolytic involved in ligninolysis such as laccases and peroxidase which break carbohydrate content and to assimilate nutrients such as iron and nitrogen (Perez-Chavez et al. 2019).

The rising global population requires surplus food supply and demands advanced production of energy, and it is a peak time for the need of novel source of energy production systems (Specht et al. 2014). The biobased-circular-economy strategy can augment the resources with efficient energy production whereas for the maintenance of an ecological sustainability. This needs a broader perceptive and innovative method for recycling nutrients from biomass resources in the production of food plus production of energy. The reports based on recent findings have recommended that a collective production of edible fungi abundant in protein as well as suitable alternative for biofuels would be an interesting strategy (Xiong et al. 2019). The mushroom cultivation, by means of local biomass resources as substrate and its implementation to delignify the substrate after as a pre-treatment at the time of cultivation and facilitating conversion of the spent substrate enzymatically for the generation of second-generation biofuels. However, the efficacy of the technique, in context of both quality and quantity of produced mushroom and lignocellulosic substrate degradation, may vary with the species of fungi, substrate composition and cultivation technique, etc. (Shirkavand et al. 2016). Biofuel production via the lignocellulosic process is one of the most promising methods as a substitute to fossil fuel which supports expansion of bioeconomy (Formas 2012). Biological conversion of lignocellulosic waste raw materials using fungal species to generate advanced biofuels like ethanol based on saccharification method, wherein cellulosic compounds are transformed into monosaccharide for further processing (Jonsson and Martin 2016). Efficient bioconversion of lignocellulosic wastes is obstructed by the feedstock recalcitrance, which is up to an immense level can be said as the unconstructive result of lignin conversion. Consequently, the process of pre-treatment for delignification is an essential step to allow enzymatic accessibility to cellulose and thus enhancing the recovery of product to improve the second-generation lignocellulosic biorefineries (Singh et al. 2015). At present, physiochemical pre-treatments are the foremost process, despite the fact that they finish up in economically high cost and may have harmful effect to the environment in addition. As mushrooms are capable of releasing powerful extracellular enzymes which degrades lignin such as peroxides and laccases biological pre-treatment of lignin and cellulose-derived biomasses could be another suitable over physiochemical pre-treatment process. Fungal pre-treatment has benefits of the low cost of capital, chemicals and less environmental impact than compared to conventional methods (Wan and Li 2012; Shirkavand et al. 2016). Furthermore, edible white-rot fungi rich in protein, for example oyster and shiitake mushrooms with a high market rate, may improve the cost-effective measure of the method due to the fruiting body production. It is a vital characteristic of fungal potential to naturally degrade lignin into its monomer units to produce bioenergy (Shirkavand et al. 2016). Studies revealed that lignin degradation is associated with yields of specific sugars in the saccharification procedure as it increases the further hydrolysis of enzymes. This new technology can make up for the investment in biomass for farmers and also the recycling of remaining agricultural waste by incorporating of fungal species, allowing in support of the accessibility of value-added products such as biofuels and food as edible mushrooms (Perez-Chavez et al. 2019).

15.3 Lignocellulosic Biomass

Lignocellulosic biomass has been entertained as a potential raw material intended for the production of biofuels such as methane, bioethanol, biohydrogen and biobutanol considering its availability, sustainable and renewable property (Foster et al. 2021). Its includes agricultural and agro-industrial residues (corn stalk, corn stover, corn bran and cobs, sugarcane bagasse, sorghum stover, rusk and leaves, wheat straw, rice straw, husk and rice bran, oat straw, etc.) wood and grass, forestry waste, etc. (Figure 15.1). Lignin and cellulose-based agro wastes are the most abundant raw material and may have complexity in its composition, primarily composed of three polymers (lignin, cellulose and hemicellulose) with ash along with other extractives (Kumar et al. 2015; Goodman 2020). Lignocellulosic biomasses have good potential for biogas production and pre-treatment required to enhance the production rate. The various pre-treatment methods including a physical, chemical, physicochemical and biological process reported for significantly improved biogas production by lignocellulosic biomass (Abraham et al. 2020).

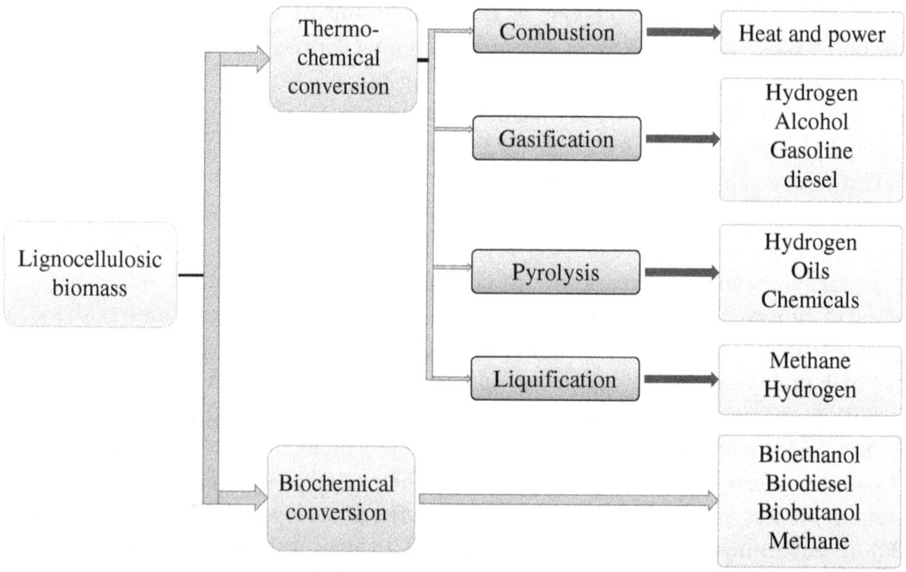

Figure 15.1 Different steps of bioenergy generation from lignocellulosic biomass.

15.3.1 Characteristics of Lignocellulosic Biomass

A typically plant cells comprise of significant units of hemicelluloses, cellulose and lignin. They are the structural unit and provide mechanical strength to plants. Agricultural wastes, grasses and woods, forestry residues and fungal biomass fall under lignocellulosic biomass. They have immense potential for their transformation into biofuel and bioenergy. Naturally, the agricultural waste biomass comprises of approximately 40–50% cellulose, 20–30% hemicelluloses and 10–25% lignin (Kumar et al. 2009; Iqbal et al. 2011). The foremost structural constituent of the plant cell wall is cellulose responsible for plant cell's mechanical strength. Simultaneously, hemicellulose is a macromolecule made up of a repeatedly bound monomeric unit of reducing sugars (pentoses or hexoses) (Anwar et al. 2014). Lignin consists of aromatic alcohols such as sinapyl alcohol, coniferyl alcohol, pcoumaryl alcohol, etc. formed by a biosynthesizing process and figuring a supporting layer around cellulose and hemicellulose (Calvo-Flores and Dobado 2010; Jiang et al. 2010; Menon and Rao 2012). Generally, the compositional property of lignocellulosic biomass is very much dependent on the source from where it is derived. It may be from grasses, hardwood, softwood or crops (Chen et al. 2020). All three components of various lignocellulosic biomasses among different sources have characteristically different chemical compositions due to the variation in the cellular system's genetic machinery. The main compositional units of lignocellulosic wastes are the following:

15.3.2 Cellulose

Cellulose is known for the supportive component providing mechanical strength to the plant cell. It is an enormously stable compound consisting of a repeating polymeric unit of glucose in linear configuration made up of 12 000 residues. (1,4)-D-glucopyranose is the main structural units linked by β-1,4 bonds having a standard molecular mass of approximately 100 000 g/mol (Himmel et al. 2007). Agro waste biomass contains cellulose molecules (40–50%) linked by intermolecular hydrogen bonds (H—H) in an inhabitant manner. Still they may have an affinity to form stable intermolecular and intramolecular hydrogen bond which enhancing rigid property and creating it extremely resistant and insoluble to a several organic solvents. Naturally, cellulose present in the form of bunches is grouped in micro-fibrils such as crystalline and amorphous regions (Taherzadeh and Karimi 2008; Iqbal et al. 2011).

15.3.3 Hemicelluloses

It is the second most abundant macromolecule comprises of the heterogeneous polymeric chains. It is mainly formed of glucomannan, glucuronoxylan and other polysaccharides in trace quantity. Galactan, xylan and arabinan are hardwood and softwood elements, while mannan is a hemicellulose present in grasses and straws (Brigham et al. 1996). These are implanted with sugar molecules as a structural unit. Among them, xylans with mannans are the generally common (Wyman et al. 2005). Arabinans and arabinogalactans along with galactans also are components of the hemicelluloses but does not possess the β-1, four linked equatorial structures. Glucuronoxylan is the principal component of hardwoods, while xylospyranose is the spinal column of the polymer chain which is connected through

β-1,4 linkages. Hemicellulose content in lignocellulosic biomass is up to 25–35% having a molecular weight of more than 30 000 g/mol. Both cellulose and hemicellulose tightly bind non-covalently to the surface micro-fibril of each cellulose. This was initially supposed to play an intermediate role in cellulose biosynthesis (Vercoe et al. 2005).

15.3.4 Lignin

It is usually the most complex plant cell that shares the least fraction of about 10–25% of the cell. It is a mixed polymer with long-chain, composed mainly of phenyl-propane units connected usually by ether bonds. It plays the role of a sticky agent by filling the space between the other two components of the cell. Since it is present in all plant-derived biomass, it is regarded as a residual product or by-product in the production of bioethanol. It is made up of complicated long-chain polymer of polyphenolic non-carbohydrate substance, phenyl-propane and methoxy groups that bind to the other components of the cell wall simultaneously (Hamelinck et al. 2005). Phenyl-propane (three carbons linked to six carbon atom rings) is the primary structural unit of lignin. This phenyl-propane is symbolized as 0, I, II methoxyl groups linked to rings with particular structure denoted as I in plants (grasses), II in wood (conifers) and III in deciduous wood (Anwar et al. 2014).

15.4 Spent Mushroom Substrate (SMS)

The remaining waste collected post harvesting of the fruiting bodies of mushroom from the cultivation bag is called compost or spent mushroom substrate (SMS) (Zhang et al. 2016; Rinker 2017). SMS is natural biomass waste generated abundantly in the mushroom farm after crop harvesting (Moon et al. 2012). For each kilogram of the fresh fruiting body, about 5 kg of SMS are formed. After harvesting mushroom, SMS residues are often discarded as waste which is a major challenge for disposal management and most often composted in many countries. Mushroom cultivation industry has been increased in the past few years, and the annual generation of solid waste is around five million tons from the SMS (Mohd Hanafi et al. 2018). In the past years, the yearly production of SMSs was calculated approximately around 170–204 billion kg (Carrasco et al. 2018). At present, the typical ways of discarding of these agro wastes are burying or dumping (Phan and Sabaratnam 2012). SMS produced after harvesting mushroom is usually a collection of straw, sawdust or mineral, gypsum and high levels of proteins, the fewer trace of carbohydrate and lipids, enzymes, etc. (Luo et al. 2018) which is potentially a great source of bio-based energy after treatment processes (Williams et al. 2001). Consequently, utilization of SMS is the best and more efficient mode of waste recycling, sustainable development and environmental protection (He et al. 2016). So far, it has been employed as a medium for the production of soil modifiers and fertilizers (Roy et al. 2015). SMS normally contains phosphorus (P), nitrogen (N)and potassium (K). Phosphorus (P) and nitrogen do not leach out at the time of decay, but potassium being more leachable element is vanished in considerable amount during decomposition (Gupta et al. 2004). Recycling of the SMS through enhancement of nutritional components to sustain mushroom production and has also been used for composting or co-composting and casing. SMS has also been used as fodder for animals as well as for production vermicompost. It is also used for the bioremediation purpose. Researchers said that

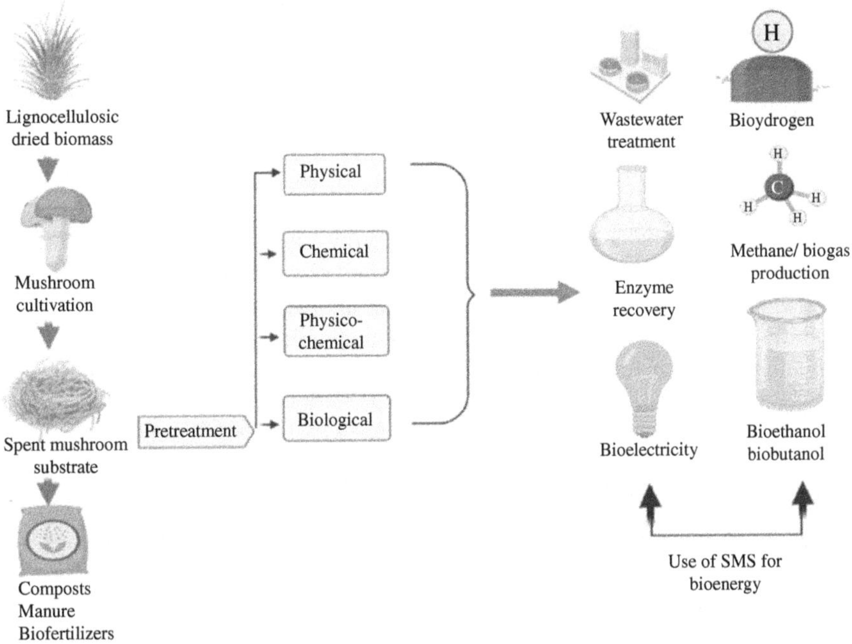

Figure 15.2 Sustainable utilization of spent mushroom substrate (SMS) for energy production.

it is a suitable alternative as it acts as biosorbent of pesticides, dyes and heavy metals, etc. as well as SMS have been efficiently employed for the recovery of hydrolytic enzymes (Phan and Sabaratnam 2012) and biofuel and bioenergy generation (Fang et al. 2017). From all bioenergy, biogas produced from the anaerobic digestion (AD) of waste is an eco-friendly fuel. It is known as the most prolonged and advantageous method for the production of bioenergy (Weiland 2010). A variety of biomass wastes have been used as substrates such as organic fraction, aquatic biomass, agricultural residues, industrial waste, municipal solid waste, forestry residues, organic waste and animal manure, etc. for biogas production. Agro wastes that are lignocellulosic are cheaply accessible raw biomass substances that can be employed for biofuels and bioenergy generation. The process is depicted in Figure 15.2. The utilization of SMS intended for the management of the agricultural wastes in the generation of recyclable energy requires appropriate surveillance on its chemical, physical and biological properties. It has different utilization in agriculture, environmental purposes and recycling process to derive energy so that it can be described as used mushroom substrate instead of SMS (Mohd Hanafi et al. 2018).

15.4.1 Biotechnological Importance of Lignocellulosic Biomass

A wide variety of agro wastes or lignocellulosic biomass are obtainable as a potential resource that can also be transformed into valuable bioproducts such as biofuels, enzymes, fertilizers, etc. A drastic improvement in applying the biotechnology field related to the utilization of lignocellulose biomass for sustainable use has appeared from the last decades. The continually rising prices of conventional fuels and their adverse effects on the environment have raised a concern to discover alternative resources which are cheaper and eco-friendly. It can be used as an approach for reduction of global warming (Iqbal et al. 2013). Environmental issues such as pollution, global warming and climate change are among the significant reasons for diverging awareness in bio-based resource as a substitute to fossil fuels. Utilization of the agro-industrial lignocellulosic-based biomass in favour of the low-cost bio-ethanol production is a potentially an excellent method to generate energy and another by-product. Carbohydrate compounds present in agro biomass are first transformed into simple sugars units (glucose) subsequently into ethanol by fermentation (Lin and Tanaka 2006). Alternative energy sources, including bioethanol, are now an area of interest and global investment. The transformation of lignocellulosic biomass into valuable products like enzyme and biochemical production, or biofuel production usually involves a multi-step processing technique including pre-treatment (chemical, biological or mechanical, etc.), enzymatic hydrolysis and fermentation process, etc. (Wyman et al. 2005; Xiao et al. 2012).

15.4.2 Applications of Spent Mushroom Substrate (SMS)

In the technology of mushroom cultivation, SMS left after harvesting fruiting bodies can also be utilized for the fabrication of various other profitable products as:

15.4.2.1 Fertilizers

The SMS consists of adequate nutrients and organic non-toxic compounds for the growth of plants which feasibly allows it for the production of biofertilizer using microbes. Ca and N ash and protein are the main components of SMS which on the incorporation of microbes can be used as biofertilizers (Lou et al. 2017a). Several investigations have shown the viability of SMS in horticultural purposes with or without the addition of other components (Nakatsuka et al. 2016). The reuse of SMSs as manure or fertilizers for improvements of soil texture and nutrients as nitrogen is present in abundance has made it the centre of attraction and can improve the soil quality (Alvarez-Martin et al. 2016). SMS is also suitable for replacing mulch because of its physicochemical characteristics and biological activity for the degradation of pesticide residues (Gao et al. 2015). Moreover, it is rich in organic components and vital phytonutrients to facilitate it to improve of soil and mulching which can extensively enhance the nitrogen composites present in the ground. For the proper growth of plants, phosphorus (P) is a chief nutrient; the function of SMS in improving soils quality by chemical addition of P can be utilized for the production of efficient of biofertilizers for crops (Zhu et al. 2012) results in enhancing soil organic matter and nutrient contents (Lou et al. 2015). Various findings showed that SMS could used as a micronutrient medium for microorganisms to manufacture fertilizer by biosorption resulting in the enhancement of soil texture and structure, soil quality and absorption capacity. SMS can be incorporated to produce new formulations for agricultural practices and the benefit of low fabrication cost and minimal environmental impact. Further study is required to discover new methods for processing and treating the biomass for the formulation of biofertilizers. This strategy leads to the formation of new micro fertilizing products which can be manufactured of low production cost (Tuhy et al. 2015). Three different SMS effectively reduces disease named *Fusarium* wilt in cucumber (*Cucumis sativus*), caused by *Fusarium oxysporum* sp. *cucumerinum* the diseases causing fungus (Wang et al. 2020).

15.4.2.2 Wastewater Treatment

Several studies showed that SMS is effectively used for the treatment of wastewater. For instance, for the removing nitrogen from wastewater, SMS is a proficient carbon source because of its good characteristics which were enhanced from 46.9 to 87.8%. Fresh SMS from the cultivated species of *Pleurotus* was worked proficiently for the elimination of fungicide such as imazalil and ortho-phenylphenol in wastewater from fruit packaging industry (Karas et al. 2016). Spent substrate from oyster mushroom utilization for bioremediation of polychlorinated biphenyls (PCBs) in contaminated water (Sredlova et al. 2019). For the treatment of dye effluents, the SMS used in batch adsorption technique is as efficient as a low-cost biosorbent. It eliminates methylene blue from wastewater with cationic dyes. It can also be employed for the bioremediation potential of SMS for water treatment by removing organic micro-pollutants (Hultberg et al. 2020). The process of pre-treatment of different types of wastewater by the application of SMS is a challenging task. In the meantime, adsorbent made from SMS was found unsuccessful for treatment of a highly polluted stream also it could not be used in refinery wastewater and industrial due to specific attributes; therefore, further research is required for new innovations and development in this area (Yan and Wang 2013).

15.4.2.3 Enzyme Recovery

Enzymes can also be produced from fungal mycelia grown in submerged culture conditions and fermentation. It may also be recovered from fermentation technology using a solid substrate such as the process of nature. The spent mushroom compost composes of various extracellular enzymes post-harvesting. Amongst the enzymes in SMS, laccase is the most abundant and common enzyme extracted from *L. edodes, P. ostreatus, Flammulina velutipes, Hericium erinaceum* and *Pleurotus erryngii* (Ko et al. 2005) and *A. bisporus* and *P. sajor-caju* (Singh et al. 2003). Lignin peroxidation process per microgram of SMS gives enzymatic activity like laccase, β-glucosidase and xylanase, etc. The range of enzymes produced by mushroom at different stages of cultivation is highly species and substrates specific (Singh et al. 2003). The most common SMS enzymes are Mn-dependant peroxidase (MnP), lignin peroxidase (LiP) and laccase (Lac), a copper-containing oxidoreductase. The particularity in variety of substrates and the enzyme releases is remarkably less allowing them more efficient for metabolism and transformation of recalcitrant organo-pollutants equivalent to lignin, for instance, polycyclic aromatic hydrocarbons (PAHs) (Pointing 2001). Lignin peroxidases enzyme is glycosylated heme proteins and is obligatory are dependent on hydrogen peroxide (H_2O_2). LiP catalyses the oxidation of a redox intermediates results in the generation of aryl cation radicals in the production of endogenously generated H_2O_2. The free radicals generated subsequently perform diversified reactions, includes oxidation of benzylic alcohol, phenol polymerization, cleavage of carbon–carbon bond, dimerization, demethylation and hydroxylatio, etc. (Reddy and D'Souza 1994).

15.4.2.4 Energy

Agricultural wastes or agro wastes biomass is lignocellulosic and renewable source of energy. It is abundantly present in nature. Utilization of spent substrate for derivation of energy provides sustainable management of agro wastes and to find an improved solution for the generation of clean energy which in turn results in lower gas emission (NOx, SOx and HCl) (Finney et al. 2009). Incorporation of SMS in sustainable energy production is a promising unconventional method to minimize the wastage of SMS production on site. SMS usage in the production of bioethanol stands for a possible alleviation solution for arising environmental issues using the mushroom cultivation industry (Ryden et al. 2017). Due to the degradation process, SMS has lower lignin content at the time of mushroom production provides an advantage for the production of energy. For instance, the combustion of SMS in the form of pellet with an amalgamation of coal was proficient (up to 91.7%) in contrast to unpelletized SMS (Durre 2007; Lee et al. 2009). It is tremendously degradable in nature and co-digestion with wheat straw efficiently enhances the production of methane or biogas (Hu and Zhu 2017). Biohydrogen can also be efficiently produced using bacteria *Clostridium thermocellum* for the degradation of lignin from SMS (Wu et al. 2013).

SMS contains several reducing sugar which on further treatment gives yields of ethanol and also it can act as a possible organic source. Some important factors regarding biomass that may affect energy production using or SMS as follows:

- Variety and species of biomass which directly influences the nature and yield of energy production,
- Species variability of mushroom also has an effect on the accessibility of lignin and cellulose part of biomass and subsequently sugar ratio and enzymatic activity,
- The optimum ratio between biomass and SMS for better processing of products,
- Co-digestion of lignocellulosic biomass with SMS enhances the variety and yield of energy produced, but the yield can be slowed as an effect of the production of some by-products during treatment (Wu et al. 2013).

15.5 Biotransformation of the Spent Mushroom Substrate (SMS) Into Energy

Global demand of energy relies primarily on fossil fuels such as coal, oil and natural gas, accounts with approximately 81% of the net primary energy contribution in 2016 worldwide (International Energy Agency 2018a). The incineration of such fossil fuels leads to the release of pollutants such as NOx, SOx, COx, ash, particulate matter and other organic compounds (Das and Veziroglu 2001), resulting in greenhouse gas emissions and air pollution that consequently contributed to climate change and hence affecting the health globally. Even though the utilization of renewable sources has risen in the last decades, still there is a need to travel a distance to lower the harmful impacts on environment from fossil fuels, as renewable match the total primary energy supply (Watts et al. 2018).

Agro wastes biomass and lignocellulosic wastes have been measured as a potential feedstock for bioenergy generation for instance ethanol, methane, biohydrogen and butanol since it is sustainable, available in bulk quantity and renewable. It includes agricultural and agro-industrial residues (cornstalk, corn stover, corn bran, corn cobs, wheat straw, rice bran, rice husk, paddy straw and oat straw, etc.) wood, grass, forestry waste and others. Agro biomass is abundantly found raw material and possess a complex structure primarily comprises of three polymers displayed in Table 15.1 (cellulose, hemicellulose and lignin), besides ash and other extractives (Kumar et al. 2015).

Table 15.1 Spent mushroom substrate (SMS) for energy generation.

Substrate	Treatment	Fuel	References
Pleurotus sp. mushroom spent	By heating and pressing	Bio-Coke solid fuel	Kamal Baharin et al. (2020)
SMS of edible mushrooms	Solid-state co-digestion (anaerobic) with wheat straw	Biogas	Lin et al. (2014)
Flammulina velutipes, Pleurotus erygii var. Tuoliensis and *P. eryngii* Substrate	Co-digestion of SMS and dairy manure	Methane	Luo et al. (2018)
Spent substrate of rice straw of *Pleurotus ostreatus*	Solid-state anaerobic digestion	Biogas, methane	Mustafa et al. (2016)
Spent substrate compost of *Agaricus bisporus*	Solid-state anaerobic fermentation	Methane	Fang et al. (2017)
Spent compost of *G. pampeanus*	Solid-state anaerobic digestion	Biogas, methane	Cordoba et al. (2016)
Spent of *P. ostreatus*	Saccharification and fermentation	Bioethanol	Ryden et al. (2017)
SMS of edible mushrooms	Anaerobic solid-state fermentation	Biohydrogen	Lin et al. (2016)
Spent mushroom substrates	Organosolv treated SMS	Butanol	Zhu et al. (2016)
Clostridium thermocellum spent	Composting	Biohydrogen	Lin et al. (2016)
SMS from Shittake mushroom	Simultaneous saccharification and fermentation (SSF)	Bioethanol	Asada (2011)
Sorghum-based SMS *P. ostreatus*	Pyrolysis	Biochar	Ryden et al. (2017)
SMS with pig manure and rice straw	Pyrolysis	Biochar	Chang et al. (2017)
SMS-based biochar	Pyrolysis	SMS-based biochar	Lou et al. (2017a)
Oil shale semi-coke SMS	Pyrolysis	Bio-oil	Jiang et al. (2017)

15.5.1 Biohydrogen Production from SMS

Hydrogen (H_2) has been calculated as a substitute to fossil fuels to undertake the environmental impacts associated with their rising utilization. H_2 can be derived from fossil fuels (natural gas, coal and nuclear energy) and renewable sources (wave, hydro, wind, solar, geothermal energy and biomass) (Abe et al. 2019). In today's world, approximately 50% of H_2 is obtained from natural gas, 18% from gasification of coals, 30% from cracking oil products and the residual is from the electrolysis of water (International Energy Agency 2018b). As known,

H_2 is generated mostly from sources of fossil fuels, which requires a high cost and are non-environmental friendly. In this context, production of biohydrogen using biological method has significant benefits over other techniques because it requires low atmospheric pressure and temperatures, requiring less dark fermentation, energy intensity photo-fermentation, direct biophotolysis, the hybrid system and indirect biophotolysis are all categories of biological processes for biohydrogen production. The dark fermentation process involves biohydrogen production by the employment of anaerobic bacteria to the substrates abundant in carbohydrate in the deficiency of light. While in photo-fermentation, a bacterium photosynthetically utilizes light for the production of biohydrogen from various food processing, agricultural wastes and organic acids (Hallenbeck and Benemann 2002). Photo-fermentation and dark fermentation are balancing methods, and both can be used sequentially in a hybrid system. In direct biophotolysis method, light energy is utilized by *Algae* and *Cyanobacteria* for the conversion of water into O_2 and biohydrogen ($2H_2O \rightarrow 2H_2O_2$). In contrast, indirect biophotolysis involves separate release of the biohydrogen and O_2 in different stages coupled via carbon dioxide (CO_2) fixation/evolution reactions (Bharathiraja et al. 2016).

For biohydrogen production, high lignin content of biomass can create hindrance to the easy and direct conversion to biohydrogen. Some of the biomass used for biohydrogen production obtained from sugarcane bagasse 8.3% of lignin and cashew apple bagasse (35.26%) had more lignin content. Even though lignin has an unconstructive impact to biohydrogen production along with other factors also influence biohydrogen production, including carbohydrate content, pre-treatment method, operational conditions, etc. Pre-treatment of biomass is a decisive aspect in the biohydrogen production via dark fermentation, due to the complexity of lignocellulosic component. Pre-treatment intends to split this recalcitrant heteropolymeric structure and enlarges the quantity of monomeric sugars required in support of microorganisms to produce bioydrogen. Figure 15.3 shows the

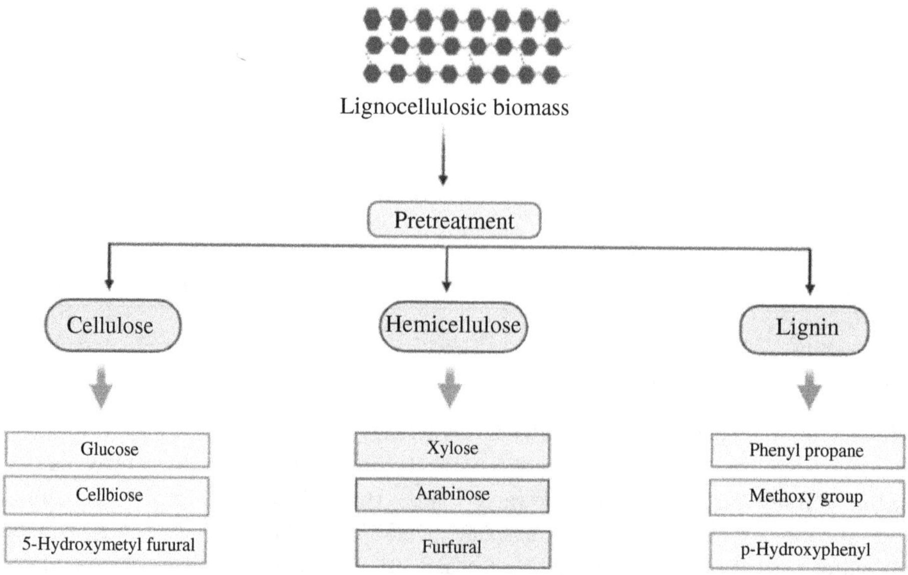

Figure 15.3 Breakdown of biomass into lignocellulosic components on pre-treatment into monomers.

major by-products achieved in the breakdown of cellulose and hemicelluloses fraction after pre-treatment into a simpler form. This breakdown of cellulose gives cellobiose, glucose and 5-hydroxymethylfurfural (5-HMF), whereas hemicellulose breaks into xylose, arabinose and furfural (Hu and Zhu 2017).

In dark fermentative process, glucose is effortlessly utilized by microorganisms than other sugar residues as xylose, which starts metabolizing when the concentration of glucose reaches up to 0% (Moodley and Gueguim Kana 2018). Pre-treatments may be categorised into physical (chipping, milling, cutting, grinding and shearing), chemical (organic solvents, acids, alkalis and ionic liquids, etc.), physico-chemical (ammonia fibre expansion, steam explosion and liquid hot water, etc.) and biological (enzymatic hydrolysis and fungal treatment, etc.) methods. The techniques when used in combination are usually more efficient and enhances the biomass digestibility and to formulate it cost-effective, pre-treatment procedure should have low operational cost, a low capital and high loading of agro wastes resulting in the recovery of most lignocellulosic compounds (Agbor et al. 2011).

15.5.2 Biogas Production from Spent Mushroom Substrate (SMS)

Biogas is generally produced from the process of AD of a broad range of agricultural waste, including lignocellulosic biomass transforming it into organic slurries and then into energy (Gupta et al. 2004). Biogas production involves a chain of chemical reactions takes place in the presence of different microbes in anaerobically digested waste biomass. Microbes will feed on the simple compounds available in the substrate contained in the medium devoid of the production of a massive amount of enzymes. Consequently, hydrolysis process stops as it is an extensive process, consumes energy and a limiting stage for AD (Perez-Chavez et al. 2019).

AD is a practice of degradation where organic substances from the substrate is biologically decomposed by a bacterial association such as *Methanothermobacter* or *Methanobacterium* for the production of methane in the form of gas and providing a good substitute way of disposing of SMS. Hydrolysis, fermentation, acetogenesis and methanogenesis are sequence of metabolic steps followed in AD of biomass (Xiao et al. 2012). It is reported in a study that at the time of AD of SMS, the biogas production is considerably higher than to untreated substrate (Bisaria et al. 1990). Though, SMS contains mycelium and extensive amounts of nutrients left over including proteins, carbohydrates and certain enzyme, etc. (Lou et al. 2015). Though, the handling of SMS in the form of substrate for aerobic digestion may still encounters numerous problems, for example, high carbon-nitrogen (C/N) ratio resulting in the inhibition of the biogas production and a highly degradable nature have negative influence on the stability of AD (Deng et al. 2012). A potential approach that might assist to conquer these difficulties is the co-digestion through a buffering substrate that is frequently applied to improve the constancy of the method. In comparison to mono-digestion process, co-digestion is profitable in terms of increased biogas production with balancing the system, easy to handle with mixed agrowastes, and so on. Co-digestion of SMS enhances the yield of methane which in turn increases biogas production by 2–16 times. It was demonstrated that the co-digestion of SMS mixed with lignocellulosic feeds enhances solid-state AD performance. Spent mushroom wastes could be used as a suitable raw material for biogas production, but the high degradability of SMS may have several opposing influence on aerobic digestion. At 3 and 5% total solids, the production of methane after co-digestion is elevated than the mono-digestion. Co-digestion method potentially augments the buffering capability of the reaction. Several studies reported that there were variations in yield of methane from AD along with the various SMSs (Luo et al. 2017a). SMS contains lower lignin content because of the extracellular secretion of lignocellulosic enzymes during formation of fruiting body is an advantage for biofuel production. The less content of lignin along with ash and high nitrogen content lead the SMS more feasible for digestion process by microbial population yielding more reducing sugars. In fact, the consequential polysaccharides produced act as an appropriate medium for hydrolysis, since the production of spent substrate itself falls under the process of pre-treatment. Moreover, the conventionally used unprocessed raw material for first-generation bioethanol production is sugarcane and corn cob, etc. will have an adverse effect on the cost of the substrates. Consequently, SMS that is a byproduct of mushroom cultivation technique is much economical and favourable. Further, biogas generated from biomass is now attaining much interest. Biogas yield by SMS of *P. florida* and *P. sajor-caju* was high as a result of the optimum C and N ratio, thus improves the vulnerability to digestion of anaerobic fermenters (Bisaria et al. 1990).

15.5.3 Bioethanol from Spent Mushroom Substrate (SMS)

Bioethanol is a biologically derived biocemical with a pleasant odour that can be extracted from natural products. It is diluted with gasoline to provide a clearer, more transparent and natural energy source. It can be a renewable fuel because it is derived from biological sources. In recent time, the demand for renewable resources has been increased for the production of biofuels and SMS may signify a high-quality renewable feedstock for biofuel including production of bioethanol. Corncob contains hemicelluloses (34.7%, without acetyl groups), cellulose (31.7%), lignin (20.3%) and acetyl groups (3.4%). Corncob-based substrate that is composed of 5% of rice bran and 1% of wheat bran, thus it can be utilized for bioethanol production (Rivas et al. 2006).

Pre-treatment of agro wastes is the crucial and vital stage towards the conversion of lignocelluloses content of biomass into ethanol. This step is essential for the breakdown and arrangement of cellulose content of biomass into its simpler forms further providing it suitable for enzymatic action (Mosier et al. 2005; Hendriks and Zeeman 2009). Even if sulphuric acid and dilute acid as chemical pre-treatments can provide higher overall yield of ethanol and sugar they obtained mixture of biochemicals contains inhibitors of the yeast that influences the growth rate of yeast performing ethanol fermentation. This mixture of chemicals consists of is hydroxymethylfurfural (HMF), acetic acid, furfural and phenolic compounds, etc. (Liu and Wyman 2005). Another method is non-acid pre-treatment's include thermal treatment and ball milling have been also applied (Hamelinck et al.2005) and reportedly generate higher production of total sugars at the same time as lowering enzyme loading. Thus, hydrolysis of the enzyme through non-acid pre-treatment is a more suitable method for raw material pre-treatment in bioethanol production. Hydrolysis of lignocellulosic biomass mainly gives glucose, xylose and hexoses fermented to form ethanol. *Candida shehatae, Pichia stipitis* and *Pachysolen tannophilus* are naturally occurring yeasts can ferment xylose to organic ethanol (Chu and Lee 2007), while *Saccharomyces cerevisiae* is not capable of utilizing xylose. *Pichia stipitis* is yeast that ferments xylose and is the most excellent producer of pure ethanol, giving a high yield (Oguri et al. 2011). Spent substrate of *P. pulmonarium* containing lignin and xylan employed as substrate for bioethanol production (Chen et al. 2020).

In the production of bioethanol, the substrate is powdered, pre-treated through the hydrothermal process, saccharified through saccharification to convert cellulose into sugars and fermented finally into alcohol. This process relies with surplus of enzyme and substrate concentration up to 10% (w/v). These methods give a better amount of yields at large scale using higher substrate concentrations as compared to the unmilled biomass. The potential method for bioethanol production by hot-air pasteurisation (HAP) using birch wood lignocellulosic substrates that employed for shiitake mushroom cultivation (Wei et al. 2020). Presently, cultivation of industries scale up of mushroom transform to more advance and sophisticated industrial scales, thus co produced many compost for mushroom cultivation. Furthermore, the potential to exploit as renewable energy source cellulosic components should not be ignored and development in the field of lignocellulosic industrial biotechnology will be expand the technology for utilization of agro wastes (Ryden et al. 2017).

15.5.4 Biobutanol from Spent Mushroom Substrate (SMS)

Butanol as a next-generation biofuel is likely to play a key role considering its highly miscible character with gasoline and high-energy density than ethanol. At the present time, bio-derived production of butanol has attracted immense attention concerning over the global oil supply and its influences on socio-economic development (Lee et al. 2008). On the other hand, due to few challenges, such as consumption of costly feedstock and low concentration of butanol in the broth as a result of toxicity of butanol, the production is still not economically feasible (Lee et al. 2008). Several attempts have been completed to encounter this obstruction so the employment of cheap biomass is one of the strategies significantly considerable. Agro wastes feedstocks rich in lignocellulosic property, like barley straw, wheat straw, wood hydrolysate, corn stover, switch grass and other agricultural by-products, may be considered as the potential substitutes (Ezeji et al. 2007). SMS as a by-product of the mushroom technology has been estimated two million tons annually in China. These spent substrates are not appropriate for animal feed as fodder; it will be more advantageous for the efficient conversion of SMS into value-added products, production of enzymes and bioenergy-like biofuels. SMS are degraded matters that potentially generate fermentable sugars (e.g. glucose and xylose), so its employment for production of biobutanol is the best alternative.

Enzymatic hydrolysis of degraded lignocellulose can attain nearly high sugars yields; however, pre-treatment processes improve the bioavailability of substrate by hydrolytic enzymes are too needed (Qureshi et al. 2007). Various pre-treatment processes have been employed and used to improve biomass fractionation of biomass, such as aqueous ammonia, dilute acid, steam explosion, alkaline, wet oxidation and liquid hot water (Larson et al. 1999). Among the above, acid pre-treatment is considered as the most developed technology. However, using acids generally involves a long term pre-treatment with a high temperature. Generation of inhibitors is not avoidable while working with fermentative microorganisms, for example, sugar deprivation products such as carboxylic acids, phenolic compounds, inorganic salts, etc. Organosolv pre-treatment has been proved to be an efficient technique for removal of the recalcitrance of lignocellulosic substances, particularly in woody biomass and for enzymatic process, i.e. saccharification. Furthermore, a modified organosolv pre-treatment method carried out under non-corrosive in mild atmosphere using only water within the washing steps has been reported for the efficient removal of hemicelluloses and lignin thus improving the efficient conversion of cellulose (Qureshi et al. 2008). It is predictable that lignin act as protective barriers in response to the enzymatic attack until effective removal is done. Also, the presence of lignin may efficiently moderate the biomass hydrolysis by means of non-definite binding of the cellulolytic enzymes (Teramoto et al. 2008). After organosolv pre-treatment, the H_2O_2 treatment greatly improves the bioavailability of biomass, as a result of its oxygen involvement in delignification process. The main objective of ethanol wash after pre-treatment (organosolv) method is to disallow the lignin precipitation. Though, the results demonstrated that it is not necessary to use ethanol wash step for an organosolv pre-treatment process involving H_2O_2 post-treatment. This may be described by the fact that H_2O_2 post-treatment method can cause oxidative degradation of the lignin component of SMS, with the precipitated one also. Besides, the second water wash can remove efficiently the residual by-products from degradation of lignin which are water-soluble thus allowing the SMS more appropriate for the consequent cellulolytic enzymes. This replacement provides a significant economic benefit to the pre-treatment process along with the expenditure of approximately 10% decrease in sugar production but two-third reduction in consumption of solvent. In the overall procedure, energy efficiency of the pre-treatment process could be improved to a great extent because of the removal of the ethanol washing steps (Zhu and Pan 2010). Through the organosolv-pre-treated SMS, the carbohydrate content cellulose of the SMS can be transformed into fermentable sugars after the enzymatic hydrolysis. Acid pre-treatment can also be used for hydrolyzing the SMS, if not hydroxymethylfurfural (HMF), little furfural and lignin-degradation products such as ferulic acid and hydroquinone can also be found in the SMS hydrolysate. After the removal of around 80% water by means of vacuum filtration, the SMS hydrolysate was concentrated to a thick syrup comprises of acetic acid, lactic acid, xylose, arabinose, glucose in g/L in increasing order. The point to be noted that evaporation method can also extensively used for the removal of the amounts of volatile compounds such as acetic acid and furfural produced from the xylose dehydration (Zhao et al. 2009).

Butanol is the main inhibiting composite in fermentation process that involves clostridial fermentation of acetone butanol-ethanol (ABE), leading to the formation of less amount of butanol and total solvent. *In situ* extractive fermentation is the most important step for minimizing the inhibition of butanol. So, the use of biodiesel can be applied as a preferential extracting method for extraction, having a high partition coefficient and low partition coefficients for butanol and acetone/ethanol, respectively, and especially does not show toxicity to the clostridial cells. The batch fermentation results in utilizing biodiesel in *in situ* extraction at a ratio of 1:1 (biodiesel: fermentation medium). This butanol enriched with biodiesel can be directly applied to the biofuel sector and also gets liberated in subsequent extraction and recycling (Ezeji et al. 2004).

15.5.5 Bio-Coke

Bio-Coke can be manufactured under very high pressure (~20 MPa) at moderate temperature ranges (120–200 °C). It exhibits unique characteristics with high calorific range between 18 and 31 MJ/kg and high density (1.4 g/cm^3). This suggests that the Bio-Coke may have an elevated heat of combustion and easy transportation and storage at a relatively low cost. It also has a relatively low tendency of retaining water and a high mechanical strength of approximately 38–149 MPa at room temperature, demonstrating that it may have the ability to bear high pressures throughout high temperature combustion (Mansor et al. 2018). This is a novel dense solid fuel that can be applied for many functions, for instance, combustion and as a heat source for boilers in generation of electricity. The conversion of waste into Bio-Coke is showed as an efficient method for proper management of the SMS and agrowaste due to high preservation of energy with cheap cost production process. In this way, 10 g of raw could transform 10 g of Bio-Coke without emission of gases including carbon dioxide and methane.

SMS has better characteristics as an unprocessed material for Bio-Coke production. It was reported that small particle size of spent substrates plays a key role in Bio-Coke production, as small particles better assist the reaction due to their larger surface area; however, it was also confirmed that when the particle size is too small the maximum compressive strength is low at room and high temperature. The production of Bio-Coke from SMS could be performed at four different ranges of temperature as 130, 150, 170 and 190 °C. The raw material is then compressed to particle sizes smaller than 1 mm. (Mizuno et al. 2014). A pressure of 21.7 MPa, molding size of 20 mm and a holding time of 6.15 mm are provided during the process. Larger molding extents may potentially cause heat permeation through thermal conduction centre of the Bio-Coke, resulting in the formation of less efficient Bio-Coke. The ideal conditions for the fabrication of Bio-Coke are the initial moisture content, particle size of the raw materials and formation temperature, which influences its apparent density. In a study, it was showed that the conversion of SMS waste to Bio-Coke, based on its characteristics, can potentially be used as a combustion fuel. The Bio-Coke produced at temperature range of 150 and 170 °C possess high potential due to its high apparent density, high mechanical strength at 25 and temperature 700 °C. High mechanical strength of Bio-Coke above temperature makes it an ideal biomass solid fuel and is unbreakable throughout transportation or storage with stable combustion at high temperature. The formation of Bio-Coke technique also shows that the conversion of SMS into Bio-Coke is influential for sustainable development as it reuses SMS and increases the efficiency of biomass waste management (Kamal Baharin et al. 2020), and it can also be employed for manufacturing the activated carbon or fuel gas by using internal flue gas and external air as activation agents (Liu et al. 2021). Also the development of integrated system to produce mushroom and the utilization of its spent substrate for biosolids production can be done using spent substrate of sugarcane bagasse, wheat bran, corn meal and calcitic limestone of *P. florida* which was dried, milled and reconverted into briquettes (De Almeida et al. 2020).

15.5.6 Electricity Generation Using Mushroom Technology

Mushroom or white rot fungus has a smal amount of simpler nutrients when it is cultivated in the solid state than in a liquid condition, making the conditions favourable for the formation of more ligninolytic enzymes, i.e. laccase outside fungal cells. Although the laccase activity that is performed by white rot fungus cultivated in the solid state generally exceeds than in a liquid cultivation state. A better laccase activity is a crucial step to formulate this technique adequately effective. The white-rot fungi being employed in the microbial fuel cell (MFC) are acquired from the mushroom cultivation technology. White rot fungus-microbial fuel cell (WRF-MFC) which uses fungus to generate electrons grown at its cathode. The addition of copper peroxidase to the solid medium containing white rot fungus stimulates the secretion of laccase, which acts as a catalyst for the oxidation and reduction process on the MFC and thus supporting the generation of electricity. The addition of copper peroxidase to *Ganoderma lucidum*-containing medium supplies the maximum laccase stimulation and increases the activity of laccase inside the chamber of WRF-MFC which increases its decolorization and its power density which is 1.77 times than that of an MFC without WRF. The enhanced decolorization process further enhances the power generation improving the performance of the WRF-MFC, representing that a laccase-catalysis-based cathode that potentially hold enormous effectiveness as a microbial fuel cell (Liu et al. 2020).

15.5.7 Solar Steam Generation Device

Solar steam generation potentially harvests solar energy for a variety of purposes such as sterilization and desalination, etc. and has established tremendous attention. Over the past decades, numerous artificial configurations have been fabricated and designed for the improvement of efficiency of conversion of energy by enhancing the heat localization, solar absorption, water supply and vapour translocation, etc. Mushrooms can be efficiently used to enable the solar steam generation because of their distinctive structure. This efficiency of solar steam generation is attributed as it has the unique natural structures, such as counting their porous framework, umbrella-like pileus and fibrous stipe. This feature of mushroom enables efficient absorption of sunlight and suppresses the heat loss components. In recent studies, the report provides the support to further development of advanced high-performance devices for thermal conversion of solar energy into heat energy. Initially, the black umbrella-shaped pileus can facilitate the proficient absorption of solar energy. The process can be further enhanced with the help of a suitable carbonization treatment. The fibrous stipe that is hydrophilic in nature could provide a sufficient water supply that is pumped by the property of capillary force. The porous framework plays the role as a suspension bridge to pump the water further into the top pileus also providing it sufficient vapour channels. The geometrical structure of mushrooms is also naturally designed for reducing all the elements of heat loss such as radiation convection and conduction. The ratio of the diameters of pileus and fibrous stipes is approximately 6 : 1, providing the mushroom an efficient thermal organization to suppress the heat conduction and loss of bulk water. The mushroom maintains its hydrophilic property before and after carbonization because of its components, which include proteins, carbohydrates and nitrogen functional groups which is present even after carbonization. The surface chemical compositions and functional groups of natural and carbonized mushrooms were recognized by X-ray photoelectron spectroscopy and Fourier Transform Infrared (Xu et al. 2017).

15.6 Challenges

Production of biofuel from biomass is not a conventional technology. Second-generation biofuel generation from lignocellulosic substances is technically difficult and uneconomic. Skilled personnel will be required to handle the process. Manufacturing processes such as hydrothermal and chemical pre-treatments, saccharification and fermentation are of high cost and require extensive capital investment and acceptance in alternative local markets results in it as a challenging task. One of the challenges is to determine the supporting logistics, which includes the strategic design of a logistics network, residue handling, a feedstock supply, distribution and a planned operation schedule (Tembo et al. 2003).The industrial supporting logistics system for the conversion of biomass-to-bio energy is different from any other industry, for instance, coal-based power plants (Najafi and Ardabili 2018). Most of the biological processes require an ambient temperature of 15–26 °C. Athough the emission of flue gas occurs at a high temperature range, it should be cooled before subjecting to a biological system. There is a necessity to seek out for suitable microbes that can perform at higher temperature as flue gas temperatures reach up to 120 °C. Moreover, pH is also a significant parameter for the growth of microbes that can negatively affect its growth as most of them grow near a favourable pH (Bhatia et al. 2019).

Another challenge is accessibility and availability, accessibility to modern energy denotes that energy must be obtainable at affordable and sustainable cost which meets the expenses of under privileged population as well as reflecting the definite costs of energy production and distribution supporting the financial capability of companies for the maintenance and development of their services. Availability deals with both the reliability and quality of delivered energy. Energy availability requires a diversified energy portfolio dependable with particular circumstances with national policies and planning. It is obvious that all energy resources along with the alternative sources will be requisite over the upcoming decades, and perhaps there will be no reason for the unacquainted exclusion of any source of energy. For this, it must be ensured that the implementation of integrated microbial processes that may fulfil the environmental-socio-economic criteria for an affordable bioderived energy production. If all these targets are accomplished, the future of integrated microbial processes with agro biomass may potentially give satisfying contribution to the emergent demand of energy sources, the alternative replacement of fossil fuels and for the manufacturing of a broad range of bio-based products to a great extent of commercial interest.

15.7 Conclusion

The rising requirement for protein-rich high nutritional food and the incapability of conventional fuels to fulfil the demand of energy resulted to investigate choices for the production of food and low-cost energy. To battle this problem, edible fungi or mushrooms are emerging as a good alternative to tackle the food and energy crisis. Mushroom cultivation technology is a labour-intensive job and therefore has the potential for a good alternative for employment opportunities, particularly for rural people within their surrounding area of agro-processing industries to develop their livelihoods. On a large-scale mushroom cultivation will thus provide extension to its availability at affordable rate, and consequently providing the population with a supplementary nutrient-rich food which in turn provide to health benefit to human. It is a profitable agri-business and can be sold in local markets and also can be exported outside country to earn foreign revenue for sustainable economic development. The mushroom industry is recently emerging as a business with production greater than 25 metric tons all over the world. In the same way, 1 kg of mushroom harvest can generate 5 kg of spent substrate residue. Management of SMS waste is a problematic for mushroom farmers in some countries. Therefore, the solution of this problem is to recycle and reuse the SMS exploring new ways for sustainable utilization which in turn increases the demand for agro wastes economically. The high organic substance of SMS has attracted the attention of scientists to research in this particular area. These multi-level systems modelling framework facilitates global learning of the food and biomass-derived energy value chain integration and explores the possible solution. SMS, which contains less lignin due to the pre-treatment and digestion procedure by lignocellulosic enzymes, presents extracellularly during the formation of mushroom provides a value point for bioenergy production such as bioethanol, biogas, biohydrogen and enzymes, etc. The lower lignin content with high nitrogen and ash content allocates the SMS more effortlessly digestible by microbial degraders to produce more amount of reducing sugars. As a result, the polysaccharides produced serves as an appropriate medium for hydrolysis, since the fabrication of SMS itself form of pre-treated substrate. SMSs produced from the mushroom industry are now no longer considered as a waste, however, as a renewable resource. It not only can be employed in several green technologies but also to generate clean energy and bioproducts. The enzymes recovered from these techniques are potentially supportive for the generation of bioenergy, remediation of pollutants biologically and other industrial purposes. The 'food and fuel' trade-off cases where innovative technological investments are basically absent and trade and subsidy policies are failing, the mushroom technology may emerges as a novel idea for fulfilment of both food and fuel. As the consequences of past agricultural policies, such technologies cannot be ruled out unfortunately due to circumstances of the previous infrastructures' setup, but it could certainly be avoided. This scenario altered significantly when crop production and biofuel technology advancements are considerably taken into account. Even though there is some uncertainty about the eventual large-scale industrial uses of cellulosic biomass conversion technologies for the generation of bioenergy, the potential benefits are well predictable and requires further research in that area.

References

Abe, J.O., Popoola, A.P.I., Ajenifuja, E. et al. (2019). Hydrogen energy, economy and storage: review and recommendation. *International Journal of Hydrogen Energy* 44 (29): 15072–15086. doi:https://doi.org/10.1016/j.ijhydene.2019.04.068

Abraham, A., Mathew, A. K., Park, H. et al. (2020). Pretreatment strategies for enhanced biogas production from lignocellulosic biomass. *Bioresource Technology* 301:122725.

Agbor, V.B., Cicek, N., Sparling, R. (2011). Biomass pretreatment: fundamentals toward application. *Biotechnology Advances* 29: 675–85. doi:https://doi.org/10.1016/j.biotechadv.2011.05.005.

Gupta, P., Indurani, C., Ahlawat, O. P. et al. (2004). Physicochemical properties of spent mushroom substrates of *Agaricus bisporus*. *Mushroom Research* 13 (2): 84–94.

Hallenbeck, P.C., Benemann, J.R. (2002). Biological hydrogen production; Fundamentals and limiting processes. *International Journal of Hydrogen Energy* 27: 1185–93. https://doi.org/10.1016/B978-008044356-0/50008-7.

Hamelinck, C.N., Hooijdonk, G.V. Andre, P.C.F. (2005). Ethanol from lignocellulosic biomass: techno-economic performance in short-, middle- and long-term. 28:(4), 384–410. doi:https://doi.org/10.1016/j.biombioe.2004.09.002

He, P., Li, F., Huang, L., et al. (2016). Chemical characterization and antioxidant activity of polysaccharide extract from spent mushroom substrate of *Pleurotus eryngii*. *Journal of the Taiwan Institute of Chemical Engineers*, 69:48–53. https://doi.org/10.1016/j.jtice.2016.10.017

Hendriks, A.T.W.M., and Zeeman, G. (2009). Pretreatments to enhance the digestibility of lignocellulosic biomass. *Bioresource Technology*, 100:10–18.

Himmel, M. E., Ding, S. Y., Johnson, D. K. et al. (2007). Biomass recalcitrance: engineering plants and enzymes for biofuel production. *Science*, 315: 804–807.

Hoekstra, A.Y., Gerbens-Leenes, P. W., Van der Meer, T. H. (2010). The water footprint of bio-energy. In: Howe, C.J., Smith, B., Henderson, J. (eds.) *Climate Change and Water: International Perspectives on Mitigation and Adaptation*. London, UK: American Water Works Association, IWA Publishing; 2010. p. 81–95.

https://www.ieabioenergy.com/wp-content/uploads/2017/01/BIOENERGY-AND-SUSTAINABLE-DEVELOPMENT-final-20170215.pdf

Hu, B. B., Zhu, M. J. (2017). Enhanced hydrogen production and biological saccharification from spent mushroom compost by *Clostridium thermocellum* 27405 supplemented with recombinant β-glucosidases. *International Journal of Hydrogen Energy* 42 (12): 7866–7874. doi:https://doi.org/10.1016/j.ijhydene.2017.01.031.

Hultberg, M., Ahrens, L., Golovko, O. (2020). Use of lignocellulosic substrate colonized by oyster mushroom (*Pleurotus ostreatus*) for removal of organic micropollutants from water. *Journal of Environmental Management* 272: 111087

International Energy Agency. (2018a). World energy balances: overview. https://doi.org/10.15713/ins.mmj.3.

International Energy Agency. (2018b). World energy outlook.

International Energy Agency. Energy Technology Perspectives. (2010). *Scenarios & Strategies to 2050*. Paris: OECD/IEA; 2010.

Iqbal, H. M. N., Asgher, M., and Bhatti, H. N. (2011). Optimization of physical and nutritional factors for synthesis of lignin degrading enzymes by a novel strain of *Trametes versicolor*. *Bio Resources*, 6:1273–1278.

Iqbal,H.M.N., Kyazze, G. & Keshavarz, T. (2013). Advances in the Valorization of Lignocellulosic Materials by Biotechnology: An Overview. *Bioresources*, 8 (2): 3157–3176.

Islam, M.Z., Rahman, M.H., Hafiz, F. (2009). Cultivation of oyster mushroom (*Pleurotus flabellatus*) on different substrates. *International Journal of Sustainable Crop Production* 4 (1): 45–48.

Jafari, M. A., Nikkhah, A., Sadeghi, A. A., and Chamani, M. (2007). The effect of *Pleurotus* spp. fungi on chemical composition and in vitro; digestibility of rice straw. *Pakistan Journal of Biological Sciences*, 10(15):2460–2464.

Jiang, G., Nowakowski, D. J., and Bridgwater, A. V. (2010). A systematic study of the kinetics of lignin pyrolysis. *Thermochimica Acta*, 498(1), 61–66.

Jiang, H., Zhang, M., Chen, J. et al. (2017). Characteristics of bio-oil produced by the pyrolysis of mixed oil shale semi-coke and spent mushroom substrate. *Fuel* 200: 218–224.

Jonsson, L., Martin, C. (2016). Pretreatment of lignocellulose: formation of inhibitory by-products and strategies for minimizing their effects. *Bioresource Technology* 199: 103–112. doi:https://doi.org/10.1016/j.biortech.2015.10.009.

Kamal Baharin, N. S., Koesoemadinata, V. C., Nakamura, S. et al. (2020). Production of Bio-Coke from spent mushroom substrate for a sustainable solid fuel. *Biomass Conversion and Biorefinery* doi:https://doi.org/10.1007/s13399-020-00844-5

Karas, P. A., Makri, S., Papadopoulou, E. S. et al. (2016). The potential of organic substrates based on mushroom substrate and straw to dissipate fungicides contained in effluents from the fruit-packaging industry-is there a role for *Pleurotus ostreatus*?. *Ecotoxicology and Environmental Safety* 124: 447–454. doi:https://doi.org/10.1016/j.ecoenv.2015.11.022.

Klein-Marcuschamer, D., Oleskowicz-Popiel, P., Simmons, B.A. et al. (2012). The challenge of enzyme cost in the production of lignocellulosic biofuels. *Biotechnology and Bioenergy* 109(4): 1083–1087.

Ko, H. G., Park, S. H., Kim, S. H. et al. (2005). Detection and recovery of hydrolytic enzymes from spent compost of four mushroom species. *Folia Microbiology*, 50:103–106.

Kopetz, H. (2015). The future role of bioenergy in the global energy system. In: Swaaij, W., Van Kersten, S., Palz, W. (eds.), *Biomass Power for the World. Pan Stanford Series on Renewable Energy*, volume 6, Pan Stanford Publishing, Singapore, pp. 141161.

Kumar, P., Barrett, D. M., Delwiche, M. J. et al. (2009). Methods for pretreatment of lignocellulosic biomass for efficient hydrolysis and biofuel production. *Industrial & Engineering Chemistry Research*, 48: 3713–3729.

Kumar, G., Bakonyi, P., Periyasamy, S. et al. (2015). Lignocellulose biohydrogen: practical challenges and recent progress. *Renewable and Sustainable Energy Reviews* 44: 728–37. doi:https://doi.org/10.1016/j.rser.2015.01.042.

Larson, S., Palmqvist, E., Hanhn-Heagerdal, B. et al. (1999). The generations of fermentation inhibitors during dilute acid hydrolysis of softwood. *Enzyme and Microbial Technology* 24: 151–159. doi:https://doi.org/10.1016/S0141-0229(98)00101-X.

Lee, S. Y., Park, J. H., Jang, S. H. et al. (2008). Fermentative butanol production by *Clostridia*. *Biotechnology and Bioenergy* 101: 209–228.

Lee, C. Y., Park, J. E., Kim, B. B. et al. (2009). Determination of mineral components in the cultivation substrates of edible mushrooms and their uptake into fruiting bodies. *Mycobiology* 37 (2): 109–113. doi:https://doi.org/10.4489/MYCO.2009.37.2.109.

Li, Y. (2012). Present development situation and tendency of edible mushroom industry in China. In: *18th Congress of the International Society for Mushroom Science*. Beijing, China, pp. 3–9.

Lin, Y., and Tanaka, S. (2006). Ethanol fermentation from biomass resources: current state and prospects. *Applied Microbiology and Biotechnology*, 69: 627–642.

Lin, Y., Ge, X., Li, Y. (2014). Solid-state anaerobic co-digestion of spent mushroom substrate with yard trimmings and wheat straw for biogas production. *Bioresource Technology* 169: 468–474. doi:https://doi.org/10.1016/j.biortech.2014.07.020.

Lin, C. Y., Lay, C. H., Chen, C. C. et al. (2016). Biohydrogen production from mushroom cultivation waste by anaerobic solid-state fermentation. *Journal of Chinese Chemical Society*, 63:199–204. https://doi.org/10.1002/jccs.201500359

Liu, C., Wyman, C. E. (2005). Partial flow of compressed-hot water through corn stover to enhance hemicellulose sugar recovery and enzymatic digestibility of cellulose. *Biotechnology Resources*, 96(18):1978–1985. doi:https://doi.org/10.1016/j.biortech.2005.01.012 .

Liu, S. H., Tsai, S. L., Guo, P.Y. et al. (2020). Inducing laccase activity in white rot fungi using copper ions and improving the efficiency of azo dye treatment with electricity generation using microbial fuel cells. *Chemosphere*, 243:125304

Liu, L., Qianb, H., Mua, L., et al. (2021). Techno-economic analysis of biomass processing with dual outputs of energy and activated carbon. *Bioresource Technology* 319: 124108.

Lou, Z., Zhu, J., Wang, Z. et al. (2015). Release characteristics and control of nitrogen, phosphate, organic matter from spent mushroom compost amended soil in a column experiment. *Process Safety and Environment Protection* 98: 417–423. doi:https://doi.org/10.1016/j.psep.2015.10.003.

Lou, Z., Sun, Y., Bian, S. et al. (2017a). Nutrient conservation during spent mushroom compost application using spent mushroom substrate derived biochar. *Chemosphere* 169: 23–31. doi:https://doi.org/10.1016/j.chemosphere.2016.11.044.

Lou, Z., Sun, Y., Zhou, X. et al. (2017b). Composition variability of spent mushroom substrates during continuous cultivation, composting process and their effects on mineral nitrogen transformation in soil. *Geoderma* 307: 30–37. doi:https://doi.org/10.1016/j.geoderma.2017.07.033.

Luo, X., Yuan, X., Wang, S. et al. (2018). Methane production and characteristics of the microbial community in the co-digestion of spent mushroom substrate with dairy manure. *Bioresource Technolnology* 250: 611–620. doi:https://doi.org/10.1016/j.biortech.2017.11.088.

Mansor, A. M., Wai, L. T., Jeng, S. L. et al. (2018). Potential commercialization of biocoke in Malaysia – a best evidence review. *Renewable and Sustainable Energy Reviews* 90: 636–649. doi:https://doi.org/10.1016/j.rser.2018.03.008.

McKendry, P. (2002). Energy production from biomass (part 1): overview of biomass. *Bioresource Technology* 83: 37–46. doi:https://doi.org/10.1016/S0960-8524(01)00118-3.

Menon, V., and Rao, M. (2012). Trends in bioconversion of lignocellulose: biofuels, platform chemicals & biorefinery concept. *Progress in Energy and Combustion Science*, 38(4): 522–550.

Mizuno, S., Hamidun, O., Ida, T. et al. (2014). Effect of crushing particle size on cold/ hot compressive strength of high-density and high-hardness solid fuel produced from raw rice husk. *Journal of Smart Processing* 3 (5): 269–275. https://doi.org/10.7791/jspmee.3.269.

Mohd Hanafi, F. H., Rezania, S., Mat Taib, S. et al. (2018). Environmentally sustainable applications of agro-based spent mushroom substrate (SMS): an overview. *Journal of Material Cycles and Waste Management* 20 (3): 1383–1396. doi:https://doi.org/10.1007/s10163-018-0739-0.

Moodley, P., Gueguim Kana, E. B. (2018). Comparative study of three optimized acid-based pretreatments for sugar recovery from sugarcane leaf waste: a sustainable feedstock for biohydrogen production. *Engineering Science and Technology, an International Journal* 21: 107–16. doi:https://doi.org/10.1016/j.jestch.2017.11.010.

Moon, Y. H., Shin, P. G., Cho, S. J. (2012). Feeding value of spent mushroom (*Pleurotus eryngii*) substrate. *Journal of Mushroom* 10 (4): 236–243.

Mosier, N., Wyman, C., Dale, B. et al. (2005). Features of promising technologies for pretreatment of lignocellulosic biomass. *Bioresource Technology*, 96: 673–686.

Mustafa, A. M., Poulsen, T. G., Sheng, K. (2016). Fungal pre-treatment of rice straw with *Pleurotus ostreatus* and *Trichoderma reesei* to enhance methane production under solid-state anaerobic digestion. *Applied Energy* 18: 661–671.

Najafi, B., Ardabili, S. F. (2018). Application of ANFIS, ANN, and logistic methods in estimating biogas production from spent mushroom compost (SMC). *Resources, Conservation and Recycling* 133: 169–178. https://doi.org/10.1016/j.resconrec.2018.02.025.

Nakatsuka, H., Oda, M., Hayashi, Y. et al. (2016). Effects of fresh spent mushroom substrate of Pleurotus ostreatus on soil micro morphology in Brazil. *Geoderma* 269: 54–60. doi:https://doi.org/10.1016/j.geoderma.2016.01.023.

Oguri, E., Takimura. O., Matsushika, A. et al. (2011). Bioethanol production by *Pichia stipitis* from enzymatic hydrolysates of corncob-based spent mushroom substrate. *Food Science and Technology Research*, 17:267–272.

Perez-Chavez, A. M., Mayer, L., Alberto, E. (2019). Mushroom cultivation and biogas production: a sustainable reuse of organic resources. *Energy for Sustainable Development* 50: 50–60. doi:https://doi.org/10.1016/j.esd.2019.03.002.

Phan, C.W., Sabaratnam, V. (2012). Potential uses of spent mushroom substrate and its associated lignocellulosic enzymes. *Applied Microbiology and Biotechnology* 96 (4): 863–73. doi:https://doi.org/10.1007/s00253-012-4446-9.

Pointing, S. B. (2001). Feasibility of bioremediation by white-rot fungi. *Applied Microbiology and Biotechnology*, 57:20–33.

Popp, J., Lakner, Z., Harangi-Rakos, M. et al. (2014). The effect of bioenergy expansion: food, energy, and environment. *Renewable and Sustainable Energy Reviews* 32: 559–578. doi:https://doi.org/10.1016/j.rser.2014.01.056.

Qureshi, N., Saha, B. C., Cotta, M. A. (2007). Butanol production from wheat straw hydrolysate using *Clostridium beijerinckii*. *Bioprocess and Biosystems Engineering* 30: 419–427. doi:https://doi.org/10.1007/s00449-007-0137-9.

Qureshi, N., Sahaa, B. C., Hector, R. E. et al. (2008). Butanol production from wheat straw by simultaneous saccharification and fermentation using *Clostridium beijerinckii*: part I e Batch fermentation. *Biomass and Bioenergy* 32: 168–175. doi:https://doi.org/10.1016/j.biombioe.2007.07.004.

Rani, P., Kalyani, N., and Prathiba, K. (2008). Evaluation of lignocellulosic wastes for production of edible mushrooms. *Applied Biochemistry and Biotechnology*, 151(2–3): 151–159. https://doi.org/10.1007/s12010-008-8162-y

Reddy, C. A., D'Souza, T. M. (1994). Physiology and molecular biology of the lignin peroxidases of *Phanerochaete chrysosporium*. *FEMS Microbiology Review*, 13:137–152.

Rinker, D. (2017). Spent mushroom substrate uses: technology and applications (pp 427–454). doi: https://doi.org/10.1002/9781119149446.ch20

Rivas, B., Torre, P., Dominguez, J. M. et al. (2006). Purification of xylitol obtained by fermentation of corncob hydrolysates. *Journal of Agriculture and Food Chemistry*, 54: 4430–4435.

Roy, S., Barman, S., Chakraborty, U. et al. (2015). Evaluation of spent mushroom substrate as biofertilizer for growth improvement of *Capsicum annuum* L. *Journal of Applied Biology & Biotechnology*, 3 (03): 022–027.

Royse, D. J., Baars, J., and Tan, Q. (2017). Current overview of mushroom production in the World. Edible and medicinal mushrooms: technology and applications. In Zied, D. C. and Pardo-Gimenez, A (Ed.), *Edible and Medicinal Mushrooms* (pp. 5–13). Wiley-Blackwell, Inc. https://doi.org/10.1002/9781119149446.ch2

Ryden, P., Efthymiou, M. N., Tindyebwa, T. A. et al. (2017). Bioethanol production from spent mushroom compost derived from chaff of millet and sorghum. *Biotechnology for Biofuels* 10 (1): 195. doi:https://doi.org/10.1186/s13068-017-0880-3.

Shirkavand, E., Baroutian, S., Gapes, D. et al. (2016). Combination of fungal and physicochemical processes for lignocellulosic biomass pretreatment – a review. *Renewable and Sustainable Energy Reviews* 54: 217–234. doi:https://doi.org/10.1016/j.rser.2015.10.003.

Singh, A. D., Abdullah, N., and Vikineswary, S. (2003). Optimization of extraction of bulk enzymes from spent mushroom compost. *Journal of Chemical Technology and Biotechnology*, 78:743–752.

Singh, S., Cheng, G., Sathitsuksanoh, N. (2015). Comparison of different biomass pretreatment techniques and their impact on chemistry and structure. *Frontiers in Energy Research* 2: 1–12. https://doi.org/10.3389/fenrg.2014.00062.

Specht, K., Siebert, R., Hartmann, I. et al. (2014). Urban agriculture of the future: an overview of sustainability aspects of food production in and on buildings. *Agriculture and Human Values* 31 (1): 33–51. doi:https://doi.org/10.1007/s10460-013-9448-4.

Sredlova, K., Skrob, Z., Filipova, A. et al. (2019). Biodegradation of PCBs in contaminated water using spent oyster mushroom substrate and a trickle-bed bioreactor. *Water Research*. doi:https://doi.org/10.1016/j.watres.2019.115274.

Suguimoto, H., Barbosa, A. M., Dekker, R. F. H. et al. (2001). Veratryl alcohol stimulates fruiting body formation in the oyster mushroom, *Pleurotus ostreatus*. *FEMS Microbiological Letter*, 194:235–238.

Taherzadeh, M. J., and Karimi, K. (2008). Pretreatment of lignocellulosic wastes to improve ethanol and biogas production: a review. *International Journal of Molecular Sciences*, 9:1621–1651.

Tembo, G., Epplin, F.M., and Huhnke. R.L. (2003). Integrative investment appraisal of a lignocellulosic biomass-to-ethanol industry. *Journal of Agricultural and Resource Economics*, 28: 611–633

Teramoto, Y., Lee, S.H., Endo, T. (2008). Pretreatment of woody and herbaceous biomass for enzymatic saccharification using sulfuric acid-free ethanol cooking. *Bioresource Technology* 99: 8856–8863. doi:https://doi.org/10.1016/j.biortech.2008.04.049.

Tuhy, L., Samoraj, M., Witkowska, Z. et al. (2015). Conversion of spent mushroom substrate into micronutrient fertilizer via biosorption in a pilot plant. *Ecological Engineering* 84: 370–374. doi:https://doi.org/10.1016/j.ecoleng.2015.09.032.

Vercoe, D., Stack, K., Blackman, A. et al. (2005). A study of the interactions leading to wood pitch deposition. In 59th Appita Annual Conference and Exhibition: Incorporating the 13th ISWFPC (International Symposium on Wood, Fibre and Pulping Chemistry): Proceedings (p. 123) (16–19 May 2005), Auckland, New Zealand.

Wan, C., Li, Y. (2012). Fungal pretreatment of lignocellulosic biomass. *Biotechnology Advances* 30 (6): 1447–1457. doi:https://doi.org/10.1016/j.biotechadv.2012.03.003.

Wang, H. W., Xu, M., Cai, X. Y., Feng, T., Xu, L. (2020). Application of spent mushroom substrate suppresses Fusarium wilt in cucumber and alters the composition of the microbial community of the cucumber rhizosphere. *European Journal of Soil Biology* 101: 103245

Watts, N., Amann, M., Arnell, N. et al. (2018). The 2018 report of the Lancet Countdown on health and climate change: shaping the health of nations for centuries to come. *Lancet* 392: 2479–514. doi:https://doi.org/10.1016/S0140-6736(18)32594-7.

Wei, M., Xiong, S., Chen, F. et al. (2020). Energy smart hot-air pasteurisation as effective as energy intense autoclaving for fungal preprocessing of lignocellulose feedstock for bioethanol fuel production. *Renewable Energy* 155: 237–247. doi:https://doi.org/10.1016/j.renene.2020.03.154

Weiland, P. (2010). Biogas production: current state and perspectives. *Applied Microbiology and Biotechnology*, 85: 849–860. https://doi.org/10.1007/s00253-009-2246-7

Williams, B. C., McMullan, J. T., and McCahey, S. (2001). An initial assessment of spent mushroom compost as a potential energy feedstock. *Bioresource Technology*, 79:227–230.

Wu, S., Lan, Y., Wu, Z. et al. (2013). Pretreatment of spent mushroom substrate for enhancing the conversion of fermentable sugar. *Bioresource Technology*, 148:596–600.

Wyman, C. E., Dale, B. E., Elander, R. T. et al. (2005). Coordinated development of leading biomass pretreatment technologies. *Bioresource Technology*, 96: 1959–1966.

Xiao, W., Wang, Y., Xia, S. et al. (2012). The study of factors affecting the enzymatic hydrolysis of cellulose after ionic liquid pretreatment. *Carbohydrate Polymers*, 87: 2019–2023.

Xiong, S., Martin, C., Eilertsen, L. et al. (2019). Energy-efficient substrate pasteurisation for combined production of shiitake mushroom (*Lentinula edodes*) and substrate: equilibrium, kinetics, and thermodynamics. *BioResources* 8 (3): 4722–4734.

Xu, N., Hu, X., Xu, W. et al. (2017). Mushrooms as efficient solar steam-generation devices. *Advanced Materials* 29**(**28**)**: 1606762. doi:10.1002/adma.201606762

Yan T, Wang, L. (2013). Adsorptive removal of methylene blue from aqueous solution by spent mushroom substrate: equilibrium, kinetics, and thermodynamics. *Bioresource Technology*, 8(3):4722–4734

Zhang, J. J., Li, J., Zhou, T. et al. (2016). Bioactivities and health benefits of mushrooms mainly from china. *Molecules*, 21:938. doi:https://doi.org/10.3390/molecules21070938

Zhao, X., Cheng, K., Liu, D. (2009). Organosolv pretreatment of lignocellulosic biomass for enzymatic hydrolysis. *Applied Microbiology and Biotechnology* 82: 815–827. doi:https://doi.org/10.1007/s00253-009-1883-1.

Zhu, J., and Pan, X. (2010). Woody biomass pretreatment for cellulosic ethanol production: technology and energy consumption evaluation. *Bioresource Technology*, 101 (13): 4992–5002.

Zhu, Y., Chang, Y., Guan, J. et al. (2016). Butanol production from organosolv treated spent mushroom substrate integrated with insitu biodiesel extraction. *Renewable Energy* 96: 656–661. doi:10.1016/j.renene.2016.04.048

Zhu, H.J., Sun, L.F., Zhang, Y.F. et al. (2012). Conversion of spent mushroom substrate to biofertilizer using a stress tolerant phosphate-solubilizing Pichia farinose FL7. *Bioresource Technology* 111: 410–416. doi:https://doi.org/10.1016/j.biortech.2012.02.042.

Index

1G biofuels 39, 234–236
2G biofuels 39, 234, 236–238, 243
3G biofuels 234, 243–244
4G biofuels 234, 245

a

accessibility, energy transition in Brazil 122
acetic acid 212
acetone butanol-ethanol (ABE) 290
acetone/lignocellulosic biomass pretreatment 211–212
acetyl groups 288
Achromobacter spp. 262
acids
  biobutanol and SMS 289
  detoxification of lignocellulosic biomass 212
  hydrolysis/lignocellulosic biomass 213
  jet fuel production 217, 218, 219
  pretreatment of lignocellulosic biomass 211
Acinetobacter baylyi 161
Actinomycetes spp. 237–238
active solar energy technology 132, 133
adenosine triphosphate (ATP) 194, 195, 197
Aditiya, H. B. 243
AFEX (ammonia fibre explosion) 237
Africa
  energy consumption 96–97
  final energy consumption 232–233
  pathways of energy transition/economic growth 110
  population increase 95
  progressive energy consumption 97
  ratio of energy consumption to GDP from 1860 to 1996 97, 98
Africa, Sahel region 8
Agaricus bisporus 276, 284, 285
Aghion, P. 123
agriculture 161–162
  India/sustainability 176
  renewable energy 142
  second generation biofuels 237
  wind energy 154
agro waste/biotransformation 274–302
Ai, C. 262
air pollution/air quality 77–78, 142, 169–181
airborne wind energy (AWE) 61
airports 86–87
Airth, R. L. 197, 198
alcohols
  butanol 39, 221, 289–290
  dehydrogenase 161
  jet fuel production 217, 218, 221

Energy: Crises, Challenges and Solutions, First Edition. Edited by Pardeep Singh, Suruchi Singh, Gaurav Kumar, and Pooja Baweja.

alcohols (*cont'd*)
  lignocellulosic biomass pretreatment 211–212
  methanol 211–212
  phenol/lignocellulosic biomass 211–212, 222–223
  see also ethanol
algae
  3G biofuels 39
  hydrogen production methods 43–44
  SMS/biohydrogen production 286
  third generation biofuels 243–244
  see also bioluminescence; microorganism-derived energy
alkalis, lignocellulosic biomass pretreatment 211
alkane isomerization and cracking, biofuel categorization 39
Allium sativum-derived C-dots 62–63
Alternative Energy Source Incentive Programme (PROINFA), Brazil 116, 117
alternative energy sources
  advances in 18–54
  bioenergy 37–39
  biomass energy 62–63
  challenges 44–46
  clean energy 160–164
  cost-effective production 46
  efficiency 44–45
  environmental impacts 64–65
  future of 46–47, 65
  geothermal energy 33–37, 61–62
  hydrogen 42–44
  hydropower 29–33, 60
  innovations 56–64
  large-scale production 45
  novel research 19–20
  ocean energy 40
  recent advances in 20–41, 55–71
  semiconductor technology 56–60
  wind energy 25–29, 61
Amazon Development Agency 118
ammonia fibre explosion (AFEX) 237
ammonia recycled percolation (ARP) 237
amorphous silicon photovoltaic cells 133
amorphous silicon-based photovoltaic cells 22, 44
anaerobic digestion of waste as eco-friendly fuel 281, 287–288
anaerobic yeast, microbial fuel cells 257
Ancanthodrilidae spp. 194
ANEEL (Brazilian Electricity Regulatory Agency) 116
angiospermic hardwood 238, 240
animal model studies in bioluminescence 201
Apergis, N. 110
a-pyrones luciferins 197–198
arabinans 279–280
arabinogalactans 279–280
arabinose 213
Arachnocampa spp. 194
Argentina, progressive energy consumption 97
Aristotle 192
Armillaria spp. 196–197, 199
aromatic compounds 210
ARP (ammonia recycled percolation) 237
array cables, offshore wind energy technologies 28–29
arsenide, active solar energy technology 133
Arxula adeninivorans 258
Asada, C. 284, 285
Asia
  final energy consumption 232–233
  population increase 95
  see also individual Asian countries...
Aspergillus campestris 261
Aspergillus spp. 200, 201, 237–238
ATP (adenosine triphosphate) 194, 195, 197
'attenuators'/offshore wave energy 40
Auricularia spp. 261
aurisin A and K (bioluminescence) 201

Australia
  energy consumption 96–97
  geothermal energy 34
aviation fuel 217–221
AWE (airborne wind energy) 61
azo bonds 264–265

b

Bacillus spp. 237–238, 262
Bacon, F. 192
bacteria
  second generation
    biofuels 237–238, 240
  third generation biofuels 244
  see also bioluminescence; microorganism-derived energy
"Banco do Brazil" 119
banks 115, 119
Bartholin 192
Baruah, J. 237
batteries design 258–261
beetles 193, 194, 195
Bennetto, H. P. 251
Benthic Unattended Generators (BUGs) 161
benzothiazoyl-thiazole 195
β-glucosidases 213, 283
β-isopropylmalate dehydrogenase (LEU2) 160–161
Betaproteo bacteria 262
Bharucha, E. 158
BHJ (bulk hetero-junction), structures and solar energy 23–24
binary plants 35
bioactive secondary metabolites/ bioluminescence 201
bioalcohols 39, 221, 289–290
bioassays of toxicity and bioluminescence 199
biobutanol 39, 289–290
  see also butanol
bio-coke 290–291
biocrude/syngas converts to renewable gasoline 39
biodiesel 38–39, 160–161, 243
  Brazil/energy sources 113
  Brazil/technological development 120
bioelectricity production and MFCs 262–263
bioenergy 19, 37–39
  2G biofuels 39
  biofuel categorization 39
  biopellets/biogas 38
  ethanol/diesel 38–39
  growth rate 20–21
  policy prioritization/Brazil 114–115
bioengineered fungi 160–161
bioethanol 38–39, 288–289
  see also ethanol
biofuels 62
  2G biodiesel 243
  categorization 39
  classification 234–245
  first generation 1G 39, 234–236
  fourth generation 4G 234, 245
  future of 144
  global investments 46, 47
  history 233–234
  India/electricity generation 159
  second generation 2G 39, 234, 236–238, 243
  third generation 3G 234, 243–244
  worldwide electricity generation 20
biogas
  advances in 20, 38
  biomass energy 62
  India/electricity generation 159
  production from SMS 287–288
  worldwide electricity generation 20
bio-gasoline 214–217
biohydrogen production 44
  from SMS 285–287
  microbial fuel cells 263
bioluminescence 191–207
  animal model studies 201
  bioactive secondary metabolites 201
  bioassays of toxicity 199–200

bioluminescence (*cont'd*)
  fungi 196–198
  glowing tree 198–199
  history of 192–193
  imaging mechanism 200
  in-vivo imaging 200–201
  molecules 194–196
  opportunities 198–201
  terrestrial organisms 193–194
biomass energy 62–63, 138
  advantages/disadvantages 140
  biofuels categorization 39
  clean energy 157–160
  developing nations 99, 100, 101, 102
  diagrammatic representation 153
  gasification/hydrogen 43
  global investments 46, 47
  global primary energy consumption 4–5
  hydrogen production methods 43–44
  India/exploitative policy 173–174
  liquid biofuel production 208–230
  polyol conversions to alkane 217
  SMS/biohydrogen production 286–287
  technologies/application/challenges 56, 57
  see also lignocellulosic biomass
bio-oil, hydrogen production methods 42–43
biopellets 38
bioremediation 264
biosensors 263
biotransformation and sustainability via mushrooms 274–302
birds, bioluminescence 192
Blazejczak, J. 110
BNDES (Brazil National Bank for Development) 115, 119
Bombay stock exchange (BSE) 84–85
Bond, D. R. 251, 262
Boyle, R. 192
Brazil
  banks 115, 119
  climate change National Policy 2009 114
  climate change/regime complex 185–186
  climate-resilient growth 121–122
  electricity production 159, 160
  electricity sector 113–115
  energy consumption 96–97
  energy policies 113–114
  energy sources and associated policies 113–114
  Federal Funding Agency 118
  federal programmes 119
  finance/innovation 120–121
  first phase reform 1990s 114
  funding organizations 117, 118
  global governance 183–184
  governmental programmes and laws 116, 117
  hydropower 153–154
  implications of climate regime 186
  market structures 115–116
  pathways of energy transition/economic growth 109–110, 111–112
  policy prioritization/choice of energy mix 114–115
  political economy 123
  progressive energy consumption 97
  reasearch foundations 119
  regime complex 184–186
  regime complex/technological innovation 182–190
  Regime for Funding Agency 118–120
  regime structure 118–120
  research foundations 119
  second reform 2004 114
  social consequences/accessibility 122
  source funding and R&D trends 120–121
  sources of finance 116–121
  state programmes 119
  technological innovations 187

Brazil National Bank for Development (BNDES) 115, 119
Brazilian Electricity Regulatory Agency (ANEEL) 116
Brazilian Institute of Environment and Renewable Natural Resources (IBAMA) 115
Brazilian Institute of Geography and Statistics (IBGE) 182
Bresadola, M. 251
brown rot fungi 212, 238–240
BSE (Bombay stock exchange) 84–85
buffers and MFCs 254
buffers and microbial fuel cells 254
BUGs (Benthic Unattended Generators) 161
'bulge wave devices'/offshore wave energy 40
bulk hetero-junction (BHJ) structures and solar energy 23–24, 59
business meagre thin-film cells 133
butanol 39, 221, 289–290

c

cadmium (CdS/CdTe) photovoltaic cells 22, 44
cadmium telluride photovoltaic cells 133
CAIT (climate analysis indicator tools) 79
Caldicellulosiruptor saccharolyticus 242
Camelina non-edible oil-producing plants 220
Canada
  electricity production 159
  energy consumption 96–97
  final energy consumption 232–233
  hydropower 153–154
Cancun Agreement 2010 82, 83
Candida spp. 200–201, 235, 237–238, 242, 257, 288–289
carbohydrate conversion to alkanes in gasoline production 216–217
Carbon Brief report 87–88
carbon emissions 6
  Brazil climate regime 186
  Brazilian political economy 123
  carbon sequestering 10–11
  Coronavirus pandemic 87–88
  energy crisis 6
  India/emissions 1960–2020 74–76
  India/reduction of 84–87
  novel research 19
  pathways of energy transition 110
  process and causes of climate change 9–10
  renewable energy 141–142
  SMS/biohydrogen production 286
  see also greenhouse gases
carbon fibre reinforced resin matrix composite materials (CFRC) 63
carbon footprints 78–80
carbon nanotubes 21–22
carbon sequestering 10–11
carbon/lithium cobalt oxide ($LiCoO_2$) cells 259
Carbonex, Bombay stock exchange 84–85
carboxymethylcellulases 213
carrier multiplication 21–22
Catal, T. 255
catalysts
  bioluminescent molecules 194
  jet fuel production 217–221
  microbial fuel cells 255–258
  see also enzymes
CBP (consolidated bioprocessing) 242–243
CDM (Clean Development Mechanism) 86–87
CEF (clean energy fund), India 173
cellobiohydrolase (CBH) 213
cellulases/lignocellulosic biomass hydrolysis 213
cellulose
  enzymatic pretreatment of biofuels 242
  mushrooms/LCB 279
  polyol conversions to alkane 217
  SMS and bioethanol production 288

cellulose–hemicellulose complex 211–212
centipedes 193
Central Electricity Regulatory Commission (CERC) 178
CERC (Central Electricity Regulatory Commission) 178
CFRC (carbon fibre reinforced resin matrix composite materials) 63
Chang, S. T. 284, 285
Chavanne, G. 235
Chile, geothermal energy 34
China
  biobutanol and SMS 289
  carbon sequestering/clean coal 10–11
  commercial energy consumption and projection 97, 98
  electricity production 159, 160
  energy consumption 96–97
  environmental issues/economic development 76, 77–78
  future of renewables 143–144
  hydropower 153–154
  as leaders in renewable energy 21
  pollution 77–78
  population increase 95
  progressive energy consumption 97
  ratio of energy consumption to GDP from 1860 to 1996 97, 98
  Three Gorges Dam 29–30
  wind energy 154
Chlorella vulgaris 262
CHP (combined heat power) 138
Christwardana, M. 256
chromophores 264–265
circadian rhythm and bioluminescence 198
Clean Development Mechanism (CDM) 86–87
clean energy 151–168
  alternative energy resources 160–164
  biodiesel/bioengineered fungi 160–161
  biomass 157–160
  challenges 164–165
  clean coal 10–11
  commercialization/scalability 165
  conventional sources 152–156
  ecological/environmental issues 164
  fuel cells 163
  hydro energy 153–154
  hydrogen as fuel 162–163
  material requirements 165
  microbial fuel cells 161
  mitigation strategies 157
  radiant energy 163–164
  social barriers 164
  waste-to-energy technology 161–162
clean energy fund (CEF), India 173
Clean Energy Tax 176–177
click beetles 194, 195
climate analysis indicator tools (CAIT) 79
climate change 1–17
  Brazil National Policy 2009 114
  Brazil/regime complex 185–186
  carbon sequestering 10–11
  clean coal 10–11
  coal alternatives 10–12
  energy crisis 1, 5–6
  energy demand 12
  environmental consequences 7–9
  hydrogen 12–13
  India and sustainable development 169–181
  IPCC 4–8, 12–13, 72, 82–83, 142
  IPCC and greenhouse gas reduction 12–13
  NAPCC-National Action Plan of India 2008 175–176
  natural gas/nuclear energy 11
  process and causes 9–10
  social consequences 7–9
  see also carbon emissions; Sustainable Development Goals
climate-resilient growth in Brazil 121–122
Clitellata spp. 194

Clostridium spp. 242–243, 284, 285
CNEN (National Nuclear Energy Commission) 116
CNPq ('National Council for Scientific and Technological Development') 117, 118
coal
  alternatives to 10–12
  clean coal 10–11
  developing nations 100, 101, 102
  global primary energy consumption 4–5
  India energy sector 170, 171
  pollution 76–77
  sustainable development 19
  taxes 176–177
Cochin International airport 86–87
coelenterates and ctenophores 195
coelenterazine luciferin 195
Cohen, B. 251
Cole, D. S. 251
Coleoptera spp. 194
combined heat power (CHP) 138
combined-cycle geothermal plants 35
combustion, biomass 138
commercial energy consumption and projection 97, 98
Companies Act 2013 82–83
concentrated solar PV (CPV) 21–22, 23, 59
concentrated solar/hydrogen production methods 42–43
CONCEPT, Brazil 116, 117
Conference of the Parties (COP15), Copenhagen 81, 82, 83, 176
Coniochaeta ligniaria 212
consolidated bioprocessing (CBP) 242–243
Constitutional Fund for financing the Centre West Region, Brazil 119
conventional crystalline arsenide, gallium or silicon cells 133
cooking fuels 99–100, 101, 102
Copenhagen accord 2009 81, 82, 83, 176
copper and bioluminescence 199
copper indium gallium selenide (CIS/CIGS)/silicon thin-film photovoltaic cells 22
copper peroxidase 292
copper-containing oxidoreductase 283
Cordoba, V. 284, 285
Coriolus versicolor 258
corncob and bioethanol production 288
Corporate Social Responsibility (CSR) 82–83, 85–86
corporate social responsibility (CSR) clauses 87
Costa Rica, geothermal energy 34
COVID pandemic 87–88, 144
  see also pandemics
CPV (concentrated solar PV) 21–22, 23, 59
CPV-T (thermal-PV hybrid systems) 23, 24–25
cracking, biofuel categorization 39
CRISPR-cas 9 in biodiesel production 160–161
crude oil, India energy sector 171
Cryptococcus spp. 200–201
crystalline solar cells 21–22, 46, 57, 59
CSR (Corporate Social Responsibility) 82–83, 85–86, 87
Cucumis sativus 282
cyanobacteria
  hydrogen production methods 42–43
  SMS/biohydrogen production 286
  third generation biofuels 244
cyclo-alkanes 210
cypridina luciferin 195, 196
cytochromes c OmcA and OmcB 254–255
Czechoslovakia, energy consumption 96–97

d

Dakya striata (snail) 193
dams 29–30, 153–154
  see also hydropower

Danuletiu, D. C. 110
Davda, R. R. 216
DC power take-off and array cables, offshore wind energy technologies 28–29
de Castro, Don J. 192
De Luce Animalium (Bartholin) 192
De Lunariis (Francis Drake) 192
Deen Dayal Upadhyay Gram Jyoti Yojana policy 179
Delhi metro rail corporation (DMRC) 86–87
Deltaproteobacterium 262
Demirbas, M. F. 244
depolymerization, lignin to hydrocarbon fuel 222–223
desalination systems, future of solar energy 46
design and materials of rotor blade 26
detoxification of lignocellulosic biomass 210–212
developing nations
  energy services 99–100
  energy supplies 100–102
  environment 102–103
Di Ilio, G. 251
3,5-dichlorophenol (3,5-DCP) 199
diesel bioenergy *see* biodiesel
digitalization, hydropower 30–31
dioxetanone 195
Diptera spp. 194
direct dry steam plants 34–35
direct liquid fuel cells (DLFCs) 63–64
D-luciferin 195
DMRC (Delhi metro rail corporation) 86–87
Doha agreement 2012 82, 83
Draft amendment to Electricity Act 2003, India 179
Drake, F. 192
drop-in fuels (HVO/HEFA) 39
drought 8
Dubois, R. 193
Dumesic, J. A. 210, 216
Durban agreement 2011 82, 83
dye decolorization 264–265
dye-sensitized solar cells (DSSCs) 21–22, 23, 59

e

earthworms 193
economic growth
  benefits of renewable energy 141
  energy development 91–107
  fluctuations in energy consumption 95–96
  impact on energy 94–95
  pathways of energy transition 108–130
EET (extracellular electron transfer) 254–255
EGS (enhanced geothermal systems) 35–37, 46, 155
Egypt
  progressive energy consumption 97
  wind energy 154
EIA (Environmental Impact Assessment) 156
El Salvador, geothermal energy 34
Elateridae spp. 194
electricity
  biomass energy 62–63
  conservation/Brazil policy 115
  developing nations 99–100, 101
  hydro energy 154–155
  hydropower and environment 64–65
  India Electricity Act 174–175
  India energy sector 171–172
  LCOE 45–46
  National Electricity Policy of India 175
  overview/Brazil 113–115
  rates and Brazil energy sources 114
  Renewable Energy Certificates 178
  spent mushroom substrate 292–293
  via proton exchange membrane 250
  worldwide generation 20
Electrobras, Brazil 115, 116, 117, 118

electrode of MFCs 252
electrodes and fungal battery design 258–261
electrogens 161
electrolysis 42–43
electrolyte resistance in MFCs 255
electrolytic hydrogen production methods 42–43
electron transfer mechanisms in fungi 255, 256
Enchytraeidae spp. 194
endo-1-4-β-glucanase 213
energy conservation/transition, Brazil policy prioritization 115
Energy Information Administration (EIA) 151
Energy Research Company (EPE) 116
enhanced geothermal systems (EGS) 35–37, 46, 155
environment
  alternative energy sources 64–65
  benefits of renewable energy 141
  clean energy 151–168
  climate-resilient growth/ Brazil 121–122
  consequences of climate change 7–9
  developing nations 102–103
  regime complex/technological innovation 182–190
  sustainable development 80–87
  see also renewable energy
environmental
  air quality 77–78
  carbon footprints 78–80
  economic development 76–77
Environmental Impact Assessment (EIA) 156
enzymes
  biobutanol and SMS 289–290
  biodiesel production 160–161
  bioethanol production 242
  bioluminescence 193, 195, 197–198
  electricity generation and SMS 291–292
  lignocellulosic biomass 213–214
  MFCs 254–255
  pretreatment of biofuels 240–243
  second generation biofuels 238
  spent mushroom substrate 283, 289–290, 291–292
  triacyl glycerols 235–236
EPE (Energy Research Company) 116
1-epi-nambinone 201
Escherichia coli 161, 262
ethanol 38–39, 288–289
  bioenergy 38
  biomass 62
  climate regime/Brazil 186
  energy sources/Brazil 113
  hydrogen production methods 42–43
  jet fuel 221
  lignocellulosic biomass 210–214
  policy prioritization/Brazil 114–115
  production from SMS 288–289
Ethiopia, geothermal energy 34
ethylene glycol, lignocellulosic biomass pretreatment 211–212
EURONAT EGS plants 37
Europe
  final energy consumption 232–233
  see also individual European countries...
European Union (EU) 182–183
  21st century development 72–73
  future of renewables 143–144
  Paris Accord 82–84, 87, 152–153
Excluding Organization of the Petroleum Exporting Countries (Non-OPEC) 97, 98
exoelectrogens 251–252
exoglucanase 213
extinction rates 8–9
extracellular electron transfer (EET) 254–255

f

facultative anaerobic yeast, microbial fuel cells 257

FADESP (Research and Development Foundation), Brazil 119
Falcucci, G. 251
Fang, W. 284, 285
Fapesp (Sao Paulo Research Foundation) 119
Fapex (Foundation to Support Research and Extensions), Brazil 119
Faster Adoption and Manufacturing of (Hybrid &) Electric Vehicles (FAME) schemes 179
fatty acid butyl esters (FABEs) 161
fatty acid ethyl esters (FAEEs) 161
fatty acid methyl esters (FAMEs) 39, 161
Fay, D. 76
FBB ("Banco do Brazil") 119
FCEVs (fuel cell electric vehicles) 63–64
FCPF (Forest Carbon Partnership Facility) 3
FDI (Foreign Direct Investment) 81
FEC (final energy consumption) 232–233
fermentation 210–211, 214, 242–243
ferricyanide reductase 257
fertilizers and spent mushroom substrate 282
FFA (free fatty acids) 235–236
filamentous fungi 257–258
final energy consumption (FEC) 232–233
financial resources, Sustainable Development Goals 81–82
FINEP ('Studies and Projects Financing Agency') 117, 118
fireflies 192–193, 194, 195
Firmicutes spp. 242–243, 262
first generation (1G) biofuels 39, 234–236
Fischer-Tropsch (FT) conversions to syngas/jet fuel production 220–221
fishes
  bioluminescence 192–193, 195–196
  fish-friendly hydropower 32–33
Flammulina velutipes 284, 285
flash power plants 35
flate plate solar collectors (FPSCs) 60
flavin mononucleotide (FMN) 195
floating foundations, offshore wind energy technologies 27–28
floating hybrid energy platforms, wind energy 61
flow control technologies, hydropower 30
fluctuations in energy consumption/ economic development 95–96
Foerster, G. E. 197, 198
Foreign Direct Investment (FDI) 81
Forest Carbon Partnership Facility (FCPF) 3
Former Soviet Union, ratio of energy consumption to GDP from 1860 to 1996 97, 98
fossil fuels
  gasoline 186, 214–217
  India energy sector 170, 171–172
  India subsidy cuts 178
  sustainable development 19
  see also biodiesel; carbon emissions; oil; petroleum
Foundation to Support Research and Extensions (Fapex), Brazil 119
Fourier Transform Infrared 292
fourth generation (4G) biofuels 234, 245
FPSCs (flate plate solar collectors) 60
France
  electricity production 159
  energy consumption 96–97
  ratio of energy consumption to GDP from 1860 to 1996 97, 98
free fatty acids (FFA) 235–236
freshwater habitat 193
Fridericia spp. 194
fuel cell electric vehicles (FCEVs) 63–64
fuel cells 163
Fukushima nuclear disaster 61
Fulgora (lanternflies) 192
fungi 160–161, 191–207
  agro waste/biotransformation 274–302
  batteries design 258–261
  bioluminescence 191–207

diversity 196–197
electron transfer mechanisms 255, 256
lignocellulosic biomass pretreatment 212
luciferin 195
mechanisms 197–198
microbial fuel cells 250–273
microbial lipases 235–236
second generation biofuels 237–238, 239–240
third generation biofuels 244
see also bioluminescence
Fusarium spp. 282
future-generation turbines technology, offshore wind energy technologies 27

g

G. pampeanus 284, 285
G. sulfurreducens 262
GaAs-based solar cells 22, 57
galactans 279–280
Galactomyces reessii 258
galactose 213
Galleria spp. 193
gallium, active solar energy technology 133
Gammaproteo spp. 262, 264
Ganoderma lucidum 258, 292
gas
Brazil/energy sources 114
developing nations 100, 101, 102
global primary energy consumption 4–5
India energy sector 170, 171
jet fuel production 217, 218, 220–221
LPG 99–100, 114
spent mushroom substrate 283–284
see also greenhouse gases; natural gas
gasification 39, 43, 138
gasoline 186, 214–217
GDP *see* gross domestic product
genes 195
Geobacter spp. 161, 262, 264
GEOPHIRES EGS plants 37
geothermal energy 19, 33–37, 61–62, 137–138
advantages/disadvantages 140
binary plants 35
clean energy 155
combined-cycle/hybrid plants 35
comparative account 158
comparison by country 34
direct dry steam plants 34
EGS 35–37
flash power plants 35
global investments 46, 47
hydrogen production methods 42–43
technologies/application/challenges 56, 57
worldwide electricity generation 20–21
geothermal heat pumps (GHPs) 62, 137
Geothrix spp. 161
Germany
21st century development 72–73
electricity production 159
energy consumption 96–97
as leaders in renewable energy 21
ratio of energy consumption to GDP from 1860 to 1996 97, 98
wind energy 154
Gerronema viridilucens 196, 197, 199
Gesner, C. 192
Ghost Fungus 196
Gieg, L. M. 245
global warming *see* climate change
glowing tree (bioluminescence) 198–199
glowworms 192–193
glucomannan 279
glucose substrate and *Hansenula anomala* 257
glucuronoxylan 279–280
glycerol, lignocellulosic biomass pretreatment 211–212
glycerol-3-phosphate dehydrogenase 1 (GPD1) 160

glycerol-3-phosphate dehydrogenase (GUT2) 160
Golueke 243
GPD1 (glycerol-3-phosphate dehydrogenase 1) 160
graphene nanoplatelets (GNPs) 23
gravity hydropower converters 32
green algae, hydrogen production methods 42–43
Green Bonds 81–82
Green Foreign Direct Investment (FDI) 81
Green India Mission 176
Green tags 178
Greenex, Bombay stock exchange 84–85
greenhouse gases (GHG)
  clean energy 151–168
  climate change 1–17
  geothermal energy 33
  hydrogen production methods 42–43
  India energy sector 172
  methane (CH_4) 6, 9–10, 11, 62
  mushrooms/LCB 284
  see also carbon emissions; climate change
gross domestic product (GDP) 73–76
  energy development 91–107
  pathways of energy transition 108–130
  see also economic growth
Grossman, G. M. 123–124
Grove, William 163
Growth Acceleration Programme (PAC-II), Brazil 116–117
Gude, V. G. 251
guiaiacol, lignin to hydrocarbon fuel 222–223
Gujarat, India 144
GUT2 (glycerol-3-phosphate dehydrogenase) 160
Guyot, R. 196–197

h

Hamlet (Shakespeare) 192
Hansenula anomala 257
HAP (hot-air pasteurisation) 289
hardwood hemicelluloses 279–280
Harvey, E. Newton 193
Hassan, S. H. 262
Hauksbee, F. 192
Hawksworth, D. L. 196
HCPV (high-concentration photovoltaic) solar cells 22
HDC (hydrodecarbonylation/ decarboxylation) 219–220
HDO (hydrodeoxygenation) 210, 219–220, 222–223
heat power and biomass 138
Helpman, E. 124
hemicellulase 213, 214
hemicellulose 212–215
  enzymatic pretreatment of biofuels 242
  mushrooms/LCB 279–280
  SMS and bioethanol production 288
high-concentration photovoltaic (HCPV) solar cells 22
high-temperature fuel cells (HTFCs) 63–64
high-temperature water splitting/hydrogen production methods 42–43
high-voltage alternating current (HVAC) infrastructure, offshore wind energy technologies 28
high-voltage direct current (HVDC) infrastructure, offshore wind energy technologies 28
HMF (hydroxymethylfurfural) 212, 219, 288, 290
HMS Challenger 193
Honey Mushrooms 196
hot-air pasteurisation (HAP) 289
'hot-carrier' flat plate devices 21–22
hot-electron extraction 21–22
Hou, Q. 262
Howitt, P. 123
HPMs (hydrostatic pressure machines) 32
HRJ (hydroprocessed renewable jet) fuel 219–220
HTFCs (high-temperature fuel cells) 63–64

Huber, G. W. 210
human capital 123–124
HVAC (high-voltage alternating current) infrastructure 28
HVDC (high-voltage direct current) infrastructure 28
hybrid and electric vehicles 179
hybrid geothermal plants 35
hybrid solar cells 21–22, 23, 24–25, 59
hybrid systems
  India/exploitative policy 173
  wind energy 61
hydrocarbon fuel 221–223
hydrochloric acid (HCl)/lignocellulosic biomass 213
hydrodecarbonylation/decarboxylation (HDC) 219–220
hydrodeoxygenation (HDO) 210, 219–220, 222–223
hydrogen 42–44
  bio-hydrogen production 44
  biomass gasification 43
  climate change 12
  as fuel 162–163
  microbial fuel cells 263
  production methods 43–44
  renewable electrolysis 43
  technologies/application/challenges 56, 57
  thermochemical water splitting 43
  see also biohydrogen production
hydrogen peroxide (H_2O_2) 283
hydroid *Obelia* spp. 195
hydrolases 235
hydrolysis
  biomass 138
  LCB to monosaccharides 215–216
  lignocellulosic biomass 210–211, 212–213
hydropower 19, 29–33, 60, 136–137
  advantages/disadvantages 139
  Brazil/energy sources 113
  Brazilian energy programmes 116, 117
  clean energy 153–154
  comparative account 158
  diagrammatic representation of types 153
  digitalization 30–31
  environmental impacts 64–65
  evolution 31–32
  fish-friendly 32–33
  flow control technologies 30
  future of 143–144
  global primary energy consumption 4–5
  gravity hydropower converters 32
  growth rate 20–21
  hydrogen production methods 42–43
  India/electricity generation 159
  India/exploitative policy 173
  PATs 32
  small-scale hydropower plants 31
  technologies/application/challenges 56, 57
hydroprocessed renewable jet (HRJ) fuel 219–220
hydrostatic pressure machines (HPMs) 32
hydrothermal processing, biomass 138
hydro-treatment/alkane isomerization and cracking, biofuel categorization 39
hydroxymethylfurfural (HMF) 212, 219, 288, 290

i

IBAMA (Brazilian Institute of Environment and Renewable Natural Resources) 115
IBGE (Brazilian Institute of Geography and Statistics) 182
ICCC (Interministerial Commission on Climate Change) 120
ice sheets in Antarctica 7–8
Iceland, geothermal energy 34
IEA (International Energy Agency) 4–5, 111
illudin S/illudinolone/illudosone hemiacetal (sesquiterpenes) 201
imidazolopyrazine 195

impurity-band devices 21–22
Incentives Program for Alternative Sources of Electric Energy (ANEEL) 116
INDC (Intended National Determined Contribution), India 176, 177
India
 21st century development 72–73
 air pollution/climate change 169–181
 carbon emission reductions 84–87
 carbon emissions 1960–2020 74–76
 coal taxes 176–177
 Copenhagen Accord 81, 82, 83, 176
 Electricity Act 174–175
 electricity production 159, 160
 energy consumption 96–97
 energy reserves 170
 energy sector 170–172
 environmental issues/economic development 76, 77–78
 financial instruments to promote renewable energy 176–179
 fossil fuels/electricity 171–172
 future of renewables 143–144
 greenhouse gases 172
 hydro energy 154
 INDC 176, 177
 innovative schooling 86
 as leaders in renewable energy 21
 mandatory CSR 85–86
 NAPCC-National Action Plan 2008 175–176
 National Electricity Policy 175
 national RE strategies 174–176
 PAT schemes 178–179
 policies to exploit renewable energy 172–174
 pollution 77–78
 problems faced with SDGs 81–82
 production of energy 170–171
 progressive energy consumption 97
 ratio of energy consumption to GDP from 1860 to 1996 97, 98
 RECs 178
 renewable energy national strategies 174–175
 renewable power potential 170, 171
 solar powered transport 86–87
 subsidy cuts on fossil fuels 178
 sustainability indices 84–85
Indian Renewable Energy Department Agency (IREDA) 174
Indian Solar Loan Programme 172
Indonesia
 geothermal energy 34
 progressive energy consumption 97
Industrial Revolution of 1750 9–10
INNOVA ENERGIA, Brazil 121
INPI (National Institute of Industrial Property), Brazil 187
insects, bioluminescence 193, 194
integrated turbine and foundation installation, offshore wind energy technologies 28
Intended National Determined Contribution (INDC), India 176, 177
Intergovernmental Panel on Climate Change (IPCC) 4–8, 12–13, 72, 82–83, 142
intermediate-band devices 21–22
Interministerial Commission on Climate Change (ICCC) 120
International Atomic Energy Agency, 2006 114–115
International Energy Agency (IEA) 4–5, 111
International Renewable Energy Agency (IRENA) 20–21, 159
International Solar Alliance 172
in-vivo imaging and bioluminescence 200–201
IPCC *see* Intergovernmental Panel on Climate Change
IREDA (Indian Renewable Energy Department Agency) 174
iso-alkanes 210

isomerization and cracking, biofuel
categorization 39
isoomphadione (sesquiterpenes) 201
Itaipu Dam, China 153–154
Italy
energy consumption 96–97
geothermal energy 34
as leaders in renewable energy 21

j
Jack-o-Lantern mushrooms 196
Jadhav, S. K. 254
Janusz, G. 241
Japan
21st century development 72–73
electricity production 159
energy consumption 96–97
Fukushima nuclear disaster 61
future of renewables 143–144
geothermal energy 34
Kyoto Protocol 82, 83, 86–87, 183, 184, 185–186
ratio of energy consumption to GDP from 1860 to 1996 97, 98
Jatropha non-edible oil-producing plants 220
Jawaharial Nehru National Solar Mission 172
jellyfish 192, 195
jet fuels 217–221
Jiang, H. 284, 285
Jim, Y. K. 79–80
Jin, S. 264
Jung, S. 262

k
Kamal Baharin, N. S. 284, 285
Kaneshiro, H. 258
Karnataka, India 144
Kaskova, Z. M. 197–198
Kaushik, A. 254
Ke, H. M. 196
Kenya
geothermal energy 34
pathways of energy transition/economic growth 110
kerosene 100
Kim, B. H. 251
Kim, J. R. 262
Kocuria spp. 262
Koók, L. 251
Koyna hydroelectric plant, India 154
Kyoto Protocol 82, 83, 86–87
Kyoto Protocol, Japan 183, 184, 185–186

l
laccase 283
lactate dehydrogenase redox protein 257
Langer, C. 163
lanternflies 192
Latia neritoides 193
Latin America
commercial energy consumption and projection 97, 98
final energy consumption 232–233
see also individual Latin American countries...
LCB (lignocellulosic biomass) 208–230
LCOE (levelized cost of electricity) 45–46
LDAs (lignin-degrading axillary enzymes) 240–241
Lentinula edodes 276
LEU2 (β-isopropylmalate dehydrogenase) 160–161
levelized cost of electricity (LCOE) 45–46
levulinic acid (LA) 219
Li 251
"Light for All" (Luz para Todos) energy programme, Brazil 116, 117
lignin
biobutanol production 289–290
bioethanol production 288–289
electricity generation 291–292
see also lignocellulosic biomass
lignin peroxidase (LiP) 283

ligninases 213
lignin-degrading axillary enzymes (LDAs) 240–241
lignin-modifying enzymes (LME) 240–241
lignin-modifying peroxidases (LMPs) 240–241
lignite, India energy sector 170, 171
lignocellulosic biomass (LCB) 208–230
  bio-gasoline 214–217
  biotechnological importance 281–282
  cellulose 279
  characteristics of 279
  conversion to hydrocarbons 221–223
  ethanol 210–214
  jet fuels 217–221
  mushrooms 278–280
  second generation biofuels 237, 238
  spent mushroom substrate 280–284
Lima, Peru agreement 2014 82, 83
Lin, C. Y. 284, 285
lipases 235–236
Liquefied Petroleum Gas (LPG) 99–100
liquid biofuels
  production from lignocellulosic biomass 208–230
  worldwide electricity generation 20
lithium-air batteries 260
lithium-based batteries, structure/composition 259
lithium-ion batteries 260
lithium-sulphur (Li-S) batteries 259–260
LME (lignin-modifying enzymes) 240–241
LMPs (lignin-modifying peroxidases) 240–241
long-chain myristic aldehyde 195
Lou, Z. 284, 285
Lovley, D. R. 251
LPG (Liquefied Petroleum Gas) 99–100, 114
Lucentipes spp. 196, 197
luciferin 193, 195, 196, 197–198
Lumbricidae spp. 194
luminescent orange juice-derived C-dots 62–63
Luo, X. 284, 285
Lutz, C. 110
lux operon genes 195

m

magnetic resonance imaging (MRI) 200–201
Magnus, A. 192
Maji, K. 110
mannans 279
mannose sugars 213
Marine, A. S. 97, 98
marine coelenterates and ctenophores 195
marine energy
  advantages/disadvantages 140
  global investments 46, 47
  worldwide electricity generation 20
  see also ocean energy
market structure and energy transition 115–116
Marrakesh Accord 82, 83
Mata, T. M. 244
Mbokou, S. F. 257
MCT (Ministry of Science and Technology) 120
MCTT (Ministry of Science, Technology and Innovation) 118
MDGs (Millennium development goals) 81
MEC (microbial electrolysis cell) 263
MEER (microbially enhanced energy recovery) 245
Megascolecidae spp. 194
MEOR (microbially enhanced oil recovery) 245
Meriam 192
metal doping on carbon nanostructures 63
methane, *see also* greenhouse gases

methane (CH_4) 6, 11, 62
natural gas composition 11
process and causes of climate change 9–10
methanol, lignocellulosic biomass pretreatment 211–212
Methanothermobacter and *Methanobacterium* 287
metro rail transport 86–87
Mexico
energy consumption 96–97
geothermal energy 34
progressive energy consumption 97
MFCS (microbial fuel cells) 161, 165, 250–273
Mg^{2+} and bioluminescence 195
Michaelseniella spp. 194
micro-computed tomography (μCT) 200–201
microbial conversion/biofuel categorization 39
microbial electrolysis cell (MEC) 263
microbial fuel cells (MFCS) 161, 165, 250–273
application 261–265
bioelectricity production 262–263
biohydrogen production 263
bioremediation 264
biosensors 263
buffers 254
Candida melibiosica 257
challenges/future perspectives 265
configuration of reactor 253–254
dual-chamber 253–254
dye decolorization 264–265
electrodes and fungal battery design 258–261
electrolyte resistance 255
fungi 255–258
general introduction to 251–253
Hansenula anomala 257
performance 253–255
research contributions to development of 251
Saccharomyces cerevisiae 256
single-chamber 253
SMS and electricity generation 291–292
substrate 254–255
wastewater treatment 264
microbially enhanced energy recovery (MEER) 245
microbially enhanced oil recovery (MEOR) 245
Microcystis aeruginosa 158
microorganism-derived energy 157–160
electron transfer mechanisms 253
microbial fuel cells 250–273
sustainability and future energy challenges 231–249
see also biomass
Middle-eastern energy consumption 232–233
midshipman fish 196
Millennium development goals (MDG) 81
millipedes 193
mindmaps, environmental issues 76, 77
Ministry of Mines and Energy 115
Ministry of New and Renewable Energy (MNRE) 173, 174, 178
Ministry of Non-Conventional Energy Sources, India 174
Ministry of Science, Technology and Innovation (MCTT) 118
Ministry of Science and Technology (MCT) 120
Mitiouchkina, T. 199
Mn-dependant peroxidase (MnP) 283
Mnemiopsis spp. 195
MNRE (Ministry of New and Renewable Energy) 173, 174, 178
Mohtasham, J. 158
mollusks 192, 193
Mond, L. 163
mono-crystalline silicon solar cells 22
monosaccharides
LCB hydrolysis 215–216
polyol conversions to alkane 216

Moon Night mushrooms 196
Moorella spp. 243
Morris, J. M. 264
MRI (magnetic resonance imaging) 200–201
multiple-rotor systems (MRS) 61
multi-wall carbon nanotubes (MWCNTs) 23
mushrooms 193, 196, 198
- agro waste/biotransformation 274–302
- challenges 293
- LCB/spent mushroom substrate 280–284
- lignocellulosic biomass 278–280
- *see also* fungi; spent mushroom substrate

Mustafa, A. M. 284, 285
mycelia 196–197
Mycena spp. 196, 199
Mycenoid spp. 196–197
myristic aldehyde 195

n

n-alkanes 210
NADH (nicotinamide adenine dinucleotide) 194, 195
NAD(P)H-dependent reductase 197
nambinones A-C 201
National Agency of Petroleum, Brazil 119
National Alcohol Programme of 1975, Brazil 186
National Biodiesel Programme, Brazil 186
National Biofuel Policy of India 174
National Biogas and Manure Management Programme, India 174
'National Council for Scientific and Technological Development' (CNPq) 117, 118
National Electricity Policy (NEP), India 175
National Energy Conservation Programme (PROCEL), Brazil 116, 117
National Institute of Industrial Property (INPI), Brazil 187
National Mission for Enhanced Energy Efficiency, India 175
National Mission on Strategic Knowledge for Climate Change, India 176
National Mission for Sustainable Agriculture, India 176
National Mission on Sustainable Habitat, India 175–176
National Mission for Sustaining the Himalayan Ecosystem, India 176
National Nuclear Energy Commission (CNEN) 116
National Offshore Wind Energy Policy of India 173
National Programme for the Rational Use of Oil and National Gas (CONCEPT), Brazil 116, 117
National Solar Mission of India 175
National Water Mission of India 176
National Wind Power Policy of India 173
natural gas 11
- Brazil policy prioritization 114
- developing nations 100, 101, 102
- global primary energy consumption 4–5
- India energy sector 170, 171
- sustainable development 19

Natural Gas and Biofuels, federal programme of Brazil 119
The Natural History (Pliny the Elder) 192
Naturalis Historia (Pliny the Elder) 192
n-butanone, lignocellulosic biomass pretreatment 211–212
Neonothopanus spp. 196, 198, 199, 201
NEP (National Electricity Policy), India 175
Netherlands, energy consumption 96–97
Neurospora crassa 261
New Zealand, geothermal energy 34

Newell, P. 110
Nicaragua, geothermal energy 34
nicotinamide adenine dinucleotide (NADH) 195
nitric acid (HNO_3)/lignocellulosic biomass 213
nitrogen in spent mushroom substrate 280–281
Norjannah, B. 236
Norway, electricity production 159
nuclear energy 11
  CNEN 116
  Fukushima disaster 61
  global primary energy consumption 4–5

o

ocean energy 19, 40
  clean energy 156
  comparative account 158
  OTEC 40, 41
  salinity gradient energy 40, 41
  tidal energy 41
  wave energy 40
  see also tidal energy; wave energy
ocean thermal energy conversion (OTEC) 40, 41
Octochaetidae spp. 194
OECD (Organization for Economic Co-operation and Development) 4–5, 81, 97, 98
offshore wind energy technologies 27–29
  DC power take-off and array cables 28–29
  floating foundations 27–28
  floating wind turbines 61
  future-generation turbines technology 27
  high-voltage direct current infrastructure 28
  India/exploitative policy 173
  integrated turbine and foundation installation 28
  other techniques 29
  repowering of sites 28
  site layout optimization 29
  worldwide electricity generation 20
oil
  crises 1973–1974 123
  developing nations 100, 101, 102
  global primary energy consumption 4–5
  hydrogen production methods 42–43
  India energy sector 171
  jet fuel production 217, 218, 219–220
  MEOR 245
  pollution 76–77
  shale semi-coke SMS 284, 285
  sustainable development 19
oleaginous yeast 160–161, 244
Oliveira, A. G. 196, 198
Omphalotus spp. 196, 201
onshore wind energy technologies 25–27
  design and materials of rotor blade 26
  India/electricity generation 159
  power electronics optimization 26
  recycling of materials 26–27
  smart wind turbines 26
  turbine size and ratings 25–26
  worldwide electricity generation 20
Operador Nacional do Sistema Elétrico (ONS) 116
organic materials and solar energy 133
organic/polymer PV cells 21–23
Organization for Economic Co-operation and Development (OECD) 4–5, 81, 97, 98
Organization of the Petroleum Exporting Countries (OPEC) 97, 98
organosolv pre-treatment and biobutanol production 289–290
'oscillating wave surge converters'/onshore wave energy 40

Ostracod luminous molecules 193
Oswald 243
OTEC (ocean thermal energy conversion) 40, 41
'overtopping devices'/onshore wave energy 40
oxidation and bioluminescence 194
oxidation reduction potentials (ORP) 255
oxygen and bioluminescence 195
oxyluciferin 195, 197–198
Ozanam, J. 192–193

p

P. eryngii 284, 285
P. florida 288
P. ostreatus 284, 285
P. pulmonarium 288–289
P. sajor-caju 288
Pachysolen tannophilus 242, 288–289
PAHs (polycyclic aromatic hydrocarbons) 283
Pakisthan, pollution 77–78
pandemics 82, 87–88, 144
PANTA RHEI models 110
Paraguay, hydro energy 154
Parapriacanthus spp. 196
Paris Accord 82–84, 87, 152–153
Park, D. H. 262
passive solar energy technology 133–134
PAT (Perform, Achieve and Trade schemes) of India 178–179
pathways of energy transition/economic growth 108–130
  climate-resilient growth 121–122
  electricity sector in Brazil 113–115
  market structure 115–116
  political economy 123
  public investment in R & D 111–112
  Research and Development 111–112
  social consequences/accessibility 122
  theoretical constructs 123–125
PATs (pump as turbines) 32
PCBs (polychlorinated biphenyls) 283
PCP (pentachlorophenol) 199
PEC (primary energy consumption) 232
Pedersen, E. 156–157
Pelagia noctiluca (purple jellyfish) 192
PEM (proton exchange membrane) 250, 252–254
Pempheris spp. 196
Penicillium capsulatum 242
pentachlorophenol (PCP) 199
Perform, Achieve and Trade (PAT) schemes of India 178–179
perovskites 133
peroxidases 240–241
Petrobras, Brazil 116, 117, 118
petroleum
  India energy sector 170, 171
  LPG 99–100, 114
phenol/lignocellulosic biomass 211–212, 222–223
[6,6]-phenyl C60 butyric acid methyl ester (PCBM) photovoltaic cells 22–23
phenyl-propane in lignin 280
Philippines, geothermal energy 34
Pholas dactylus 192
Pholas luminous molecules 193
phosphate buffer solution (PBS) 254
phosphoric acid (H_3PO_4)/lignocellulosic biomass 213
phosphorus in bioluminescence 192–193
phosphorus in spent mushroom substrate 280–281
photobiological water splitting/hydrogen production methods 42–43
photochemical water splitting/hydrogen production methods 42–43
photolytic hydrogen production methods 42–43
photovoltaic (PV) cells 21–24, 133
  clean energy 155–156
  cost-effective energy production 46
  efficiency 44–45
  future of 46–47
  hydrogen production 43

large-scale energy production 45
schematic representation 57, 58
solar photovoltaic energy 20, 21–24, 159
types/schematic representation 57, 58
see also solar photovoltaic energy
photovoltaic/thermal (PV/T) collectors 24–25
phytoene desaturase 160–161
phytoene synthase 160–161
Pichia stipitis 242, 288–289
pig manure 284, 285
platform molecules/jet fuel production 217, 219
Pleurotus spp. 265, 276–277, 283, 284, 285
Pliny the Elder 192
Poland, energy consumption 96–97
polar ice caps melting 7–8
political benefits of renewable energy 141
pollution 76–77, 142
India/sustainable development 169–181
mushrooms/LCB 284
see also environment; greenhouse gases
poly[3-hexylthiophene] (P3HT) photovoltaic cells 22–23
polychlorinated biphenyls (PCBs) 283
polycrystalline, GaAs materials in photovoltaic cells 57, 59
poly-crystalline silicon solar cells 22
polycyclic aromatic hydrocarbons (PAHs) 283
polyols 215–217
Pontié, M. 257–258
population increase *see* economic growth
population increases, clean energy/alternative energy 160–161
potassium in spent mushroom substrate 280–281
Potter, M. C. 251, 256
poverty 142
power electronics optimization, onshore wind energy technologies 26
'pressure retarded osmosis'/OTEC technologies 41
pretreatment/detoxification of lignocellulosic biomass 210–212
price of energy 97–99
primary energy consumption, global 2017 4–5
primary energy consumption (PEC) 232
PROCEL, Brazil 116, 117
product recovery of lignocellulosic biomass 210–211, 214
PROELICA 2001 programme, Brazil 121
Progex (Technological Export Programme for exports), Brazil 119
PROINFA, Brazil 116, 117, 120, 121
Proteus hauseri 265
proton exchange fuel cells 163
proton exchange membrane (PEM) 250, 252–254
Pseudomonas spp. 161, 235, 265
pump as turbines (PATs) 32
pumped storage power plant (PSPP) 31
Punch, J. 156–157
purple jellyfish 192
Purtov, K. V. 197
Pyphorus luminous molecules 193
Pyrearinuster mililluminans 194
pyrolysis, biomass 138
Pyrophorus augustus 194
pyruvate decarboxylase 161

q

quantum dots (QD) 21–22, 60, 133

r

radiant energy 163–164
railroad worms 194, 195
raney nickel 215–216
ratio of energy consumption to GDP from 1860 to 1996 97, 98
recycling of materials, onshore wind energy technologies 26–27
Reduce, Recycle and Reuse project, India 87

reforming of renewable liquid fuels/hydrogen production methods 42–43
Regan, J. M. 262
regime complex/technological innovation in Brazil 182–190
Regime for Funding Agency, Brazil 118–120
Remazol dye 265
renewable electrolysis/hydrogen production methods 43
Renewable Energy Certificates (RECs) 178
renewable energy (RE) 131–150
  advantages/disadvantages 139–140
  Brazil policy prioritization 114
  clean energy 151–168
  current status/future perspectives 143–144
  global investments 46, 47
  hydrogen production 42–44
  importance of 140–141
  India energy sector 170, 171, 174–175
  India/electricity generation 159
  limitations 143
  municipal waste 159
  overview 132
  societal benefits 141–142
  sources 132–139
  sustainable development 3–5
  sustainable development goals 142
  see also biomass; geothermal energy; hydropower; solar energy; tidal energy; wind energy
renewable water supplies 8
repowering of offshore wind sites 28
Research and Development Foundation (FADESP), Brazil 119
Research and Development (R&D) 42–43, 111–112, 119, 120–121, 123
'reverse electro dialysis'/OTEC technologies 41
rhizomorphs 196–197
Rhizopus spp. 235
Rhodobacter capsulatus 199
Rhodosporidium torulides oleaginous yeast 160–161
rice straw 284, 285
Romania, energy consumption 96–97
Romer, P. M. 123
Rossi, R. 254
'rotating mass converters'/offshore wave energy 40
Royal Society/HMS Challenger 193
Russian Federation, electricity production 159
Ryden, P. 284, 285

s

saccharification and co-fermentation (SSCF) 242–243
Saccharomyces cerevisiae 161, 256, 262, 288–289
Sahel region of Africa 8
salinity gradient energy 40, 41, 156
Sao Paulo Research Foundation (Fapesp) 119
SAPO (silicoaluminophosphate) 220
'Satat Bharat – Sanatan Bharat' project, India 87
SCDnezzy hybrid systems 61
Scedosporium dehoogi 257–258
Scenedesmus obliquus 158
Schrattenholzer, L. 97, 98
SDGs *see* Sustainable Development Goals
sea pansy Renilla spp. 195
second generation (2G) biofuels 39, 234, 236–238, 243
semiconductors, hydrogen production methods 42–43
sesquiterpenes 201
Shakespeare, William 192
Shewanella spp. 161, 258, 262
Shittake mushrooms 284, 285
silicoaluminophosphate (SAPO) 220
silicon-based solar technology 22, 44, 133

Simões, M. F. 265
Simsek, H. A. & N. 111
Sindhuja, M. 262
single-walled carbon nanotubes 63
Sinopyrophorinae spp. 194
site layout optimization for offshore wind 29
small-scale hydropower plant (SHP) 31, 46, 47, 136
smart wind onshore wind turbines 26
smart windows 60
SMS (spent mushroom substrate) 280–284
snails 193
softwood of coniferous ecosystems 238, 240
softwood hemicelluloses 279–280
solar energy 19, 132, 133–134
  advantages/disadvantages 139
  classification of technologies 22–23
  clean energy 155–156
  comparative account 158
  diagrammatic representation of types 153
  environmental impacts 64
  future of 46–47
  global investments 46, 47
  global primary energy consumption 4–5
  growth rate 20–21
  hydrogen production methods 42–43
  India/caps and tariffs 179
  India/exploitative policy 172
  pollution 76–77
  recent advances in 21–25
  semiconductor technology 56–60
  technologies/application/challenges 56, 57
  transportation systems 86–87
  wind/hybrid policy of India 173
solar photovoltaic energy 20, 21–24, 159
solar power generation 24
solar steam generation devices 292
solar thermal technology 20, 133
solid biofuels
  India/electricity generation 159
  worldwide electricity generation 20
solid thermal energy, India/electricity generation 159
sorghum-based SMS *P. ostreatus* 284, 285
South Africa
  energy consumption 96–97
  progressive energy consumption 97
South America
  geothermal energy 34
  see also Latin America
South Korea, progressive energy consumption 97
Spain
  energy consumption 96–97
  as leaders in renewable energy 21
spent mushroom substrate (SMS) 280–284
  applications 282–284
  biobutanol 289–290
  bio-coke 290–291
  bioethanol 288–289
  biogas production 287–288
  biohydrogen production 285–287
  biotransformation into energy 284–292
  electricity generation 192–193
  energy 283–284
  enzyme recovery 283
  fertilizers 282
  solar steam generation devices 292
  wastewater treatment 283
SRB (sulfate-reducing bacteria) 161
SSCF (saccharification and co-fermentation) 242–243
standards of living and economic development 94
Streptomyces spp. 237–238
'Studies and Projects Financing Agency' (FINEP) 117, 118
Subha, C. 262

subsidy cuts on fossil fuels 178
substrate and microbial fuel cells 254–255
succinate dehydrogenace (complex II) 254–255
sugarcane, Brazil/energy sources 113
sugars
  glucose substrate and *Hansenula anomala* 257
  jet fuels 217, 218, 219
  see also monosaccharides
sulfate-reducing bacteria (SRB) 161
sulphuric acid (H_2SO_4)/lignocellulosic biomass 213
Superintendence for the Development of the North East 118
supply and demand of energy 97–99
sustainability
  21st century development 72–90
  advances in alternative energy sources 18–54
  agro waste/biotransformation 274–302
  indices/India 84–85
  microorganism-derived energy 231–249
Sustainable Development Goals (SDGs) 80–87
  Brazil regime complex 185
  carbon emissions/sustainability indices 84–87
  clean energy 151–168
  environment 80–87
  financial resources 81–82
  illegal activities 82
  natural calamities and pandemics 82
  Paris Accord 82–84
  problems faced 81–82
  renewable energy 3–5, 142
  social issues 82
syringol, lignin to hydrocarbon fuel 222–223

t

TAG (triacyl glycerides) 244
Talaromyces emersonii 242
Tamil Nadu Development Agency (TEDA) 154
Tamil Nadu, India 144, 154
tandem cells 21–22
technological benefits of renewable energy 141
Technological Development Program for Biodiesel, Brazil 120
Technological Export Programme for exports (Progex), Brazil 119
technological innovation in Brazil 182–190
Tesla, Nikola 163–164
tetrahydro furfuryl alcohol, lignocellulosic biomass pretreatment 211–212
TGs (triacyl glycerols) 235–236
thermal hydrogen production methods 42–43
thermal-PV hybrid systems (CPV-T) 23, 24–25, 46
thermochemical gasification, biofuel categorization 39
thermochemical water splitting 43
thermoelectric systems (CSP) 23–24, 43
Thermomonospora fusca 242
thin film photovoltaic cells 21–22
third generation (3G) biofuels 234, 243–244
third-generation PV cells 21–22
Three Gorges Dam 29–30
tidal current energy 156
tidal energy 41, 138–139, 140, 153
toxicity and bioluminescence 199
Tractebel Energia 116
Tradable Renewable Certificates (TRCs) 178
Trametes versicolor 258
transesterification/biofuel categorization 39
transformation technologies 138, 274–302
transportation systems, solar energy 86–87
triacyl glycerides (TAG) 244
triacyl glycerols (TGs) 235–236

triacylglycerol acylhydrolases 235
Trichoderma reesei 212
trifluoroacetic acids (TFA)/lignocellulosic biomass 213
trimethylene glycol, lignocellulosic biomass pretreatment 211–212
Tuoliensis spp. 284, 285
turbine size and ratings in wind energy 25–26
turbine water wheels (TWWs) 32
turbines *see* wind energy
Turkey, geothermal energy 34

u

UCA (Unnat Chulah Abhiyan), India 173–174
UN General Assembly, renewable energy 142
UNCED (United Nations Conference on Environment and Development) 183, 184
underwater pumped hydro energy storage (UPHES) 31
Union of Soviet Socialist Republics (U.S.S.R.)
- commercial energy consumption and projection 97, 98
- energy consumption 96–97

United Kingdom
- 21st century development 72–73
- energy consumption 96–97
- ratio of energy consumption to GDP from 1860 to 1996 97, 98

United Nations Conference on Environment and Development (UNCED) 183, 184
United Nations Conference on Trade and Development (UNCTAD) 81
United Nations Development Programme (UNDP) 80–81
United Nations Environment Program (UNEP) 172
United Nations Framework Convention on Climate Change (UNFCCC) 82, 83, 185
United Nations (UN), carbon emission reductions 84–87
United States
- 21st century development 72–73
- commercial energy consumption and projection 97, 98
- electricity production 159
- energy consumption 96–97
- final energy consumption 232–233
- future of renewables 143–144
- geothermal energy 34
- hydropower 153–154
- as leaders in renewable energy 21
- ratio of energy consumption to GDP from 1860 to 1996 97, 98
- wind energy 154

Unnat Chulah Abhiyan (UCA), India 173–174
UPHES (underwater pumped hydro energy storage) 31
US Department of Energy (USDE) 244
US Energy Information Administration (USEIA) 151, 208

v

γ-valerolactone (GVL) 219
Vargula hilgendorfi (Ostracod luminous molecules) 193
Venezuela, progressive energy consumption 97
Verma, D. 220
Vibrio spp. 264
Vietnam War 198
Vlahinić-Dizdarević, N. 124

w

wafer-based cells 133
waste-to-energy technology 47, 161–162
wastewater treatment 264, 283
water gas shift (WGS) 220–221
water splitting/hydrogen production methods 42–43
water supplies 8
wave energy 40, 153, 156

white rot fungi (WRF) 212, 238, 239, 240, 291–292
wind energy 19, 25–29, 61, 134–136
  advantages/disadvantages 139
  Brazilian energy programmes 116, 117
  clean energy 154
  comparative account 158
  development and impacts 134, 135
  environmental impacts 65
  global investments 46, 47
  global primary energy consumption 4–5
  growth rate 20–21
  hydrogen production methods 42–43
  India/electricity generation 159
  India/exploitative policy 172–173
  offshore wind energy technologies 27–29
  onshore wind energy technologies 25–27
  pollution 76–77
  solar/hybrid policy of India 173
  technologies/application/challenges 56, 57
  turbine rotor evolution 134, 135
  worldwide electricity generation 20
World Bank, population increase 95
World Energy Conference 97
World Energy Council 2013 155
World Health Organization (WHO), air quality 77–78
World Intellectual Property Organization (WIPO) 183
Wu, X. 251

x

Xenorhabdus spp. 193
X-ray photoelectron spectroscopy 292
xylan 214–215, 279, 288–289
xylanase 283
xylose 213, 288–289
xylospyranose 280
Xystocheir bistipita 194

y

Yarrowia lipolytica yeast 160
yeast
  biodiesel production 160–161
  first generation biofuels 235–236
  microbial fuel cells 257
  SMS and bioethanol production 288–289
  third generation biofuels 244
Youssef, N. 245
Yu, Y. 254

z

Zeikus, J. G. 262
zeolites 63
Zhang, Y. 262
Zhou, M. 262
Zhu 284, 285
Žiković, S. 124
zinc and bioluminescence 199
Zymomonas mobilis 161, 242

www.ingramcontent.com/pod-product-compliance
Lightning Source LLC
LaVergne TN
LVHW062206200926
848484LV00006B/1334
9781119741442